Introduction to Environmental Impact Assessment

D0542345

Reviews of earlier editions

'This book should join a limited number of publications that provide the essential first stages on a rapid EIA learning curve ... Well written and referenced and should provide an invaluable introduction to EIA for a wide range of people including students, practitioners, developers and decision makers.' *Environmental Assessment.*

'This book gives the student a good introduction to the history and evolution of environmental impact assessment. It covers the existing legislation well and provides interesting case studies and examples along the way to keep the reader interested ... Many local authorities when faced with their first EIA would welcome such a review.' *Built Environment.*

Introduction to environmental impact assessment: 3rd edition

A comprehensive, clearly structured and readable overview of the subject, *Introduction to Environmental Impact Assessment* quickly established itself as the leading introduction to EIA. The second edition developed many issues of growing importance in this fast-moving subject area, and reinforced the success of the book. In this third edition, the major updates include, in particular,

- experience of the implementation of the revised EC and UK EIA Directive
- best practice in the EIA process
- a new case-studies section which explores some key issues in the process
- comparative EIA system worldwide
- changing prospects for EIA
- the development of SEA legislation and practice.

The book has comprehensive appendices, with a wealth of important reference material, including key websites. Written by three authors with extensive research, training and practical experience of EIA, this book brings together the most up-to-date information from many sources.

Introduction to Environmental Impact Assessment 3rd Edition not only provides a complete introductory text but will also support further studies. Students on undergraduate and postgraduate planning programmes will find it essential as a course text, as will students of environmental management/policy, environmental sciences/ studies, geography and built environment. Planners, developers and decision-makers in government and business will also welcome this new edition as a very effective means of getting to grips with this important subject.

John Glasson is Professor of Environmental Planning, Research Director of the Impacts Assessment Unit (IAU) and Co-Director of the Oxford Institute for Sustainable Development (OISD) at Oxford Brookes University. He is also Visiting Professor at Curtin University in Western Australia. **Riki Therivel** is Visiting Professor at Oxford Brookes University, a Senior Research Associate in the IAU, and partner in Levett-Therivel sustainability consultants. **Andrew Chadwick** is Senior Research Associate in the IAU.

The Natural and Built Environment Series

Editor: Professor John Glasson, Oxford Brookes University

Methods of Environmental Impact Assessment: 2nd Edition
Peter Morris and Riki Therivel

Introduction to Environmental Impact Assessment: 3rd Edition
John Glasson, Riki Therivel and Andrew Chadwick

Public Transport: 4th Edition
Peter White

Urban Planning and Real Estate Development: 2nd Edition
John Ratcliffe and Michael Stubbs

Landscape Planning and Environmental Impact Design: 2nd Edition
Tom Turner

Controlling Development
Philip Booth

Partnership Agencies in British Urban Policy
Nicholas Bailey, Alison Barker and Kelvin MacDonald

Planning, the Market and Private House-Building
Glen Bramley, Will Bartlett and Christine Lambert

British Planning Policy in Transition
Mark Tewdwr-Jones

Development Control
Keith Thomas

Forthcoming:

Urban Regeneration: A critical perspective
Sue Brownill and Neil McInroy

Strategic Planning and Regional Development in the UK
Harry Dimitriou and Robin Thompson

Introduction to Environmental Impact Assessment
Third Edition

John Glasson, Riki Therivel and
Andrew Chadwick

Routledge
Taylor & Francis Group

LONDON AND NEW YORK

First published 2005
by Routledge
2 Park Square, Milton Park, Abingdon, Oxon OX14 4RN

Simultaneously published in the USA and Canada
by Routledge
270 Madison Avenue, New York, NY 10016

Reprinted 2006 (twice)

Routledge is an imprint of the Taylor & Francis Group, an informa business

© 2005 John Glasson, Riki Therivel, Andrew Chadwick

Typeset in Goudy by
Integra Software Services Pvt. Ltd, Pondicherry, India
Printed and bound in Great Britain by
MPG Books Ltd, Bodmin

All rights reserved. No part of this book may be reprinted or reproduced
or utilised in any form or by any electronic, mechanical, or other means,
now known or hereafter invented, including photocopying and recording,
or in any information storage or retrieval system, without permission in
writing from the publishers.

British Library Cataloguing in Publication Data
A catalogue record for this book is available from the British Library

Library of Congress Cataloging in Publication Data
A catalog record for this book has been requested

ISBN 10: 0–415–33836–0 (hbk)
ISBN 10: 0–415–33837–9 (pbk)
ISBN 13: 978–415–33836–3 (hbk)
ISBN 13: 978–415–33837–0 (pbk)

Dedicated to our families

Contents

Preface to the first edition   xi
Preface to the third edition   xiii
Acknowledgements   xv
Abbreviations   xvii

PART 1
Principles and procedures   1

1   **Introduction and principles**   3

    1.1   Introduction   3
    1.2   The nature of environmental impact assessment   3
    1.3   The purposes of environmental impact assessment   8
    1.4   Changing perspectives on EIA   13
    1.5   Projects, environment and impacts   15
    1.6   Current issues in environmental impact
          assessment   21
    1.7   An outline of subsequent parts and chapters   23
    Note   24
    References   24

2   **Origins and development**   28

    2.1   Introduction   28
    2.2   The National Environmental Policy Act and subsequent
          US systems   28
    2.3   The worldwide spread of EIA   36
    2.4   Development in the UK   37
    2.5   EC Directive 85/337   40
    2.6   EC Directive 85/337, as amended by Directive
          97/11/EC   44
    2.7   An overview of EIA systems in the EU – divergent
          practice in a converging system?   46
    2.8   Summary   50
    Notes   50
    References   51

3   UK agency and legislative context   54

    3.1   *Introduction*   54
    3.2   *The principal actors*   54
    3.3   *EIA regulations: an overview*   58
    3.4   *The Town and Country Planning (Environmental Impact Assessment)*
           (England and Wales) Regulations 1999 (S1 293)   62
    3.5   *Other EIA regulations*   77
    3.6   *Summary and conclusions on changing legislation*   80
    Notes   81
    References   82

PART 2
Process   85

4   Starting up; early stages   87

    4.1   *Introduction*   87
    4.2   *Managing the EIA process*   87
    4.3   *Project screening – is an EIA needed?*   89
    4.4   *Scoping – which impacts and issues to consider?*   91
    4.5   *The consideration of alternatives*   93
    4.6   *Understanding the project/development action*   96
    4.7   *Establishing the environmental baseline*   100
    4.8   *Impact identification*   107
    4.9   *Summary*   123
    Notes   123
    References   123

5   Impact prediction, evaluation and mitigation   126

    5.1   *Introduction*   126
    5.2   *Prediction*   126
    5.3   *Evaluation*   137
    5.4   *Mitigation*   149
    5.5   *Summary*   153
    References   154

6   Participation, presentation and review   157

    6.1   *Introduction*   157
    6.2   *Public consultation and participation*   158
    6.3   *Consultation with statutory consultees*   167
    6.4   *EIA presentation*   168
    6.5   *Review of EISs*   172
    6.6   *Decisions on projects*   176
    6.7   *Summary*   181
    Notes   182
    References   182

7   Monitoring and auditing: after the decision   185

   7.1   *Introduction*   185
   7.2   *The importance of monitoring and auditing in the EIA process*   185
   7.3   *Monitoring in practice*   187
   7.4   *Auditing in practice*   191
   7.5   *A UK case study: monitoring and auditing the local socio-economic*
       impacts of the Sizewell B PWR construction project   194
   7.6   *Summary*   204
   Note   205
   References   205

PART 3
Practice   209

8   An overview of UK practice to date   211

   8.1   *Introduction*   211
   8.2   *Number, type and location of EISs and projects*   212
   8.3   *The pre-submission EIA process*   217
   8.4   *EIS quality*   222
   8.5   *The post-submission EIA process*   227
   8.6   *Costs and benefits of EIA*   233
   8.7   *Summary*   236
   Notes   236
   References   236

9   Case studies of EIA in practice   240

   9.1   *Introduction*   240
   9.2   *Wilton power station case study – project definition in EIA*   241
   9.3   *M6–M56 motorway link road – the treatment of alternatives and*
       indirect effects   248
   9.4   *N21 link road, Republic of Ireland – assessing impacts on an EU*
       priority wildlife habitat   254
   9.5   *Portsmouth incinerator – new approaches to public participation in EIA* 261
   9.6   *Cairngorm mountain railway – mitigation in EIA*   267
   9.7   *Humber Estuary projects – assessment of cumulative effects*   270
   9.8   *UK offshore wind energy development – strategic environmental*
       assessment   275
   9.9   *Pre-assessment of the impact of refugees in Guinea*   284
   9.10 *Summary*   287
   Note   287
   References   287

10 Comparative practice   289

   10.1 *Introduction*   289
   10.2 *EIA status worldwide*   290
   10.3 *Benin*   296

10.4	Peru	297
10.5	China	299
10.6	Poland	301
10.7	The Netherlands	303
10.8	Canada	305
10.9	Australia and Western Australia	307
10.10	International bodies	312
10.11	Summary	314
	Note	314
	References	314

PART 4

Prospects   319

11 Improving the effectiveness of project assessment   321

11.1	Introduction	321
11.2	Perspectives on change	321
11.3	Possible changes in the EIA process: the future agenda	323
11.4	Extending EIA to project operations: environmental management systems and environmental audits	334
11.5	Summary	337
	Notes	338
	References	338

12 Widening the scope: strategic environmental assessment   341

12.1	Introduction	341
12.2	Strategic environmental assessment (SEA)	341
12.3	SEA worldwide	344
12.4	SEA in the UK	351
12.5	Carrying out SEA under the SEA Directive	356
12.6	Summary	363
	Notes	364
	References	364

Appendices

1	The text of Council Directive 97/11/EC	366
2	Directive 2001/42/EC	382
3	The Lee and Colley review package	394
4	Environmental impact statement review package (IAU, Oxford Brookes University)	395
5	Key EIA journals and websites	408

| Author index | 413 |
| Subject index | 418 |

Preface to the first edition

There has been a remarkable and refreshing interest in environmental issues over the past few years. A major impetus was provided by the 1987 Report of the World Commission on the Environment and Development (the Brundtland Report); the Rio Summit in 1992 sought to accelerate the impetus. Much of the discussion on environmental issues and on sustainable development is about the better management of current activity in harmony with the environment. However, there will always be pressure for new development. How much better it would be to avoid or mitigate the potential harmful effects of future development on the environment at the planning stage. Environmental impact assessment (EIA) assesses the impacts of planned activity on the environment in advance, thereby allowing avoidance measures to be taken: prevention is better than cure.

Environmental impact assessment was first formally established in the USA in 1969. It has spread worldwide and received a significant boost in Europe with the introduction of an EC Directive on EIA in 1985. This was implemented in the UK in 1988. Subsequently there has been a rapid growth in EIA activity, and over three hundred environmental impact statements (EISs) are now produced in the UK each year. EIA is an approach in good currency. It is also an area where many of the practitioners have limited experience. This text provides a comprehensive introduction to the various dimensions of EIA. It has been written with the requirements of both undergraduate and postgraduate students in mind. It should also be of considerable value to those in practice – planners, developers and various interest groups. EIA is on a rapid "learning curve"; this text is offered as a point on the curve.

The book is structured into four parts. The first provides an introduction to the principles of EIA and an overview of its development and agency and legislative context. Part 2 provides a step-by-step discussion and critique of the EIA process. Part 3 examines current practice, broadly in the UK and in several other countries, and in more detail through selected UK case studies. Part 4 considers possible future developments. It is likely that much more of the EIA iceberg will become visible in the 1990s and beyond. An outline of important and associated developments in environmental auditing and in strategic environmental assessment concludes the text.

Although the book has a clear UK orientation, it does draw extensively on EIA experience worldwide, and it should be of interest to readers from many countries. The book seeks to highlight best practice and to offer enough insight to methods, and to supporting references, to provide valuable guidance to the practitioner. For information on detailed methods for assessment of impacts in particular topic areas (e.g. landscape, air quality, traffic impacts), the reader is referred to the complementary volume, *Methods of environmental impact assessment* (Morris & Therivel, 1995, London, UCL Press).

John Glasson
Riki Therivel
Andrew Chadwick
Oxford Brookes University

Preface to the third edition

The aims and scope of the third edition are unchanged from those of the first edition. However, as noted in the preface to the first edition, EIA continues to evolve and adapt, and any commentary on the subject must be seen as part of a continuing discussion. The worldwide spread of EIA is becoming even more comprehensive. In the European Union there is now over 15 years' experience of the implementation of the pioneering EIA Directive, including 5 years' experience of the important 1999 amendments. There has been considerable interest in the development of the EIA process, in strengthening perceived areas of weakness, in extending the scope of activity and also in assessing effectiveness. Reflecting such changes, this fully revised edition updates the commentary by introducing and developing a number of issues which are seen as of growing importance to both the student and the practitioner of EIA.

The structure of the first edition has been retained, plus much of the material from the second edition, but variations and additions have been made to specific sections. In Part 1 (Principles and Procedures), the importance of an adaptive EIA is addressed further. In the EU context, the implementation of the amended EIA Directive is discussed more fully, including the divergent practice across the Member States, and the specific regulations and procedures operational in the UK. In Part 2 (discussion of the EIA process), many elements have been updated, including screening and scoping, alternatives, prediction, participation, mitigation and monitoring and auditing.

We have made major changes to Part 3 (overview of practice), drawing on the findings of important reviews of EIA effectiveness and operation in practice. Chapter 9 is completely new and focuses on case studies of EIA in practice. Most of the case studies are UK-based and involve EIA at the individual project level, although an example of SEA is also discussed and there are two non-UK studies. Whilst it is not claimed that the selected case studies all represent best examples of EIA practice, they do include some novel and innovative approaches towards particular issues in EIA, such as new methods of public participation and the treatment of cumulative effects. They also draw attention to some of the limitations of the process in practice. Chapter 10 (Comparative Practice) has also had a major revision, reflecting, for example, growing experience in African countries, China and countries in transition, and major reviews for some well-established EIA systems in, for example, Canada and Australia.

Part 4 of the book (Prospects) has also been substantially revised to reflect some of the changing prospects for EIA including, for example, more consideration of cumulative impacts, socio-economic impacts, public participation and possible shifts towards integrated assessment. Chapter 11 and other parts of the book draw on some of the findings of the 2003 Five Year Review of the operation of the amended EC EIA Directive, undertaken by the Impacts Assessment Unit (IAU) at Oxford Brookes University, for the European Commission. Chapter 12 is a largely new chapter, reflecting the evolution of strategic environmental assessment (SEA), and in particular the introduction of an EU-wide SEA Directive, operational from July 2004. The chapter includes a brief guide to carrying out SEA under the new Directive, as set out in recent UK government guidance. The appendices now include the full versions of both the amended EIA Directive and the SEA Directive, a revised IAU EIS-review package, and a guide to key EIA websites worldwide.

John Glasson
Riki Therivel
Andrew Chadwick
Oxford 2005

Acknowledgements

Our grateful thanks are due to many people without whose help this book would not have been produced. We are particularly grateful to Carol Glasson and Maureen Millard, who typed and retyped several drafts to tight deadlines and to high quality, and who provided invaluable assistance in bringing together the disparate contributions of the three authors. Our thanks also go to Rob Woodward for his production of many of the illustrations. In addition, Helen Ibbotson of Taylor and Francis, and Satishna Gokuldas of Integra, have provided vital contributions in turning the manuscript into the published document. We are very grateful to our consultancy clients and research sponsors, who have underpinned the work of the Impacts Assessment Unit in the School of Planning at Oxford Brookes University (formerly Oxford Polytechnic). In particular we wish to record the support of UK government departments (variously DoE, DETR and ODPM), the EC Environment Directorate, the Economic and Social Research Council (ESRC), the Royal Society for the Protection of Birds (RSPB), many local and regional authorities, and especially the various branches of the UK energy industry which provided the original impetus to and continuing positive support for much of our EIA research and consultancy.

Our students at Oxford Brookes University on both undergraduate and postgraduate programmes have critically tested many of our ideas. In this respect we would like to acknowledge, in particular, the students on the MSc course in Environmental Assessment and Management. The editorial and presentation support for the third edition by the staff at Taylor and Francis is very gratefully acknowledged. We have benefited from the support of colleagues in the Schools of Planning and Biological and Molecular Sciences, and from the wider community of EIA academics, researchers and consultants, who have helped to keep us on our toes.

We are also grateful for permission to use material from the following sources:

- British Association of Nature Conservationists (cartoons: Parts 2 and 3)
- Rendel Planning (Figure 4.3)
- UNEP Industry and Environment Office (Figure 4.5 and Table 4.2)
- University of Manchester, Department of Planning and Landscape, EIA Centre (Tables 5.8, 8.3, Figure 8.5, Appendix 3)
- John Wiley & Sons (Tables 6.1, 6.2)
- Baseline Environmental Consulting, West Berkeley, California (Figure 7.2)
- UK Department of Environment (Table 6.4)
- UK Department of Environment, Transport and the Regions (Tables 3.3, 3.4 and 3.5, Figures 4.1 and 5.11)
- *Planning* newspaper (cartoon: Part 4)

- Beech Tree Publishing (Figure 7.7)
- European Commission (Figure 4.7, Table 11.1)
- West Yorkshire County Council (Figure 4.17)
- West Australian Environmental Protection Agency (Table 10.2, Figure 10.5)
- Office of the Deputy Prime Minister (Box 12.2 and 12.3)

Abbreviations

ADB	Asian Development Bank
AEE	assessment of environmental effects
ANZECC	Australia and New Zealand Environment and Conservation Council
AONB	area of outstanding natural beauty
BATNEEC	best available technique not entailing excessive costs
BPEO	best practicable environmental option
CBA	cost–benefit analysis
CC	county council
CEA	cumulative effects assessment
CEAA	Canadian Environmental Assessment Agency
CEC	Commission of the European Communities
CEGB	Central Electricity Generating Board
CEPA	Commonwealth Environmental Protection Agency (Australia)
CEQ	Council on Environmental Quality (US)
CEQA	California Environmental Quality Act
CHP	combined heat and power
CIE	community impact evaluation
CPO	compulsory purchase order
CPRE	Campaign to Protect Rural England
CVM	contingent valuation method
DC	district council
DETR	Department of Environment, Transport and the Regions
DG	Directorate General (CEC)
DMRB	Design Manual for Roads and Bridges
DoE	Department of the Environment
DoEn	Department of Energy
DoT	Department of Transport
DfT	Department for Transport
DTI	Department of Trade and Industry
EA	environmental assessment
EEA	European Environment Agency
EBRD	European Bank for Reconstruction and Development
EC	European Community
EES	environmental evaluation system
EIA	environmental impact assessment
EIB	European Investment Bank
EIR	environmental impact report

EIS	environmental impact statement
EMAS	eco-management and audit scheme (CEC)
EMS	environmental management system
EN	English Nature
ENDS	Environmental Data Services
EPA	Environmental Protection Act
ERMP	Environmental Review and Management Programme
ES	environmental statement
ESI	electricity supply industry
ESRC	Economic and Social Research Council
EU	European Union
FEARO	Federal Environmental Assessment Review Office
FGD	flue gas desulphurization
FoE	Friends of the Earth
FONSI	finding of no significant impact
GAM	goals achievement matrix
GIS	geographical information system
GNP	gross national product
ha	hectares
HSE	Health and Safety Executive
HMIP	Her Majesty's Inspectorate of Pollution
HMSO	Her Majesty's Stationery Office
IAIA	International Association for Impact Assessment
IAU	Impacts Assessment Unit (Oxford Brookes)
IEA	Institute of Environmental Assessment
IEMA	Institute of Environmental Management and Assessment
IFI	International Funding Institution
IPC	integrated pollution control
ISO	International Standards Office
JNCC	Joint Nature Conservancy Council
km	kilometre
LCA	life cycle assessment
LCP	large combustion plant
LI	Landscape Institute
LPA	local planning authority
LULU	locally unacceptable land use
MAFF	Ministry of Agriculture, Fisheries and Food
MAUT	multi-attribute utility theory
MEA	Manual of environmental appraisal
MW	megawatts
NEPA	National Environmental Policy Act (US)
NEPP	National Environmental Policy Plan (Netherlands)
NGC	National Grid Company
NGO	non-government organization
NIMBY	not in my back yard
NRA	National Rivers Authority
ODPM	Office of Deputy Prime Minister
OECD	Organisation for Economic Co-operation and Development
PADC	project appraisal for development control

PBS	planning balance sheet
PER	Public Environmental Review
PPG	Planning Policy Guidance
PPS	Planning Policy Statement
PPPs	policies, plans and programmes
PWR	pressurized water reactor
QOLA	quality of life assessment
RA	risk assessment
RMA	Resource Management Act
RSPB	Royal Society for the Protection of Birds
RTPI	Royal Town Planning Institute
SACTRA	Standing Advisory Committee on Trunk Road Assessment
SDD	Scottish Development Department
SEA	strategic environmental assessment
SEDD	Scottish Executive Development Department
SEERA	South East England Regional Assembly
SIA	social impact assessment
SoS	Secretary of State
SPA	Special Protection Area
SSSI	site of special scientific interest
T&CP	town and country planning
UK	United Kingdom
UNCED	United Nations Conference on Environment and Development
UNECE	United Nations Economic Commission for Europe
UNEP	United Nations Environment Programme
US	United States
WRAM	Water Resources Assessment Method

Part 1

Principles and procedures

1 Introduction and principles

1.1 Introduction

Over the last three decades there has been a remarkable growth of interest in environmental issues – in sustainability and the better management of development in harmony with the environment. Associated with this growth of interest has been the introduction of new legislation, emanating from national and international sources, such as the European Commission, that seeks to influence the relationship between development and the environment. Environmental impact assessment (EIA) is an important example. EIA legislation was introduced in the USA over 35 years ago. A European Community (EC) directive in 1985 accelerated its application in EU Member States and, since its introduction in the UK in 1988, it has been a major growth area for planning practice. The originally anticipated 20 environmental impact statements (EIS) per year in the UK have escalated to over 600, and this is only the tip of the iceberg. The scope of EIA continues to widen and grow.

It is therefore perhaps surprising that the introduction of EIA met with strong resistance from many quarters, particularly in the UK. Planners argued, with partial justification, that they were already making such assessments. Many developers saw it as yet another costly and time-consuming constraint on development, and central government was also unenthusiastic. Interestingly, initial UK legislation referred to environmental assessment (EA), leaving out the apparently politically sensitive, negative-sounding reference to impacts. The scope of the subject continues to evolve. This chapter therefore introduces EIA as a process, the purposes of this process, types of development, environment and impacts and current issues in EIA.

1.2 The nature of environmental impact assessment

1.2.1 Definitions

Definitions of EIA abound. They range from the oft-quoted and broad definition of Munn (1979), which refers to the need "to identify and predict the impact on the environment and on man's health and well-being of legislative proposals, policies, programmes, projects and operational procedures, and to interpret and communicate information about the impacts", to the narrow UK DoE (1989) operational definition: "The term 'environmental assessment' describes a technique and a process by which information about the environmental effects of a project is collected, both by the developer and from other sources, and taken into account by the planning authority in forming their judgements on whether the development should go ahead." The UNECE (1991) has an altogether more succinct and pithy definition: "an assessment of the impact of a planned activity on the environment".

1.2.2 Environmental impact assessment: a process

In essence, EIA is *a process*, a systematic process that examines the environmental consequences of development actions, in advance. The emphasis, compared with many other mechanisms for environmental protection, is on prevention. Of course, planners have traditionally assessed the impacts of developments on the environ-ment, but invariably not in the systematic, holistic and multidisciplinary way required by EIA. The process involves a number of steps, as outlined in Figure 1.1.

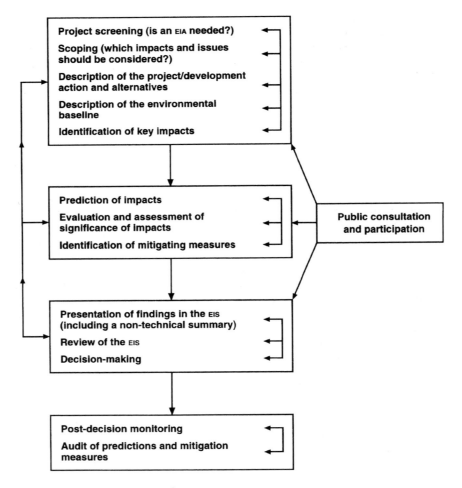

Figure 1.1 Important steps in the EIA process.
Note: EIA should be a cyclical process with considerable interaction between the various steps. For example, public participation can be useful at most stages of the process; monitoring systems should relate to parameters established in the initial project and baseline descriptions.

These are briefly described below, pending a much fuller discussion in Chapters 4–7. It should be noted at this stage that, although the steps are outlined in a linear fashion, EIA should be a cyclical activity, with feedback and interaction between the various steps. It should also be noted that practice can and does vary considerably from the process illustrated in Figure 1.1. For example, until recently UK EIA legislation did not require some of the steps, including the consideration of alternatives, and still does not require post-decision monitoring (DETR 2000). The order of the steps in the process may also vary.

- *Project screening* narrows the application of EIA to those projects that may have significant environmental impacts. Screening may be partly determined by the EIA regulations operating in a country at the time of assessment.
- *Scoping* seeks to identify at an early stage, from all of a project's possible impacts and from all the alternatives that could be addressed, those that are the crucial, significant issues.
- *The consideration of alternatives* seeks to ensure that the proponent has considered other feasible approaches, including alternative project locations, scales, processes, layouts, operating conditions and the "no action" option.
- *The description of the project/development action* includes a clarification of the purpose and rationale of the project, and an understanding of its various characteristics – including stages of development, location and processes.
- *The description of the environmental baseline* includes the establishment of both the present and future state of the environment, in the absence of the project, taking into account changes resulting from natural events and from other human activities.
- *The identification of the main impacts* brings together the previous steps with the aim of ensuring that all potentially significant environmental impacts (adverse and beneficial) are identified and taken into account in the process.
- *The prediction of impacts* aims to identify the magnitude and other dimensions of identified change in the environment with a project/action, by comparison with the situation without that project/action.
- *The evaluation and assessment of significance* assesses the relative significance of the predicted impacts to allow a focus on the main adverse impacts.
- *Mitigation* involves the introduction of measures to avoid, reduce, remedy or compensate for any significant adverse impacts.
- *Public consultation and participation* aim to ensure the quality, comprehensiveness and effectiveness of the EIA, and that the public's views are adequately taken into consideration in the decision-making process.
- *EIS presentation* is a vital step in the process. If done badly, much good work in the EIA may be negated.
- *Review* involves a systematic appraisal of the quality of the EIS, as a contribution to the decision-making process.
- *Decision-making* on the project involves a consideration by the relevant authority of the EIS (including consultation responses) together with other material considerations.
- *Post-decision monitoring* involves the recording of outcomes associated with development impacts, after a decision to proceed. It can contribute to effective project management.

- *Auditing* follows from monitoring. It can involve comparing actual outcomes with predicted outcomes, and can be used to assess the quality of predictions and the effectiveness of mitigation. It provides a vital step in the EIA learning process.

1.2.3 Environmental impact statements: the documentation

The EIS documents the information and estimates of impacts derived from the various steps in the process. Prevention is better than cure; an EIS revealing many significant unavoidable adverse impacts would provide valuable information that could contribute to the abandonment or substantial modification of a proposed development action. Where adverse impacts can be successfully reduced through mitigation measures, there may be a different decision. Table 1.1 provides an example of the content of an EIS for a project.

The *non-technical summary* is an important element in the documentation; EIA can be complex, and the summary can help to improve communication with the various parties involved. Reflecting the potential complexity of the process, a *methods statement*, at the beginning, provides an opportunity to clarify some basic information (e.g. who the developer is, who has produced the EIS, who has been consulted and how, what methods have been used, what difficulties have been encountered and what the limitations of the EIA are). A *summary statement of key issues*, upfront, can also help to improve communications. A more enlightened EIS would also include a *monitoring programme*, either here or at the end of the document. The *background to the proposed development* covers the early steps in the EIA process, including clear descriptions of a project, and baseline conditions (including relevant planning policies and plans). Within each of the *topic areas* of an EIS there would normally be a discussion of existing conditions, predicted impacts, scope for mitigation and residual impacts.

Table 1.1 An EIS for a project – example or contents

Non-technical summary

Part 1: Methods and key issues
  1. Methods statement
  2. Summary of key issues; monitoring programme statement

Part 2: Background to the proposed development
  3. Preliminary studies: need, planning, alternatives and site selection
  4. Site description, baseline conditions
  5. Description of proposed development
  6. Construction activities and programme

Part 3: Environmental impact assessment – topic areas
  7. Land use, landscape and visual quality
  8. Geology, topography and soils
  9. Hydrology and water quality
  10. Air quality and climate
  11. Ecology: terrestrial and aquatic
  12. Noise
  13. Transport
  14. Socio-economic impact
  15. Interrelationships between effects

Environmental impact assessment and EIS practices vary from study to study, from country to country, and best practice is constantly evolving. An early UN study of EIA practice in several countries advocated changes in the process and documentation (UNECE 1991). These included giving a greater emphasis to the socio-economic dimension, to public participation, and to "after the decision" activity, such as monitoring. A recent review of the operation of the amended EC Directive (CEC 2003) raised similar, and other emerging, issues a decade later (see Chapter 2). Sadler (1996) provided a wider agenda for change based on a major international study of the effectiveness of EIA (see Chapter 11).

1.2.4 Other relevant definitions

Development actions may have impacts not only on the physical environment but also on the social and economic environment. Typically, employment opportunities, services (e.g. health, education) and community structures, lifestyles and values may be affected. *Socio-economic impact assessment* or *social impact assessment* (SIA) is regarded here as an integral part of EIA. However, in some countries it is (or has been) regarded as a separate process, sometimes parallel to EIA, and the reader should be aware of its existence (Carley & Bustelo 1984, Finsterbusch 1985, IAIA 1994, Vanclay 2003).

Strategic environmental assessment (SEA) expands EIA from projects to policies, plans and programmes (PPPs). Development actions may be for a project (e.g. a nuclear power station), for a programme (e.g. a number of pressurized water reactor (PWR) nuclear power stations), for a plan (e.g. in the town and country planning (T&CP) system in England and Wales, for local plans and structure plans) or for a policy (e.g. the development of renewable energy). EIA to date has generally been used for individual projects, and that role is the primary focus of this book. But EIA for programmes, plans and policies, otherwise known as SEA, is currently being introduced in the European Union (EU) and beyond (Therivel 2004, Therivel & Partidario 1996, Therivel et al. 1992). SEA informs a higher, earlier, more strategic tier of decision-making. In theory, EIA should be carried out first for policies, then for plans, programmes, and finally for projects.

Risk assessment (RA) is another term sometimes found associated with EIA. Partly in response to events such as the chemicals factory explosion at Flixborough (UK), and nuclear power station accidents at Three Mile Island (USA) and Chernobyl (Ukraine), RA has developed as an approach to the analysis of risks associated with various types of development. The major study of the array of petrochemicals and other industrial developments at Canvey Island in the UK provides an example of this approach (Health and Safety Commission 1978). Calow (1997) gives an overview of the growing area of environmental RA and management and Flyberg (2003) a critique of risk assessment in practice.

Vanclay & Bronstein (1995) and others note several other relevant definitions, based largely on particular foci of specialization and including demographic impact assessment, health impact assessment, climate impact assessment, gender impact assessment, psychological impact assessment and noise impact assessment. Other more encompassing definitions include policy assessment, technology assessment and economic assessment. There is a semantic explosion which requires some clarification. As a contribution to the latter, Sadler (1996) suggests that we should view "EA as the generic process that includes EIA of specific projects, SEA of PPPs, and their relationships to a larger set of impact assessment and planning-related tools".

1.3 The purposes of environmental impact assessment

1.3.1 An aid to decision-making

Environmental impact assessment is a process with several important purposes. It is an aid to decision-making. For the decision-maker, for example a local authority, it provides a systematic examination of the environmental implications of a proposed action, and sometimes alternatives, before a decision is taken. The EIS can be considered by the decision-maker along with other documentation related to the planned activity. EIA is normally wider in scope and less quantitative than other techniques, such as cost–benefit analysis (CBA). It is not a substitute for decision-making, but it does help to clarify some of the trade-offs associated with a proposed development action, which should lead to more rational and structured decision-making. The EIA process has the potential, not always taken up, to be a basis for negotiation between the developer, public interest groups and the planning regulator. This can lead to an outcome that balances the interests of the development action and the environment.

1.3.2 An aid to the formulation of development actions

Many developers no doubt see EIA as another set of hurdles to jump before they can proceed with their various activities; the process can be seen as yet another costly and time-consuming activity in the permission process. However, EIA can be of great benefit to them, since it can provide a framework for considering location and design issues and environmental issues in parallel. It can be an aid to the formulation of development actions, indicating areas where a project can be modified to minimize or eliminate altogether its adverse impacts on the environment. The consideration of environmental impacts early in the planning life of a development can lead to environmentally sensitive development; to improved relations between the developer, the planning authority and the local communities; to a smoother planning permission process; and sometimes, as argued by developers such as British Gas, to a worthwhile financial return on the extra expenditure incurred (Breakell & Glasson 1981). O'Riordan (1990) links such concepts of negotiation and redesign to the important environmental themes of "green consumerism" and "green capitalism". The emergence of a growing demand by consumers for goods that do no environmental damage, plus a growing market for clean technologies, is generating a response from developers. EIA can be the signal to the developer of potential conflict; wise developers may use the process to negotiate "green gain" solutions, which may eliminate or offset negative environmental impacts, reduce local opposition and avoid costly public inquiries.

1.3.3 An instrument for sustainable development

Underlying such immediate purposes is of course the central and ultimate role of EIA as one of the instruments to achieve sustainable development: development that does not cost the Earth! Existing environmentally harmful developments have to be managed as best as they can. In extreme cases, they may be closed down, but they can still leave residual environmental problems for decades to come. How much better it would be to mitigate the harmful effects in advance, at the planning stage, or in some cases avoid the particular development altogether. Prevention is better than cure.

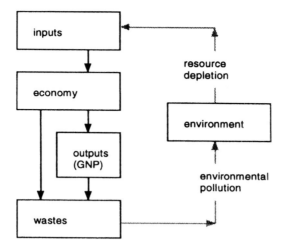

Figure 1.2 The economic development process in its environmental context. (Adapted from Boulding 1966.)

Economic development and social development must be placed in their environmental contexts. Boulding (1966) vividly portrays the dichotomy between the "throughput economy" and the "spaceship economy" (Figure 1.2). The economic goal of increased *gross national product* (GNP), using more inputs to produce more goods and services, contains the seeds of its own destruction. Increased output brings with it not only goods and services but also more waste products. Increased inputs demand more resources. The natural environment is the "sink" for the wastes and the "source" for the resources. Environmental pollution and the depletion of resources are invariably the ancillaries to economic development.

The interaction of economic and social development with the natural environment and the reciprocal impacts between human actions and the biophysical world have been recognized by governments from local to international levels. Attempts have been made to manage the interaction better, but the EC report, *Towards Sustainability* (CEC 1992), revealed disquieting trends that could have devastating consequences for the quality of the environment. Such EU trends included a 25 per cent increase in energy consumption by 2010 if there was no change in current energy demand growth rates; a 25 per cent increase in car ownership and a 17 per cent increase in miles driven by 2000; a 13 per cent increase in municipal waste between 1987 and 1992, despite increased recycling; a 35 per cent increase in the EU's average rate of water withdrawal between 1970 and 1985; and a 60 per cent projected increase in Mediterranean tourism between 1990 and 2000. Such trends are likely to be even more pronounced in developing countries, where, because population growth is greater and current living standards lower, there will be more pressure on environmental resources. The revelation of the state of the environment in many central and eastern European countries, and worldwide, added weight to the urgency of the situation.[1]

The 1987 Report of the UN World Commission on Environment and Development (usually referred to as the Brundtland Report, after its chairwoman) defined sustainable development as "development which meets the needs of the present generation without compromising the ability of future generations to meet their own needs"

(UN World Commission on Environment and Development 1987). Sustainable development means handing down to future generations not only "man-made capital", such as roads, schools and historic buildings, and "human capital", such as knowledge and skills, but also "natural/environmental capital", such as clean air, fresh water, rain forests, the ozone layer and biological diversity. The Brundtland Report identified the following chief characteristics of sustainable development: it maintains the quality of life, it maintains continuing access to natural resources and it avoids lasting environmental damage. It means living on the Earth's income rather than eroding its capital (DoE et al. 1990). In addition to a concern for the environment and the future, Brundtland also emphasizes participation and equity, thus highlighting both inter- and intra-generational equity.

There is, however, a danger that "sustainable development" may become a weak catch-all phrase; there are already many alternative definitions. Holmberg & Sandbrook (1992) found over 70 definitions of sustainable development. Redclift (1987) saw it as "moral convictions as a substitute for thought"; to O'Riordan (1988) it was "a good idea which cannot sensibly be put into practice". But to Skolimowski (1995), sustainable development

> struck a middle ground between more radical approaches which denounced all development, and the idea of development conceived as business as usual. The idea of sustainable development, although broad, loose and tinged with ambiguity around its edges, turned out to be palatable to everybody. This may have been its greatest virtue. It is radical and yet not offensive.

Readers are referred to Reid (1995) and Kirkby et al. (1995) for an overview of the concept, debate and responses.

Turner & Pearce (1992) and Pearce (1992) have drawn attention to alternative interpretations of maintaining the capital stock. A policy of conserving the whole capital stock (man-made, human and natural) is consistent with running down any part of it as long as there is substitutability between capital degradation in one area and investment in another. This can be interpreted as a "weak sustainability" position. In contrast, a "strong sustainability" position would argue that it is not acceptable to run down environmental assets, for several reasons: uncertainty (we do not know the full consequences for human beings), irreversibility (lost species cannot be replaced), life support (some ecological assets serve life-support functions) and loss aversion (people are highly averse to environmental losses). The "strong sustainability" position has much to commend it, but institutional responses have varied.

Institutional responses to meet the goal of sustainable development are required at several levels. Issues of global concern, such as ozone-layer depletion, climate change, deforestation and biodiversity loss, require global political commitments to action. The United Nations Conference on Environment and Development (UNCED) held in Rio de Janeiro in 1992 was an example not only of international concern, but also of the problems of securing concerted action to deal with such issues. Agenda 21, an 800-page action plan for the international community into the twenty-first century, sets out what nations should do to achieve sustainable development. It includes topics such as biodiversity, desertification, deforestation, toxic wastes, sewage, oceans and the atmosphere. For each of 115 programmes, the need for action, the objectives and targets to be achieved, the activities to be undertaken, and the means of implementation are all outlined. Agenda 21 offers policies and programmes to achieve a sustainable balance between consumption, population and the Earth's life-supporting capacity. Unfortunately it is not legally binding. It relies on national governments, local governments and others to implement most of the programmes. The Rio Conference called for a Sustainable Development Commission to be established to progress the implementation of Agenda 21. The Commission met for the first time in 1993 and reached agreement on a thematic programme of work for 1993–97. This provided the basis for an appraisal of Agenda 21 in preparation for a special session of the UN in 1997. The Johannesburg Earth Summit of 2002 re-emphasized the difficulties of achieving international commitment on environmental issues. Whilst there were some positive outcomes – for example, on water and sanitation (with a target to halve the number without basic sanitation – about 1.2 billion – by 2015), on poverty, health, sustainable consumption and on trade and globalization – many other outcomes were much less positive. Delivering the Kyoto protocol on legally enforceable reductions of greenhouse gases continues to be difficult, as does progress on safeguarding biodiversity and natural resources, and on delivering human rights in many countries. Such problems severely hamper progress on sustainable development.

Within the EU, four Community Action Programmes on the Environment were implemented between 1972 and 1992. These gave rise to specific legislation on a wide range of topics, including waste management, the pollution of the atmosphere, the protection of nature and EIA. The Fifth Programme, "Towards sustainability" (1993–2000), was set in the context of the completion of the Single European Market. The latter, with its emphasis on major changes in economic development resulting from the removal of all remaining fiscal, material and technological barriers between Member States, could pose additional threats to the environment. The Fifth Programme recognized the need for the clear integration of performance targets – in relation to environmental protection – for several sectors, including manufacturing, energy, transport and tourism. EU policy on the environment will be based on the "precautionary principle" that preventive action should be taken, that environmental damage should be rectified at source and that the polluter should pay. Whereas previous EU programmes relied almost exclusively on legislative instruments, the Fifth Programme advocates a broader mixture, including "market-based instruments", such as the internalization of environmental costs through the application of fiscal measures, and "horizontal, supporting instruments", such as improved baseline and statistical data and improved spatial and sectoral planning. Figure 1.3 illustrates the interdependence of resources, sectors and policy areas. EIA has a clear role to play.

The Sixth Programme, "Our future, our choice" (2001–10), builds on the broader approach introduced in the previous decade. It recognizes that sustainable development has social and economic as well as physical environmental dimensions, although the focus is on four main priority issues: tackling climate change, protecting nature and biodiversity, reducing human health impacts from environmental pollution and ensuring the sustainable management of natural resources and waste. It also recognizes the importance of empowering citizens and changing behaviour, and of "greening land-use planning and management decisions". "The Community directive on EIA and proposal on SEA, which aim to ensure that the environmental implications of planned infrastructure projects and planning are properly addressed, will also help ensure that the environmental considerations are better integrated into planning decisions" (CEC 2001).

In the UK, the publication of *This common inheritance: Britain's environmental strategy* (DoE et al. 1990) provided the country's first comprehensive White Paper on the environment. The report includes a discussion of the greenhouse effect, town and

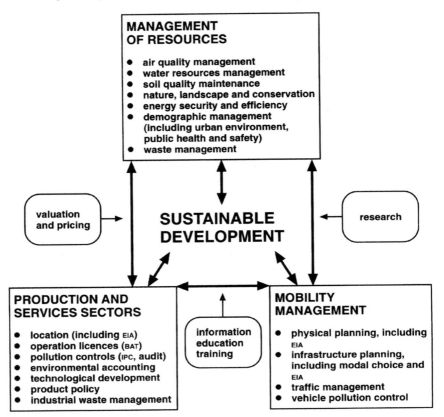

**MANAGEMENT
OF RESOURCES**

- air quality management
- water resources management
- soil quality maintenance
- nature, landscape and conservation
- energy security and efficiency
- demographic management
  (including urban environment,
  public health and safety)
- waste management

valuation
and pricing

**SUSTAINABLE
DEVELOPMENT**

research

**PRODUCTION AND
SERVICES SECTORS**

- location (including EIA)
- operation licences (BAT)
- pollution controls (IPC, audit)
- environmental accounting
- technological development
- product policy
- industrial waste management

information
education
training

**MOBILITY
MANAGEMENT**

- physical planning, including
  EIA
- infrastructure planning,
  including modal choice and
  EIA
- traffic management
- vehicle pollution control

Figure 1.3 An EC framework for sustainable development. (*Source:* CEC 1992.)

country, pollution control, and awareness and organization with regard to environmental issues. Throughout it emphasizes that responsibility for our environment should
be shared between the government, business and the public. The range of policy
instruments advocated includes legislation, standards, planning and economic measures. The last, building on work by Pearce et al. (1989), includes charges, subsidies,
market creation and enforcement incentives. The report also notes, cautiously, the
recent addition of EIA to the "toolbox" of instruments. Subsequent UK government
reports, such as *Sustainable development: the UK strategy* (HMG 1994), recognize the role
of EIA in contributing to sustainable development and raise the EIA profile among
key user groups. The UK government reports also reflect the extension of the scope of
sustainable development to include social, economic and environmental factors. This
is reflected in the UK Strategy for Sustainable Development, *A Better Quality of Life*
(DETR 1999a), with its four objectives of:

1.   social progress which recognizes the needs of everyone;
2.   effective protection of the environment;
3.   prudent use of natural resources; and
4.   maintenance of high and stable levels of economic growth and employment.

To measure progress, the UK government has published a set of sustainable development indicators, including a set of 15 key headline indicators (DETR 1999b). It also required a high-level sustainable development framework to be produced for each English region (see, for example, A *Better Quality of Life in the South East* (SEERA 2001)).

1.4 Changing perspectives on EIA

The arguments for EIA vary in time, in space and according to *the perspective of those involved*. From a minimalist defensive perspective, developers, and possibly also some parts of government, might see EIA as a necessary evil, an administrative exercise, something to be gone through that might result in some minor, often cosmetic, changes to a development that would probably have happened anyway. For the "deep ecologists" or "deep Greens", EIA cannot provide total certainty about the environmental consequences of development proposals; they feel that any projects carried out under uncertain or risky circumstances should be abandoned. EIA and its methods must straddle such perspectives, partly reflecting the previous discussion on weak and strong sustainability. EIA can be, and is now often, seen as a positive process that seeks a harmonious relationship between development and the environment. The nature and use of EIA will change as relative values and perspectives also change. EIA must adapt, as O'Riordan (1990) noted:

> One can see that EIA is moving away from being a defensive tool of the kind that dominated the 1970s to a potentially exciting environmental and social betterment technique that may well come to take over the 1990s...If one sees EIA not so much as a technique, rather as a process that is constantly changing in the face of shifting environmental politics and managerial capabilities, one can visualize it as a sensitive barometer of environmental values in a complex environmental society. Long may EIA thrive.

EIA must also be re-assessed in its *theoretical context*, and in particular in the context of decision-making theory (see Lawrence 1997, Weston 2000). EIA had its origins in a climate of a rational approach to decision-making in the USA in the 1960s. The focus was on the systematic process, objectivity, a holistic approach, a consideration of alternatives and an approach often seen as primarily linear. This rational approach is assumed to rely on a scientific process in which facts and logic are pre-eminent. In the UK this rational approach was reflected in planning in the writings of, *inter alia*, Faludi (1973), McLoughlin (1969), and Friend & Jessop (1977).

However, other writings on the theoretical context of EIA have recognized the importance of the subjective nature of the EIA process. Kennedy (1988) identified EIA as both a "science" and an "art", combining political input and scientific process. More colourfully, Beattie (1995), in an article entitled "Everything you already know about EIA, but don't often admit", reinforces the point that EIAs are not science; they are often produced under tight deadlines and data gaps and simplifying assumptions are the norm under such conditions. They always contain unexamined and unexplained value judgements, and they will always be political. They invariably deal with controversial projects, and they have distributional effects. EIA professionals should therefore not be surprised, or dismayed, when their work is selectively used by various parties in the process.

In the context of decision-making theory, this recognition of the political, the subjective, is reflected most fully in a variety of behavioural/participative theories. Braybrooke & Lindblom (1963), for example, saw decisions as incremental adjustments, with a process that is not comprehensive, linear and orderly, and is best characterized as "muddling through". Lindblom (1980) further developed his ideas through the concept of "disjointed incrementalism", with a focus on meeting the needs and objectives of society, often politically defined. The importance of identifying and confronting trade-offs, a major issue in EIA, is clearly recognized. The participatory approach includes processes for open communication among all affected parties.

The recognition of multiple parties and the perceived gap between government and citizens have stimulated other theoretical approaches, including communicative and collaborative planning (Healey 1996, 1997). This approach draws upon the work of Habermas (1984), Forester (1989) and others. Much attention is devoted to consensus-building, co-ordination and communication, and the role of government in promoting such actions as a means of dealing with conflicting stakeholder interests to come to collaborative action.

It is probably now realistic to place the current evolution of EIA somewhere between the rational and behavioural approaches – reflecting elements of both. It does include strands of rationalism, but there are many participants, and many decision points – and politics and professional judgement are often to the fore. This tends to fit well with the classic concept of "mixed scanning" advocated by Etzioni (1967), utilizing rational techniques of assessment, in combination with more intuitive value judgements, based upon experience and values. The rational-adaptive approach of Kaiser et al. (1995) also stresses the importance of a series of steps in decision-making, with both (scientific-based) rationality and (community-informed) participation, moderating the selection of policy options and desired outcomes.

Environmental impact assessment must also be seen in the context of *other environmental management decision tools*. Petts (1999) provides a good overview of the recent evolution. These tools are additional to the family of assessment approaches discussed in Section 1.2, and include, for example, life cycle assessment (LCA), CBA, and environmental auditing. LCA differs from EIA in its focus not on a particular site or facility, but on a product or system and the cradle-to-grave environmental effects of that product or system (see White et al. 1995). In contrast, CBA focuses on economic impacts of a development, but taking a wide and long view of those impacts. It involves as far as possible the monetization of all the costs and benefits of a proposal. It came to the fore in the UK in relation to major transport projects in the 1960s, but is enjoying a new lease of life (see Hanley & Splash 1993, Lichfield 1996). Environmental auditing is the systematic, periodic and documented evaluation of the environmental performance of facility operations and practices, and this area has seen the development of procedures, such as the International Standard 14001 (ISO 14001). But in general, these other tools have been much less internalized into decision-making procedures and legislation than EIA, and now SEA. They also tend to be more technocentric, and with less attention paid to process and the wider stakeholder environment. However, they can be seen as complementary tools to EIA. Thus Chapter 5 explores the potential role of CBA approaches in EIA evaluation, and in Chapter 11 the role of environmental auditing is explored further, in relation to environmental management systems (EMSs).

This brief discussion on perspectives, theoretical context, associated tools and processes emphasizes the need to continually re-assess the role and operation of EIA and the importance of an adaptive EIA.

1.5 Projects, environment and impacts

1.5.1 *The nature of major projects*

As noted in Section 1.2, EIA is relevant to a broad spectrum of development actions, including policies, plans, programmes and projects. The focus here is on projects, reflecting the dominant role of project EIA in practice. The SEA of the "upper tiers" of development actions is considered further in Chapter 12. The scope of projects covered by EIA is widening, and is discussed further in Chapter 4. Traditionally, project EIA has applied to major projects; but what are major projects, and what criteria can be used to identify them? One could take Lord Morley's approach to defining an elephant: it is difficult, but you easily recognize one when you see it. In a similar vein, the acronym LULU (locally unacceptable land uses) has been applied in the USA to many major projects, such as in energy, transport and manufacturing, clearly reflecting the public perception of the negative impacts associated with such developments. There is no easy definition, but it is possible to highlight some important characteristics (Table 1.2).

Most large projects involve considerable investment. In the UK context, "megaprojects" such as the Sizewell B PWR nuclear power station (budgeted to cost about £2 billion), the Channel Tunnel (about £6 billion) and the proposed Severn Barrage (about £8 billion) constitute one end of the spectrum. At the other end may be industrial estate developments, small stretches of road, various waste-disposal facilities, with considerably smaller, but still substantial, price tags. Such projects often cover large areas and employ many workers, usually in construction, but also in operation for some projects. They also invariably generate a complex array of inter- and intra-organizational activity during the various stages of their lives. The developments may have wide-ranging, long-term and often very significant impacts on the environment. The definition of significance with regard to environmental effects is an important issue in EIA. It may relate, *inter alia*, to scale of development, to sensitivity of location and to the nature of adverse effects; it will be discussed further in later chapters. Like a large stone thrown into a pond, a major project can create major ripples with impacts spreading far and wide. In many respects such projects tend to be regarded as exceptional, requiring special procedures. In the UK, these procedures have included public inquiries, hybrid bills that have to be passed through parliament (for example, for the Channel Tunnel) and EIA procedures.

Major projects can also be defined according to type of activity. They include manufacturing and extractive projects, such as petrochemicals plants, steelworks, mines and quarries; services projects, such as leisure developments, out-of-town shopping centres, new settlements and education and health facilities; and utilities and infrastructure,

Table 1.2 Characteristics of major projects

- Substantial capital investment
- Cover large areas; employ large numbers (construction and/or operation)
- Complex array of organizational links
- Wide-ranging impacts (geographical and by type)
- Significant environmental impacts
- Require special procedures
- Extractive and primary (including agriculture); services; infrastructure and utilities
- Band, point

such as power stations, roads, reservoirs, pipelines and barrages. An EC study adopted a further distinction between band and point infrastructures. Point infrastructure would include, for example, power stations, bridges and harbours; band or linear infrastructure would include electricity transmission lines, roads and canals (CEC 1982).

A major project also has a planning and development life cycle, including a variety of stages. It is important to recognize such stages because impacts can vary considerably between them. The main stages in a project's life cycle are outlined in Figure 1.4. There may be variations in timing between stages, and internal variations within each stage, but there is a broadly common sequence of events. In EIA, an important distinction is between "before the decision" (stages A and B) and "after the decision" (stages C, D and E). As noted in Section 1.2, the monitoring and auditing of the implementation of a project following approval are often absent from the EIA process.

Projects are initiated in several ways. Many are responses to market opportunities (e.g. a holiday village, a subregional shopping centre, a gas-fired power station); others

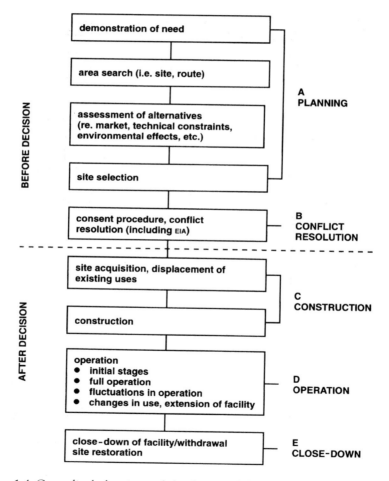

Figure 1.4 Generalized planning and development life cycle for major projects (with particular reference to impact assessment on host area). (Adapted from Breese et al. 1965.)

may be seen as necessities (e.g. the Thames Barrier); others may have an explicit prestige role (e.g. the programme of Grands Travaux in Paris including the Bastille Opera, Musée d'Orsay and Great Arch). Many major projects are public-sector initiatives, but with the move towards privatization in many countries, there has been a move towards private sector funding, exemplified by such projects as the North Midlands Toll Road and the Channel Tunnel. The initial planning stage A may take several years, and lead to a specific proposal for a particular site. It is at stage B that the various control and regulatory procedures, including EIA, normally come into play. The construction stage can be particularly disruptive, and may last up to 10 years for some projects. Major projects invariably have long operational lives, although extractive projects can be short compared with infrastructure projects. The environmental impact of the eventual close-down/decommissioning of a facility should not be forgotten; for nuclear power facilities it is a major undertaking. Figure 1.5 shows how the stages in the life cycles of different kinds of project may vary.

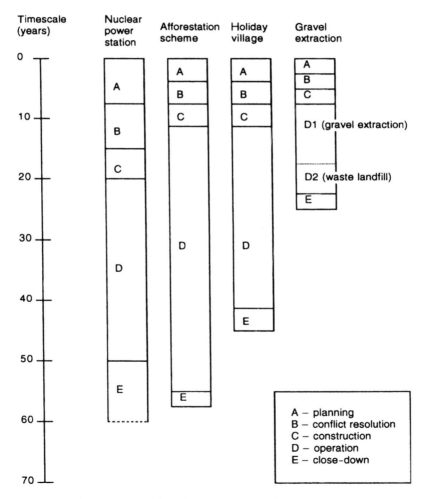

Figure 1.5 Broad variations in life cycle stages between different types of project.

1.5.2 Dimensions of the environment

The environment can be structured in several ways, including components, scale/space and time. A narrow definition of environmental components would focus primarily on the biophysical environment. For example, the UK Department of the Environment (DoE) used the term to include all media susceptible to pollution, including air, water and soil; flora, fauna and human beings; landscape, urban and rural conservation and the built heritage (DoE 1991). The DoE checklist of environmental components is outlined in Table 1.3. However, as already noted in Section 1.2, the environment has important economic and socio-cultural dimensions. These include economic structure, labour markets, demography, housing, services (education, health, police, fire, etc.), lifestyles and values, and these are added to the checklist in Table 1.3. This wider definition is more in tune with an Australian definition, "For the purposes of EIA, the meaning of environment incorporates physical, biological, cultural, economic and social factors" (ANZECC 1991).

The environment can also be analysed at various scales (Figure 1.6). Many of the spatial impacts of projects affect the local environment, although the nature of "local" may vary according to the aspect of environment under consideration and to the stage in a project's life. However, some impacts are more than local. Traffic noise, for example, may be a local issue, but changes in traffic flows caused by a project may have a regional impact, and the associated CO_2 pollution contributes to the global greenhouse problem. The environment also has a time dimension. Baseline data on the state of the environment are needed at the time a project is being considered. This in itself may be a daunting request. In the UK, local development plans and national statistical sources, such as the Digest of Environmental Protection and

Table 1.3 Environmental components

Physical environment (adapted from DoE 1991)	
Air and atmosphere	Air quality
Water resources and water bodies	Water quality and quantity
Soil and geology	Classification, risks (e.g. erosion, contamination)
Flora and fauna	Birds, mammals, fish, etc.; aquatic and terrestrial vegetation
Human beings	Physical and mental health and well-being
Landscape	Characteristics and quality of landscape
Cultural heritage	Conservation areas; built heritage; historic and archaeological sites
Climate	Temperature, rainfall, wind, etc.
Energy	Light, noise, vibration, etc.
Socio-economic environment	
Economic base – direct	Direct employment; labour market characteristics; local and non-local trends
Economic base – indirect	Non-basic and services employment; labour supply and demand
Demography	Population structure and trends
Housing	Supply and demand
Local services	Supply and demand of services: health, education, police, etc.
Socio-cultural	Lifestyles, quality of life; social problems (e.g. crime); community stress and conflict

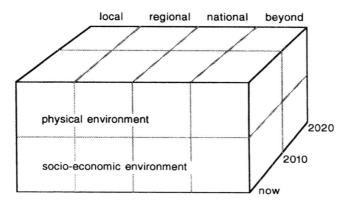

Figure 1.6 Environment: components, scale and time dimensions.

Water Standards, may provide some relevant data. However, tailor-made state-of-the-environment reports and audits are still in limited supply (see Chapter 11 for further information). Even more limited are time-series data highlighting trends in environmental quality. The environmental baseline is constantly changing, irrespective of any development under consideration, and it requires a dynamic rather than a static analysis.

1.5.3 The nature of impacts

The environmental impacts of a project are those resultant changes in environmental parameters, in space and time, compared with what would have happened had the project not been undertaken. The parameters may be any of the type of environmental receptors noted previously: air quality, water quality, noise, levels of local unemployment and crime, for example. Figure 1.7 provides a simple illustration of the concept.

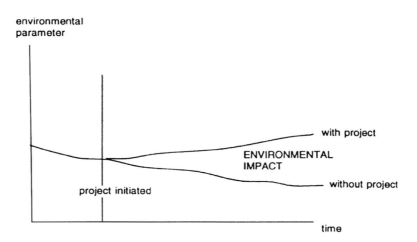

Figure 1.7 The nature of an environmental impact.

Table 1.4 Types of impact

- Physical and socio-economic
- Direct and indirect
- Short-run and long-run
- Local and strategic (including regional, national and beyond)
- Adverse and beneficial
- Reversible and irreversible
- Quantitative and qualitative
- Distribution by group and/or area
- Actual and perceived
- Relative to other developments

Table 1.4 provides a summary of some of the types of impact that may be encountered in EIA. The biophysical and socio-economic impacts have already been noted. These are often seen as synonymous with adverse and beneficial. Thus, new developments may produce harmful wastes but also produce much needed jobs in areas of high unemployment. However, the correlation does not always apply. A project may bring physical benefits when, for example, previously polluted and derelict land is brought back into productive use; similarly the socio-economic impacts of a major project on a community could include pressure on local health services and on the local housing market, and increases in community conflict and crime. Projects may also have immediate and direct impacts that give rise to secondary and indirect impacts later. A reservoir based on a river system not only takes land for the immediate body of water but also may have severe downstream implications for flora and fauna and for human activities such as fishing and sailing.

The direct and indirect impacts may sometimes correlate with short-run and long-run impacts. For some impacts the distinction between short-run and long-run may also relate to the distinction between a project's construction and its operational stage; however, other construction-stage impacts, such as change in land use, are much more permanent. Impacts also have a spatial dimension. One distinction is between local and strategic, the latter covering impacts on areas beyond the immediate locality. These are often regional, but may sometimes be of national or even international significance.

Environmental resources cannot always be replaced; once destroyed, some may be lost for ever. The distinction between reversible and irreversible impacts is a very important one, and the irreversible impacts, not susceptible to mitigation, can constitute particular significant impacts in an EIA. It may be possible to replace, compensate for or reconstruct a lost resource in some cases, but substitutions are rarely ideal. The loss of a resource may become more serious later, and valuations need to allow for this. Some impacts can be quantified, others are less tangible. The latter should not be ignored. Nor should the distributional impacts of a proposed development be ignored. Impacts do not fall evenly on affected parties and areas. Although a particular project may be assessed as bringing a general benefit, some groups and/or geographical areas may be receiving most of any adverse effects, the main benefits going to others elsewhere. There is also a distinction between actual and perceived impacts. Subjective perceptions of impacts may significantly influence the responses and decisions of people towards a proposed development. They constitute an important source of information, to be considered alongside more objective predictions of impacts. Finally, all impacts should be compared with the "do-nothing" situation, and the state of the environment predicted without the project. This can be widened to include comparisons with anticipated impacts from alternative development scenarios for an area.

We conclude on a semantic point: the words "impact" and "effect" are widely used in the literature and legislation on EIA, but it is not always clear whether they are interchangeable or should be used only for specifically different meanings. In the United States, the regulations for implementing the National Environmental Policy Act (NEPA) expressly state that "effects and impacts as used in these regulations are synonymous". This interpretation is widespread, and is adopted in this text. But there are other interpretations relating to timing and to value judgements. Catlow & Thirlwall (1976) make a distinction between effects which are "... the physical and natural changes resulting, directly or indirectly, from development" and impacts which are "... the consequences or end products of those effects represented by attributes of the environment on which we can place an objective or subjective value". In contrast, an Australian study (CEPA 1994) reverses the arguments, claiming that "there does seem to be greater logic in thinking of an impact resulting in an effect, rather than the other way round". Other commentators have introduced the concept of value judgement into the differentiation. Preston & Bedford (1988) state that "the use of the term 'impacts' connotes a value judgement". This view is supported by Stakhiv (1988), who sees a distinction between "scientific assessment of facts (effects), and the evaluation of the relative importance of these effects by the analyst and the public (impacts)". The debate continues!

1.6 Current issues in environmental impact assessment

Although EIA now has over 30 years of history in the USA, elsewhere the development of concepts and practice is more recent. Development is moving apace in many other countries, including the UK and the other EU Member States. Such progress has not been without its problems, and a number of the current issues in EIA are highlighted here and will be discussed more fully in later chapters.

1.6.1 Scope of the assessment

Whereas legislators may seek to limit coverage, best practice may lead to its widening. For example, project EIA may be mandatory only for a limited set of major projects. In practice many others have been included. But which projects should have assessments? In the UK, case law is now building up, but the criteria for the inclusion or exclusion of a project for EIA are still developing. In a similar vein, there is a case for widening the dimensions of the environment under consideration to include socio-economic impacts more fully. The trade-off between the adverse biophysical impacts of a development and its beneficial socio-economic impacts often constitutes the crucial dilemma for decision-makers. Coverage can also be widened to include other types of impacts only very partially covered to date. Distributional impacts would fall into this category. Lichfield and others are seeking to counter this problem (see Lichfield 1996).

1.6.2 *The nature of methods of assessment*

As noted in Section l.2, some of the main steps in the EIA process (e.g. the consideration of alternatives, monitoring) may be missing from many studies. There may also be problems with the steps that are included. The prediction of impacts raises various conceptual and technical problems. The problem of establishing the environmental baseline position has already been noted. It may also be difficult to establish the dimensions and development stages of a project clearly. Further conceptual problems include establishing what would have happened in the relevant environment without a project, clarifying the complexity of interactions of phenomena, and making trade-offs in an integrated way (i.e. assessing the trade-offs between economic apples, social oranges and physical bananas). Other technical problems are the general lack of data and the tendency to focus on the quantitative, and often single, indicators in some areas. There may also be delays and discontinuities between cause and effect, and projects and policies may discontinue. The lack of auditing of predictive techniques limits the feedback on the effectiveness of methods. Nevertheless, innovative methods are being developed to predict impacts, ranging from simple checklists and matrices to complex mathematical models. These methods are not neutral, in the sense that the more complex they are, the more difficult it becomes for the general public to participate in the EIA process.

1.6.3 *The relative roles of participants in the process*

The various "actors" in the EIA process – the developer, the affected parties, the general public and the regulators at various levels of government – have differential access to the process, and their influence on the outcome varies. Many would argue that in countries such as the UK the process is too developer-orientated. The developer or the developer's consultant carries out the EIA and prepares the EIS, and is unlikely to predict that the project will be an environmental disaster. Notwithstanding this, developers themselves are concerned about the potential delays associated with the requirement to submit an EIS. They are also concerned about cost. Details about costs are difficult to obtain. Clark (1984) estimates EIA costs of 0.5–2.0 per cent of a project's value. Hart (1984) and Wathern (1988) suggest figures of a similar order. Estimates by Coles et al. (1992) suggest a much wider range, from 0.000025 to 5 per cent, for EISs in the UK. The UK DETR (1997) suggested £35,000 as an appropriate median figure for the cost of undertaking an EIA under the new regulations.

Procedures for and the practice of public participation in the EIA process vary between, and sometimes within, countries, from the very comprehensive to the very partial and largely cosmetic. An important issue is the stages in the EIA process to which the public should have access. Government roles in the EIA process may be conditioned by caution at extending systems, by limited experience and expertise in this new and rapidly developing area, and by resource considerations. A central government may offer limited guidance on best practice, and make inconsistent decisions. A local government may find it difficult to handle the scope and complexity of the content of EISs.

1.6.4 *The quality of assessments*

Many EISs fail to meet even minimum standards. For example, a survey by Jones et al. (1991) of the EISs published under UK EIA regulations highlighted some shortcomings.

They found that "one-third of the EISs did not appear to contain the required non-technical summary, that, in a quarter of the cases, they were judged not to contain the data needed to assess the likely environmental effects of the development, and that in the great majority of cases, the more complex, interactive impacts were neglected". An update by Glasson et al. (DoE 1996) suggests that although there has been some learning from experience, many EISs in the UK are still unsatisfactory (see Chapter 8 for further discussion). Quality may vary between types of project. It may also vary between countries supposedly operating under the same legislative framework.

1.6.5 Beyond the decision

Many EISs are for one-off projects, and there is little incentive for developers to audit the quality of the assessment predictions and to monitor impacts as an input to a better assessment for the next project. EIA up to and no further than the decision on a project is a very partial linear process, with little opportunity for a cyclical learning process. In some areas of the world (e.g. California, Western Australia), the monitoring of impacts is mandatory, and monitoring procedures must be included in an EIS. The extension of such approaches constitutes another significant current issue in the largely project-based EIA process.

1.6.6 Beyond project assessment

As noted in Section 1.2, the SEA of PPPs represents a logical extension of project assessment. SEA can cope better with cumulative impacts, alternatives and mitigation measures than project assessment. SEA systems already exist in California and The Netherlands, and to a lesser extent in Canada, Germany and New Zealand. Following the Fifth Community Action Programme on the Environment which stated: "Given the goal of achieving sustainable development, it seems only logical, if not essential, to apply an assessment of the environmental implications of all relevant policies, plans and programmes" (CEC 1992), an EU SEA Directive is now in place, to be implemented from 2004 (see Therivel 2004, and Chapter 12).

1.7 An outline of subsequent parts and chapters

This book is in four parts. The first establishes the context of EIA in the growth of concern about environmental issues and in relevant legislation, with particular reference to the UK. Following from the first chapter, which provides an introduction to EIA and an overview of principles, Chapter 2 focuses on the origins of EIA under the US NEPA of 1969, on interim developments in the UK, and on the subsequent introduction of EC Directive 85/337 and subsequent amendments (CEC 97/11). The details of the UK legislative framework for EIA, under T&CP and other legislation, are discussed in Chapter 3.

Part 2 provides a rigorous step-by-step approach to the EIA process. This is the core of the text. Chapter 4 covers the early start-up stages, establishing a management framework, clarifying the type of developments for EIA, and outlining approaches to scoping, the consideration of alternatives, project description, establishing the baseline and identifying impacts. Chapter 5 explores the central issues of prediction, the assessment of significance and the mitigation of adverse impacts. The approach draws out broad principles affecting prediction exercises, exemplified with reference to particular cases. Chapter 6 provides coverage of an important issue identified above: participation in the EIA process. Communication in the EIA process, EIS presentation and EIA review are also covered in this chapter. Chapter 7 takes the process beyond the decision on a project and examines the importance of, and approaches to, monitoring and auditing in the EIA process.

Part 3 exemplifies the process in practice. Chapter 8 provides an overview of UK practice to date, including quantitative and qualitative analyses of the EISs prepared. Chapter 9 provides a review of EIA practice in several key sectors, including energy, transport, waste management and tourism. A feature of the chapter is the provision of a set of case studies of recent and topical EIA studies from the UK and overseas, illustrating particular features of and issues in the EIA process. Chapter 10 draws on comparative experience from a number of developed countries (The Netherlands, Canada and Australia) and from a number of countries from the developing and emerging economies (Peru, China, Benin and Poland) – presented to highlight some of the strengths and weaknesses of other systems in practice; the important role of international agencies in EIA practice – such as the UN and the World Bank – is also discussed in this chapter.

Part 4 looks to the future. It illuminates many of the issues noted in Section 1.5. Chapter 11 focuses on improving the effectiveness of the current system of project assessment. Particular emphasis is given to the development of environmental auditing to provide better baseline data, to various procedural developments and to achieving compatibility for EIA systems in Europe. Chapter 12 discusses the extension of assessment to PPPs, concluding full circle with a further consideration of EIA, SEA and sustainable development.

A set of appendices provide details of legislation and practice not considered appropriate to the main text. A list of further reading is included there.

Note

1. A comprehensive up-to-date overview of the state of the environment in Europe is provided in European Environment Agency, 2003. *Europe's environment: the third assessment.* Copenhagen: EEA.

References

ANZECC (Australia and New Zealand Environment and Conservation Council) 1991. *A national approach to EIA in Australia.* Canberra: ANZECC.

Beattie, R. 1995. Everything you already know about EIA, but don't often admit. *Environmental Impact Assessment Review* **15**.

Boulding, K. 1966. The economics of the coming Spaceship Earth. In *Environmental quality to a growing economy*, H. Jarrett (ed.), 3–14. Baltimore: Johns Hopkins University Press.

Braybooke, C. & D. Lindblom 1963. *A strategy of decision.* New York: Free Press of Glencoe.

Breakell, M. & J. Glasson (eds) 1981. *Environmental impact assessment: from theory to practice.* Oxford School of Planning, Oxford Polytechnic.

Breese, G. et al. 1965. *The impact of large installations on nearby urban areas.* Los Angeles: Sage.

Calow, P. (ed.) 1997. *Handbook of environmental risk assessment and management.* Oxford: Blackwell Science.

Carley, M. J. & E. S. Bustelo 1984. *Social impact assessment and monitoring: a guide to the literature*. Boulder: Westview Press.

Catlow, J. & C. G. Thirlwall 1976. *Environmental impact analysis*. London: DoE.

CEC (Commission of the European Communities) 1982. *The contribution of infrastructure to regional development*. Brussels: CEC.

CEC 1992. *Towards sustainability: a European Community programme of policy and action in relation to the environment and sustainable development*, vol. II. Brussels: CEC.

CEC 2001. *Our future, our choice. The sixth EU environment action programme 2001–10*. Brussels: CEC.

CEC 2003. (Impacts Assessment Unit, Oxford Brookes University) *Five Years' Report to the European Parliament and the Council on the Application and Effectiveness of the EIA Directive*. website of EC DG Environment: http://europa.eu.int/comm/environment/eia/home.htm.

CEPA (Commonwealth Environmental Protection Agency) 1994. *Assessment of cumulative impacts and strategic assessment in EIA*. Canberra: CEPA.

Clark, B. D. 1984. Environmental impact assessment (EIA): scope and objectives. In *Perspectives on environmental impact assessment*, B. D. Clark et al. (eds). Dordrecht: Reidel.

Coles, T., K. Fuller, M. Slater 1992. *Practical experience of environmental assessment in the UK*. East Kirkby, Lincolnshire: Institute of Environmental Assessment.

DETR (Department of Environment, Transport and the Regions) 1997. *Consultation paper: implementation of the EC Directive (97/11/EC) – determining the need for environmental assessment*. London: DETR.

DETR 1999a. *UK strategy for sustainable development: A better quality of life*. London: Stationery Office.

DETR 1999b. *Quality of life counts – Indicators for a strategy for sustainable development for the United Kingdom: a baseline assessment*. London: Stationery Office.

DETR 2000. *Environmental impact assessment: a guide to the procedures*. Tonbridge: Thomas Telford Publishing.

DoE (Department of the Environment) 1989. *Environmental assessment: a guide to the procedures*. London: HMSO.

DoE 1991. *Policy appraisal and the environment*. London: HMSO.

DoE 1996. *Changes in the quality of environmental impact statements*. London: HMSO.

DoE et al. 1990. *This common inheritance: Britain's environmental strategy* (Cmnd 1200). London: HMSO.

Etzioni, A. 1967. Mixed scanning: a "third" approach to decision making. *Public Administration Review* 27(5), 385–92.

Faludi, A. (ed.) 1973. *A reader in planning theory*. Oxford: Pergamon.

Finsterbusch, K. 1985. State of the art in social impact assessment. *Environment and Behaviour* 17, 192–221.

Flyberg, B. 2003. *Megaprojects and risk: on anatomy of risk*. Cambridge: Cambridge University Press.

Forester, J. 1989. *Planning in the face of power*. Berkeley, CA: University of California Press.

Friend, J. & N. Jessop 1977. *Local government and strategic choice*, 2nd edn. Toronto: Pergamon Press.

Habermas, J. 1984. *The Theory of Communicative Action*, vol. 1, *Reason and the rationalisation of society*. London: Polity Press.

Hanley, N. D. & C. Splash 1993. *Cost–benefit analysis and the environment*. Aldershot: Edward Elgar.

Hart, S. L. 1984. The costs of environmental review. In *Improving impact assessment*, S. L. Hart et al. (eds). Boulder: Westview Press.

Healey, P. 1996. The communicative turn in planning theory and its implication for spatial strategy making. *Environment and Planning B: Planning and Design* 23, 217–34.

Healey, P. 1997. *Collaborative planning: Shaping places in fragmented societies*. Basingstoke: Macmillan.

Health and Safety Commission 1978. *Canvey: an investigation of potential hazards from the operations in the Canvey Island/Thurrock area*. London: HMSO.

HMG, Secretary of State for the Environment 1994. *Sustainable development: the UK strategy.* London: HMSO.

Holmberg, J. & R. Sandbrook 1992. Sustainable development: what is to be done? In *Policies for a small planet*, J. Holmberg (ed.), 19–38. London: Earthscan.

IAIA (International Association for Impact Assessment) 1994. Guidelines and principles for social impact assessment. *Impact Assessment* 12(2).

Jones, C. E., N. Lee, C. M. Wood 1991. UK environmental statements 1988–1990: an analysis. Occasional Paper 29, Department of Planning and Landscape, University of Manchester.

Kaiser, E., D. Godshalk, S. Chapin 1995. *Urban land use planning*, 4th edn. Chicago: University of Illinois Press.

Kennedy, W. V. 1988. Environmental impact assessment in North America, Western Europe: what has worked where, how and why? *International Environmental Reporter* 11(4), 257–62.

Kirkby, J., P. O'Keefe, L. Timberlake 1995. *The earthscan reader in sustainable development*. London: Earthscan.

Lawrence, D. 1997. The need for EIA theory building. *Environmental Impact Assessment Review* 17(2), 79–107.

Lichfield, N. 1996. *Community impact evaluation*. London: UCL Press.

Lindblom, E. C. E. 1980. *The Policy Making Process*, 2nd edn. New Jersey: Englewood Cliffs.

McLoughlin, J. B. 1969. *Urban and regional planning – a systems approach*. London: Faber & Faber.

Munn, R. E. 1979. *Environmental impact assessment: principles and procedures*, 2nd edn. New York: Wiley.

O'Riordan, T. 1988. The politics of sustainability. In *Sustainable environmental management: principles and practice*, R. K. Turner (ed.). London: Belhaven.

O'Riordan, T. 1990. EIA from the environmentalist's perspective. *VIA* 4, March, 13.

Pearce, D. W. 1992. *Towards sustainable development through environment assessment*. Working Paper PA92–11, Centre for Social and Economic Research in the Global Environment, University College London.

Pearce, D., A. Markandya, E. Barbier 1989. *Blueprint for a Green economy*. London: Earthscan.

Petts, J. 1999. Environmental impact assessment versus other environmental management decision tools. In *Handbook of environmental impact assessment*, J. Petts (ed.), vol. 1. Oxford: Blackwells Science.

Preston, D. & B. Bedford 1988. Evaluating cumulative effects on wetland functions: a conceptual overview and generic framework. *Environmental Management* 12(5).

Redclift, M. 1987. *Sustainable development: exploring the contradictions*. London: Methuen.

Reid, D. 1995. *Sustainable development: an introductory guide*. London: Earthscan.

Sadler, B. 1996. *Environmental assessment in a changing world: evaluating practice to improve performance*. International study on the effectiveness of environmental assessment. Ottawa: Canadian Environmental Assessment Agency.

SEERA 2001. *A better quality of life in the south east*. Guildford: South East England Regional Assembly.

Skolimowski, P. 1995. Sustainable development – how meaningful? *Environmental Values* 4.

Stakhiv, E. 1988. An evaluation paradigm for cumulative impact analysis. *Environmental Management* 12(5).

Therivel, R. 2004. *Strategic environmental assessment in action*. London: Earthscan.

Therivel, R. & M. R. Partidario 1996. *The practice of strategic environmental assessment*. London: Earthscan.

Therivel, R., E. Wilson, S. Thompson, D. Heaney, D. Pritchard 1992. *Strategic environmental assessment*. London: RSPB/Earthscan.

Turner, R. K. & D. W. Pearce 1992. *Sustainable development: ethics and economics*. Working Paper PA92–09, Centre for Social and Economic Research in the Global Environment, University College London.

UNECE (United Nations Economic Commission for Europe) 1991. *Policies and systems of environmental impact assessment*. Geneva: United Nations.

UN World Commission on Environment and Development 1987. *Our common future*. Oxford: Oxford University Press.

Vanclay, F. 2003. International principles for social impact assessment. *International Assessment and Project Appraisal* **21**(1), 5–12.

Vanclay, F. & D. Bronstein (eds) 1995. *Environment and social impact assessments*. London: Wiley.

Wathern, P. (ed.) 1988. *Environmental impact assessment: theory and practice*. London: Unwin Hyman.

Weston, J. 2000. EIA, Decision-making Theory and Screening and Scoping in UK Practice. *Journal of Environmental Planning and Management* **43**(2), 185–203.

White, P. R., M. Franke, P. Hindle 1995. *Integrated solid waste management: a lifecycle inventory*. London: Chapman & Hall.

2 Origins and development

2.1 Introduction

Environmental impact assessment was first formally established in the USA in 1969 and has since spread, in various forms, to most other countries. In the UK, EIA was initially an *ad hoc* procedure carried out by local planning authorities and developers, primarily for oil- and gas-related developments. A 1985 European Community directive on EIA (Directive 85/337) introduced broadly uniform requirements for EIA to all EU Member States and significantly affected the development of EIA in the UK. However, 10 years after the Directive was agreed, Member States were still carrying out widely diverse forms of EIA, contradicting the Directive's aim of "levelling the playing field". Amendments of 1997 aimed to improve this situation. The nature of EIA systems – e.g. mandatory or discretionary, level of public participation, types of action requiring EIA – and their implementation in practice vary widely from country to country. However, the rapid spread of the concept of EIA and its central role in many countries' programmes of environmental protection attest to its universal validity as a proactive planning tool.

This chapter first discusses how the system of EIA evolved in the US. The present status of EIA worldwide is then briefly reviewed (Chapter 10 will consider a number of countries' systems of EIA in greater depth). EIA in the UK and the EU are then discussed. Finally, we review the various systems of EIA in the EU Member States.

2.2 The National Environmental Policy Act and subsequent US systems

The US National Environmental Policy Act (NEPA) of 1969, also known as NEPA, was the first legislation to require EIAs. Consequently it has become an important model for other EIA systems, both because it was a radically new form of environmental policy and because of the successes and failures of its subsequent development. Since its enactment, NEPA has resulted in the preparation of well over 25,000 full and partial EISs, which have influenced countless decisions and represent a powerful base of environmental information. On the other hand, NEPA is unique. Other countries have shied away from the form it takes and the procedures it sets out, not least because they are unwilling to face a situation like that in the USA, where there has been extensive litigation over the interpretation and workings of the EIA system.

This section covers NEPA's legislative history, i.e. the early development before it became law, the interpretation of NEPA by the courts and the Council on Environmental Quality (CEQ), the main EIA procedures arising from NEPA, and likely future developments. The reader is referred to Anderson et al. (1984), Bear (1990), Canter (1996), Mandelker (2000), Orloff (1980) and the annual reports of the CEQ for further information.

2.2.1 Legislative history

The National Environmental Policy Act is in many ways a fluke, strengthened by what should have been amendments weakening it, and interpreted by the courts to have powers that were not originally intended. The legislative history of NEPA is interesting not only in itself but also because it explains many of the anomalies of its operation and touches on some of the major issues involved in designing an EIA system. Several proposals to establish a national environmental policy were discussed in the US Senate and House of Representatives in the early 1960s. All these proposals included some form of unified environmental policy and the establishment of a high-level committee to foster it. In February 1969, Bill S1075 was introduced in the Senate; it proposed a programme of federally funded ecological research and the establishment of a CEQ. A similar bill, HR6750, introduced in the House of Representatives, proposed the formation of a CEQ and a brief statement on national environmental policy. Subsequent discussions in both chambers of Congress focused on several points:

- the need for a declaration of national environmental policy (now Title I of NEPA);
- a proposed statement that "each person has a fundamental and inalienable right to a healthful environment" (which would put environmental health on a par with, say, free speech). This was later weakened to the statement in §101(c) that "each person should enjoy a healthful environment";
- action-forcing provisions similar to those then being proposed for the Water Quality Improvement Act, which would require federal officials to prepare a detailed statement concerning the probable environmental impacts of any major action; this was to evolve into NEPA's §102(2)(C) which requires EIA. The initial wording of the Bill had required a "finding", which would have been subject to review by those responsible for environmental protection, rather than a "detailed statement" subject to inter-agency review. The Senate had intended to weaken the Bill by requiring only a detailed statement. Instead, the "detailed assessment" became the subject of external review and challenge; the public availability of the detailed statements became a major force shaping the law's implementation in its early years. NEPA became operational on 1 January 1970. Table 2.1 summarizes its main points.

2.2.2 An interpretation of NEPA

The National Environmental Policy Act is a generally worded law that required substantial early interpretation. The CEQ, which was set up by NEPA, prepared guidelines to assist in the Act's interpretation. However, much of the strength of NEPA came from early court rulings. NEPA was immediately seen by environmental activists as a significant vehicle for preventing environmental harm, and the early 1970s saw a series of influential lawsuits and court decisions based on it. These lawsuits were of three broad types, as described by Orloff (1980):

Table 2.1 Main points of NEPA

NEPA consists of two titles. Title I establishes a national policy on the protection and restoration of environmental quality. Title II sets up a three-member Council on Environmental Quality (CEQ) to review environmental programmes and progress, and to advise the president on these matters. It also requires the president to submit an annual "Environmental Quality Report" to Congress. The provisions of Title I are the main determinants of EIA in the USA, and they are summarized here.

Section 101 contains requirements of a substantive nature. It states that the Federal Government has a continuing responsibility to "create and maintain conditions under which man and nature can exist in productive harmony, and fulfil the social, economic and other requirements of present and future generations of Americans". As such the government is to use all practicable means, "consistent with other essential considerations of national policy", to minimize adverse environmental impact and to preserve and enhance the environment through federal plans and programmes. Finally, "each person should enjoy a healthful environment", and citizens have a responsibility to preserve the environment.

Section 102 requirements are of a procedural nature. Federal agencies are required to make full analyses of all the environmental effects of implementing their programmes or actions. Section 102(1) directs agencies to interpret and administer policies, regulations and laws in accordance with the policies of NEPA. Section 102(2) requires federal agencies

- to use "a systematic and interdisciplinary approach" to ensure that social, natural and environmental sciences are used in planning and decision-making;
- to identify and develop procedures and methods so that "presently unquantified environmental amenities and values may be given appropriate consideration in decision-making along with traditional economic and technical considerations";
- to "include in every recommendation or report on proposals for legislation and other *major Federal actions significantly affecting the quality of the human environment, a detailed statement* by the responsible official" on:

  - the environmental impact of the proposed action;
  - any adverse environmental effects which cannot be avoided should the proposal be implemented;
  - alternatives to the proposed action;
  - the relationship between local short-term uses of man's environment and the maintenance and enhancement of long-term productivity;
  - any irreversible and irretrievable commitments of resources which would be involved in the proposed action should it be implemented (authors' emphases).

Section 103 requires federal agencies to review their regulations and procedures for adherence to NEPA, and to suggest any necessary remedial measures.

1. Challenging an agency's decision not to prepare an EIA. This generally raised issues such as whether a project was major, federal, an "action", or had significant environmental impacts (see NEPA §102(2)(C)). For instance, the issue of whether an action is federal came into question in some lawsuits concerning the federal funding of local government projects.[1]
2. Challenging the adequacy of an agency's EIS. This raised issues such as whether an EIS adequately addressed alternatives, and whether it covered the full range of significant environmental impacts. A famous early court case concerned the Chesapeake Environmental Protection Association's claim that the Atomic Energy Commission did not adequately consider the water quality impacts of its proposed nuclear power plants, particularly in the EIA for the Calvert Cliffs

power plant.[2] The Commission argued that NEPA merely required the consideration of water quality standards; opponents argued that it required an assessment beyond mere compliance with standards. The courts sided with the opponents.

3.   Challenging an agency's substantive decision, namely its decision to allow or not to allow a project to proceed in the light of the contents of its EIS. Another influential early court ruling[3] laid down guidelines for the judicial review of agency decisions, noting that the court's only function was to ensure that the agency had taken a "hard look" at environmental consequences, not to substitute its judgement for that of the agency.

The early proactive role of the courts greatly strengthened the power of environmental movements and caused many projects to be stopped or substantially amended. In many cases the lawsuits delayed construction for long enough to make them economically infeasible or to allow the areas where projects would have been sited to be designated as national parks or wildlife areas (Turner 1988). More recent decisions have been less clearly pro-environment than the earliest decisions. The flood of early lawsuits, with the delays and costs involved, was a lesson to other countries in how *not* to set up an EIA system. As will be shown later, many countries carefully distanced their EIA systems from the possibility of lawsuits.

The CEQ was also instrumental in establishing guidelines to interpret NEPA, producing interim guidelines in 1970, and guidelines in 1971 and 1973. Generally the courts adhered closely to these guidelines when making their rulings. However, the guidelines were problematic: they were not detailed enough, and were interpreted by the federal agencies as being discretionary rather than binding. To combat these limitations, President Carter issued Executive Order 11992 in 1977, giving the CEQ authority to set enforceable regulations for implementing NEPA. These were issued in 1978 (CEQ 1978) and sought to make the NEPA process more useful for decision-makers and the public, to reduce paperwork and delay and to emphasize real environmental issues and alternatives.

2.2.3 A summary of NEPA procedures

The process of EIA established by NEPA, and developed further in the CEQ regulations, is summarized in Figure 2.1. The following citations are from the CEQ regulations (CEQ 1978).

> [The EIA process begins] as close as possible to the time the agency is developing or is presented with a proposal... The statement shall be prepared early enough so that it can serve practically as an important contribution to the decision-making process and will not be used to rationalize or justify decisions already made. (§1502.5)

A "*lead agency*" is designated that co-ordinates the EIA process. The lead agency first determines whether the proposal requires the preparation of a full EIS, no EIS at all, or a "finding of no significant impact" (FONSI). This is done through a series of tests. A first test is whether a federal action is likely to individually or cumulatively have a significant environmental impact. All federal agencies have compiled lists of "categorical exclusions" which are felt not to have such impacts. If an action is on such a list, then no further EIA action is generally needed. If an action is not categorically

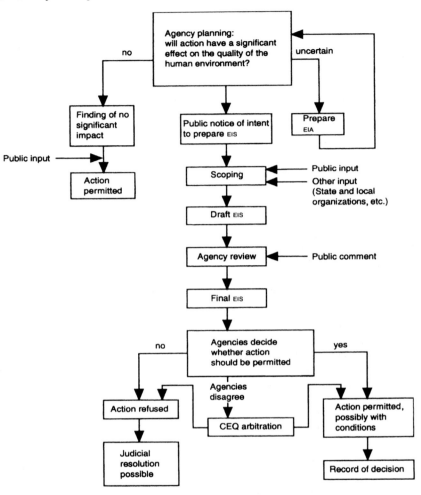

Figure 2.1 Process of EIA under NEPA. (Adapted from Legore 1984.)

excluded, an "environmental assessment" is carried out to determine whether a full EIS or a FONSI is needed. A FONSI is a public document which explains why the action is not expected to have a significant environmental impact.

If a FONSI is prepared, then a permit would usually be granted following public discussion. If a full EIS is found to be needed, the lead agency publishes a "Notice of Intent", and the *process of scoping begins*. The aim of the scoping exercise is to determine the issues to be addressed in the EIA: to eliminate insignificant issues, focus on those that are significant and identify alternatives to be addressed. The lead agency invites the participation of the proponent of the action, affected parties and other interested persons. "[The alternatives] section is the heart of the environmental impact statement...[It] should present the environmental impacts of the proposal and the alternatives in comparative form, thus sharply defining the issues and providing a clear basis for choice..." (§1502.14)

Table 2.2 Typical format for an EIS under NEPA

(a) Cover sheet

- list of responsible agencies
- title of proposed action
- contact persons at agencies
- designation of EIS as draft, final or supplement
- abstract of EIS
- date by which comments must be received

(b) Summary (usually 15 pages or less)

- major conclusions
- areas of controversy
- issues to be resolved

(c) Table of contents
(d) Purpose of and need for action
(e) Alternatives, including proposed action
(f) Affected environment
(g) Environmental consequences

- environmental impacts of alternatives, including proposed action
- adverse environmental effects which cannot be avoided if proposal is implemented
- mitigation measures to be used and residual effects of mitigation
- relation between short-term uses of the environment and maintenance and enhancement of long-term productivity
- irreversible or irretrievable commitments of resources if proposal is implemented discussion of:

    - direct and indirect effects and their significance
    - possible conflicts between proposed action and objectives of relevant land-use plans, policies and controls
    - effects of alternatives, including proposed action
    - energy requirements and conservation potential of various alternatives and mitigation measures
    - natural or depletable resource requirements and conservation of various alternatives and mitigation measures
    - effects on urban quality, historic and cultural resources, and built environment
    - means to mitigate adverse impacts

(h) List of preparers
(i) List of agencies, etc. to which copies of EIS are sent
(j) Index
(k) Appendices, including supporting data

A *draft EIS is then prepared*, and is reviewed and commented on by the relevant agencies and the public. These comments are taken into account in the subsequent preparation of a final EIS. An EIS is normally presented in the format shown in Table 2.2. In an attempt to be comprehensive, early EISs tended to be so bulky as to be virtually unreadable. The CEQ guidelines consequently emphasize the need to concentrate only on important issues and to prepare readable documents: "The text of final environmental impact statements shall normally be less than 150 pages…Environmental impact statements shall be written in plain language…" (§1502.7–8)

The public is involved in this process, both at the scoping stage and after publication of the draft and final EISs:

> Agencies shall: (a) Make diligent efforts to involve the public in preparing and implementing NEPA procedures...(b) Provide public notice of NEPA-related hearings, public meetings and the availability of environmental documents... (c) Hold or sponsor public hearings...whenever appropriate...(d) Solicit appropriate information from the public. (e) Explain in its procedures where interested persons can get information or status reports...(f) Make environmental impact statements, the comments received, and any underlying documents available to the public pursuant to the provisions of the Freedom of Information Act...(§1506.6)

Finally, a decision is made about whether the proposed action should be permitted:

> Agencies shall adopt procedures to ensure that decisions are made in accordance with the policies and purposes of the Act. Such procedures shall include but not be limited to: (a) Implementing procedures under section 102(2) to achieve the requirements of sections 101 and 102(1)...(e) Requiring that...the decision-maker consider the alternatives described in the environmental impact statement. (§1505.1)
>
> Where all relevant agencies agree that the action should not go ahead, permission is denied, and a judicial resolution may be attempted. Where agencies agree that the action can proceed, permission is given, possibly subject to specified conditions (e.g. monitoring, mitigation). Where the relevant agencies disagree, the CEQ acts as arbiter (§1504). Until a decision is made, "no action concerning the proposals shall be taken which could: (1) have an adverse environmental impact; or (2) limit the choice of reasonable alternatives..." (§1506.1).

2.2.4 Recent trends

During the first 10 years of NEPA's implementation, about 1,000 EISs were prepared annually. Subsequently, negotiated improvements to the environmental impacts of proposed actions have become increasingly common during the preparation of "environmental assessments". This has led to many "mitigated findings of no significant impact" (no perfect acronym exists for this), reducing the number of EISs prepared: whereas 1,273 EISs were prepared in 1979, only 456 were prepared in 1991 and the annual number has been approximately 500 in recent years (EPA website: es.epa.gov/oeca/ofa/; and CEQ 1993). This trend can be viewed positively, since it means that environmental impacts are considered earlier in the decision-making process, and since it reduces the costs of preparing EISs. However, the fact that this abbreviated process allows less public participation causes some concern. Of the 456 EISs prepared in 1991, 145 were filed by the Department of Agriculture (primarily for forestry and range management) and 87 were filed by the Department of Transportation (primarily for road construction). Between 1979 and 1991, the number of EISs filed by the Department of Housing and Urban Development fell from 170 to 7! The number of legal cases filed against federal departments and agencies on the basis of NEPA also fell slightly, from 139 in 1979 to 128 in 1991. The most common complaints were "no EIS when one should have been prepared" (41 cases in 1991) and "inadequate EIS" (26 cases in 1991).

The National Environmental Policy Act's twentieth year of operation, 1990, was marked by a series of conferences on the Act and the presentation to Congress of a bill of NEPA amendments. Under the Bill (HR1113), which was not passed, federal actions that take place outside the USA (e.g. projects built in other countries with US federal assistance) would have been subject to EIA, and all EISs would have been required to consider global climatic change, the depletion of the ozone layer, the loss of biological diversity and transboundary pollution. This latter amendment was controversial: although the need to consider the global impacts of programmes was undisputed, it was felt to be infeasible at the level of project EIA. Finally, the Bill would have required all federal agencies to survey a statistically significant sample of EISs to determine whether mitigation measures promised in the EIS had been implemented and, if so, whether they had been effective.

The context of EIA has also become a matter of concern. EIA is only one part of a broader environmental policy (NEPA), but the procedural provisions set out in NEPA's §102(2)(C) have overshadowed the rest of the Act. It has been argued that mere compliance with these procedures is not enough, and that greater emphasis should be given to the environmental goals and policies stated in §101. EIA must also be seen in the light of other environmental legislation. In the USA, many laws dealing with specific aspects of the environment were enacted or strengthened in the 1970s, including the Clean Water Act and the Clean Air Act. These laws have in many ways superseded NEPA's substantive requirements and have complemented and buttressed its procedural requirements. Compliance with these laws does not necessarily imply compliance with NEPA. However, the permit process associated with these other laws has become a primary method for evaluating project impacts, reducing NEPA's importance except for its occasional role as a focus of debate on major projects (Bear 1990).

The scope of EIA, and in particular the recognition of the social dimension of the environment, has been another matter of concern. After long campaigning by black and ethnic groups, particularly about inequalities in the distribution of hazardous waste landfills and incinerators, a working group was set up within the Environmental Protection Act (EPA) to make recommendations for dealing with environmental injustice (Hall 1994). The outcome was the Clinton "Executive Order on Federal Actions to Address Environmental Justice in Minority Populations and Low-Income Populations" (White House 1994). Under this Order, each federal agency must analyse the environmental effects, including human health, economic and social effects, of federal actions, including effects on minority and low-income communities, when such analysis is required under NEPA. Mitigation measures, wherever feasible, should also address the significant and adverse environmental effects of federal actions on the same communities. In addition, each federal agency must provide opportunities for communities to contribute to the NEPA process, identifying potential effects and mitigation measures in consultation with affected communities and improving the accessibility of meetings and crucial documents.

Other issues remain, and Canter (1996) highlights four areas for which NEPA requirements need further elaboration:

1. how much an agency should identify and plan mitigation before issuing an EIS;
2. ways to assess the cumulative impacts of proposed developments;

3.   ways to conduct "reasonable foreseeability" (or worst-case) analyses; and
4.   the monitoring and auditing of impact predictions.

2.2.5 Little NEPAs and the particular case of California

Many state-level EIA systems have been established in the USA in addition to NEPA. Fifteen of the USA's states[4] have so-called "little NEPAs", which require EIA for state actions (actions that require state funding or permission) and/or projects in sensitive areas. Other states[5] have no specific EIA regulations, but have EIA requirements in addition to those of NEPA.

Of particular interest is the Californian system, established under the California Environmental Quality Act (CEQA) of 1973, and subsequent amendments. This is widely recognized as one of the most advanced EIA systems in the world. The legislation applies not only to government actions but also to the activities of private parties that require the approval of a government agency. It is not merely a procedural approach but one that requires state and local agencies to protect the environment by adopting feasible mitigation measures and alternatives in environmental impact reviews (EIRs). The legislation extends beyond projects to higher levels of actions, and an amendment in 1989 also added mandatory mitigation, monitoring and reporting requirements to CEQA. Annual guidance on the California system is provided in an invaluable publication by the State of California, which sets out the CEQA Statutes and Guidelines in considerable detail (State of California 1992).

2.3 The worldwide spread of EIA

Since the enactment of NEPA, EIA systems have been established in various forms throughout the world, beginning with more developed countries – e.g. Canada in 1973, Australia in 1974, West Germany in 1975, France in 1976 – and later also in the less developed countries. The approval of a European Directive on EIA in 1985 stimulated the enactment of EIA legislation in many European countries in the late 1980s. The formation of new countries after the break-up of the Soviet Union in 1991 led to the enactment of EIA legislation in many of these countries in the early to mid-1990s. The early 1990s also saw a large growth in the number of EIA regulations and guidelines established in Africa and South America. By 1996, more than 100 countries had EIA systems (Sadler 1996). Now, at least 120 countries have EIA systems and Figure 2.2 summarizes the present state of EIA systems worldwide, to the best of the authors' knowledge.

These EIA systems vary greatly. Some are in the form of *mandatory regulations*, *acts* or *statutes*; these are generally enforced by the authorities' requiring the preparation of an adequate EIS before permission is given for a project to proceed. In other cases, EIA *guidelines* have been established. These are not enforceable but generally impose obligations on the administering agency. Other legislation allows government officials to require EIAs to be prepared at their *discretion*. Elsewhere, EIAs are prepared in an *ad hoc* manner, often because they are required by funding bodies (e.g. the World Bank, USAID) as part of a funding approval process. However, these classifications are not necessarily indicative of how thoroughly EIA is carried out. For instance, the EIA regulations of Brazil and the Philippines are not well carried out or enforced in practice (Glasson & Salvador 2000, Moreira 1988), whereas Japan's

Figure 2.2 EIA systems worldwide (the authors apologize for any omissions or inaccuracies).

guidelines are thoroughly implemented, and some very good *ad hoc* EIAs have been prepared in the UK.

Another important distinction between types of EIA system is that sometimes the actions that require EIA are given as *a definition* (e.g. the USA's definition of "major federal actions significantly affecting the quality of the human environment"), sometimes as *a list of projects* (e.g. roads of more than 10 km in length). Most countries use a list of projects, in part to avoid legal wrangling such as that surrounding NEPA's definition. Another distinction asks whether EIA is required for *government projects only* (as in NEPA), for *private projects only* or for both.

Finally, some international development and funding agencies have set up EIA guidelines, including the European Bank for Reconstruction and Development (1992), Overseas Development Administration (1996), UNEP (1997) and World Bank (1992, 1995, 1999).

2.4 Development in the UK

The UK has had formal legislation for EIA since 1988, in the form of several laws that implement European Community Directive 85/337/EEC (CEC 1985) and subsequent amendments. It is quite possible that without pressure from the European Commission such legislation would have been enacted much more slowly, since the UK government felt that its existing planning system more than adequately controlled environmentally unsuitable developments. However, this does not mean that the UK had no EIA system at all before 1988; many EIAs were prepared voluntarily or at the request of local authorities, and guidelines for EIA preparation were drawn up.

2.4.1 Limitations of the land-use planning system

The UK's statutory land-use planning system has since 1947 required local planning authorities (LPAs) to anticipate likely development pressures, assess their significance, and allocate land, as appropriate, to accommodate them. Environmental factors are a fundamental consideration in this assessment. Most developments require planning consent, so environmentally harmful developments can be prevented by its denial. This system resulted in the accumulation of considerable planning expertise concerning the likely consequences of development proposals.

After the mid-1960s, however, the planning system began to seem less effective at controlling the impacts of large developments. The increasing scale and complexity of developments, the consequently greater social and physical environmental impacts and the growing internationalization of developers (e.g. oil and chemicals companies) all outstripped the capability of the development control system to predict and control the impacts of developments. In the late 1960s, public concern about environmental protection also grew considerably, and the relation between statutory planning controls and the development of large projects came under increasing scrutiny. This became particularly obvious in the case of the proposed third London Airport. The Roskill Commission was established to select the most suitable site for an airport in south-east England, with the mandate to prepare a CBA of alternative sites. The resulting analysis (HMSO 1971) focused on socio-economic rather than physical environmental impacts; it led to an understanding of the difficulties of expanding CBA to impacts not easily measured in monetary terms, and to the realization that other assessment methods were needed to achieve a balance between socio-economic and physical environmental objectives.

2.4.2 North Sea oil- and gas-related EIA initiatives

The main impetus towards the further development of EIA, however, was the discovery of oil and gas in the North Sea. The extraction of these resources necessitated the construction of large developments in remote areas renowned for their scenic beauty and distinctive ways of life (e.g. the Shetlands, the Orkneys and the Highlands region). Planning authorities in these areas lacked the experience and resources needed to assess the impacts of such large developments. In response, the Scottish Development Department (SDD) issued a technical advice note to LPAs (SDD 1974). *Appraisal of the impact of oil-related development* noted that these developments and other large and unusual projects need "rigorous appraisal", and suggested that LPAs should commission an impact study of the developments if needed. This was the first government recognition that major developments needed special appraisal. Some EIAs were carried out in the early 1970s, mostly for oil and gas developments. Many of these were sponsored by the SDD and LPAs, and were prepared by environmental consultants, but some (e.g. for the Flotta Oil Terminal and Beatrice Oilfield) were commissioned by the developers. Other early EIAs concerned a coal mine in the Vale of Belvoir, a pumped-storage electricity scheme at Loch Lomond and various motorway and trunk road proposals (Clark & Turnbull 1984).

In 1973, the Scottish Office and DoE commissioned the University of Aberdeen's Project Appraisal for Development Control (PADC) team to develop a systematic procedure for planning authorities to make a balanced appraisal of the environmental, economic and social impacts of large industrial developments. PADC produced an interim report, *The assessment of major industrial applications – a manual* (Clark et al. 1976), which was issued free of charge to all LPAs in the UK and "commended by central government for use by planning authorities, government agencies and developers". The PADC procedure was designed to fit into the existing planning framework, and was used to assess a variety of (primarily private sector) projects. An extended and updated version of the manual was issued in 1981 (Clark et al. 1981).

In 1974, the Secretaries of State for the Environment, Scotland and Wales commissioned two consultants, J. Catlow and C. G. Thirwall, to investigate the "desirability of introducing a system of impact analysis in Great Britain, the circumstances in which a system should apply, the projects it should cover and the way in which it might be incorporated into the development control system" (Catlow & Thirwall 1976). The resulting report made recommendations about who should be responsible for preparing and paying for EIAs, what legislative changes would be needed to institute an EIA system, and similar issues. The report concluded that about 25–50 EIAs per year would be needed, for both public and private sector projects. EIA was given additional support by the Dobry Report on the development control system (Dobry 1975), which advocated that LPAs should require developers to submit impact studies for particularly significant development proposals. The report outlined the main topics such a study should address, and the information that should be required from developers. Government reactions to the Dobry Report were mixed: the Royal Commission on Environmental Pollution endorsed the report, but the Stevens Committee (1976) on Mineral Workings recommended that a comprehensive standard form for mineral applications should be introduced, arguing that such a form would make EIAs for mineral workings unnecessary.

2.4.3 Department of the Environment scepticism

However, overall the DoE remained sceptical about the need, practicality and cost of EIA. In fact, the government's approach to EIA has been described as being "from the outset grudging and minimalist" (CPRE 1991). In response to the Catlow and Thirwall report, the DoE stated: "Consideration of the report by local authorities should not be allowed to delay normal planning procedures and any new procedures involving additional calls on central or local government finance and manpower are unacceptable during the present period of economic restraint" (DoE 1977). A year later, after much deliberation, the DoE was slightly more positive:

> We fully endorse the desirability . . . of ensuring careful evaluation of the possible effects of large developments on the environment . . . The approach suggested by Thirwall/Catlow is already being adopted with many [projects] . . . The sensible use of this approach [should] improve the practice in handling these relatively few large and significant proposals. (DoE 1978)

The government's foreword to the PADC manual of 1981 also emphasized the need to minimize the costs of EIA procedures: "It is important that the approach suggested in the report should be used selectively to fit the circumstances of the proposed development and with due economy" (Clark et al. 1981). As will be seen in later chapters, the government remained sceptical for some time about the value of EIA, and about extending its remit, as suggested by the EC.

By the early 1980s, more than 200 studies on the environmental impacts of projects in the UK had been prepared on an *ad hoc* basis. These are listed by Petts & Hills (1982). Many of these studies were not full EIAs, but focused on only a few impacts. However, large developers such as British Petroleum, British Gas, the Central Electricity Generating Board and the National Coal Board were preparing a series of increasingly comprehensive statements. In the case of British Gas, these were shown to be a good investment, saving the company £30 million in 10 years (House of Lords 1981a).

2.5   EC Directive 85/337

The development and implementation of Directive 85/337 greatly influenced the EIA systems of the UK and other EU Member States. In the UK, central government research on a UK system of EIA virtually stopped after the mid-1970s, and attention focused instead on ensuring that any future Europe-wide system of EIA would fully incorporate the needs of the UK for flexibility and discretion. Other Member States were eager to ensure that the Directive reflected the requirements of their own more rigorous systems of EIA. Since the Directive's implementation, EIA activity in all the EU Member States has increased dramatically.

2.5.1   Legislative history

The EC had two main reasons for wanting to establish a uniform system of EIA in all its Member States. First, it was concerned about the state of the physical environment and eager to prevent further environmental deterioration. The EC's First Action Programme on the Environment of 1973 (CEC 1973) advocated the prevention of environmental harm: "the best environmental policy consists of preventing the creation of pollution or nuisances at source, rather than subsequently trying to counteract their effects", and, to that end, "effects on the environment should be taken into account at the earliest possible stage in all technical planning and decision-making processes". Further Action Programmes of 1977, 1983, 1987, 1992 and 2001 have reinforced this emphasis. Land-use planning was seen as an important way of putting these principles into practice, and EIA was viewed as a crucial technique for incorporating environmental considerations into the planning process.

Second, the EC was concerned to ensure that no distortion of competition should arise through which one Member State could gain unfair advantage by permitting developments that, for environmental reasons, might be refused by another. In other words, it considered environmental policies necessary for the maintenance of a level economic playing field. Further motivation for EC action included a desire to encourage best practice across Member States. In addition, pollution problems transcend territorial boundaries (witness acid rain and river pollution in Europe), and the EC can contribute at least a subcontinental response framework.

The EC began to commission research on EIA in 1975. Five years later and after more than 20 drafts, the EC presented a draft directive to the Council of Ministers (CEC 1980); it was circulated throughout the Member States. The 1980 draft attempted to reconcile several conflicting needs. It sought to benefit from the US experience with NEPA, but to develop policies appropriate to European need. It also sought to make EIA applicable to all actions likely to have a significant environmental impact, but to ensure that procedures would be practicable. Finally, and perhaps most challenging, it sought to make EIA requirements flexible enough to adapt to the needs and institutional arrangements of the various Member States, but uniform enough to prevent problems arising from widely varying interpretations of the procedures. The harmonization of the types of project to be subject to EIA, the main obligations of the developers and the contents of the EIAs were considered particularly important (Lee & Wood 1984, Tomlinson 1986).

As a result, the draft directive incorporated a number of important features. First, planning permission for projects was to be granted only after an adequate EIA had been completed. Second, LPAs and developers were to co-operate in providing information on the environmental impacts of proposed developments. Third, statutory bodies responsible for environmental issues, and other Member States in cases of trans-frontier effects, were to be consulted. Finally, the public was to be informed and allowed to comment on issues related to project development.

In the UK the draft directive was examined by the House of Lords Select Committee on the European Commission, where it received widespread support:

> The present draft Directive strikes the right kind of balance: it provides a frame-
> work of common administrative practices which will allow Member States with
> effective planning controls to continue with their system...while containing
> enough detail to ensure that the intention of the draft cannot be evaded...The
> Directive could be implemented in the United Kingdom in a way which would
> not lead to undue additions delay and costs in planning procedures and which
> need not therefore result in economic and other disadvantages. (House of Lords
> 1981a)

However, the Parliamentary Undersecretary of State at the DoE dissented. Although accepting the general need for EIA, he was concerned about the bureaucratic hurdles, delaying objections and litigation that would be associated with the proposed direct-ive (House of Lords 1981b). The UK Royal Town Planning Institute (RTPI) also commented on several drafts of the directive. Generally the RTPI favoured it, but was concerned that it might cause the planning system to become too rigid:

> The Institute welcomes the initiative taken by the European Commission to
> secure more widespread use of EIA as it believes that the appropriate use of EIA
> could both speed up and improve the quality of decisions on certain types of
> development proposals. However, it is seriously concerned that the proposed
> Directive, as presently drafted, would excessively codify and formalize procedures
> of which there is limited experience and therefore their benefits are not yet
> proven. Accordingly the Institute recommends the deletion of Article 4 and
> annexes of the draft. (House of Lords 1981a)

More generally, slow progress in the implementation of EC legislation was symptom-atic of the wide range of interest groups involved, of the lack of public support for increasing the scope of town planning and environmental protection procedures, and of the unwillingness of Member States to adapt their widely varying planning systems and environmental protection legislation to those of other countries (Williams 1988). In March 1982, after considering the many views expressed by the Member States, the Commission published proposed amendments to the draft directive

(CEC 1982). Approval was expected in November 1983. However, this was delayed by the Danish Government, which was concerned about projects authorized by Acts of Parliament. On 7 March 1985, the Council of Ministers agreed on the proposal; it was formally adopted as a directive on 27 June 1985 (CEC 1985) and became operational on 3 July 1988.

Subsequently, the EC's Fifth Action Programme, *Towards sustainability* (CEC 1992), stressed the importance of EIA, particularly in helping to achieve sustainable development, and the need to expand the remit of EIA:

> Given the goal of achieving sustainable development it seems only logical, if not essential, to apply an assessment of the environmental implications of all relevant policies, plans and programmes. The integration of environmental assessment within the macro-planning process would not only enhance the protection of the environment and encourage optimization of resource management but would also help to reduce those disparities in the international and inter-regional competition for new development projects which at present arise from disparities in assessment practices in the Member States...

In response to a (belated) five-year review of the Directive (CEC 1993), amendments to the Directive were agreed in 1997. Appendix 1 gives the complete consolidated version of the amended Directive.

The reader is referred to Clark & Turnbull (1984), Lee & Wood (1984), O'Riordan & Sewell (1981), Swaffield (1981), Tomlinson (1986), Williams (1988) and Wood (1981, 1988) for further discussions on the development of EIA in the UK and EC.

2.5.2 *Summary of EC Directive 85/337 procedures*

The Directive differs in important respects from NEPA. It requires EIAs to be prepared by both public agencies and private developers, whereas NEPA applies only to federal agencies. It requires EIA for a specified list of projects, whereas NEPA uses the definition "major federal actions..." It specifically lists the impacts that are to be addressed in an EIA, whereas NEPA does not. Finally, it includes fewer requirements for public consultation than does NEPA.

Under the provisions of the European Communities Act of 1972, Directive 85/337 is the controlling document, laying down rules for EIA in Member States. Individual states enact their own regulations to implement the Directive and have considerable discretion. According to the Directive, EIA is required for two classes of project, one mandatory (Annex I) and one discretionary (Annex II):

> projects of the classes listed in Annex I shall be made subject to an assessment... for projects listed in Annex II, the Member States shall determine through: (a) a case-by-case examination; or (b) thresholds or criteria set by the Member State whether the project shall be made subject to an assessment... When [doing so], the relevant selection criteria set out in Annex III shall be taken into account. (Article 4)

Table 2.3 summarizes the projects listed in Annexes I and II. The EC (CEC 1995) also published guidelines to help Member States determine whether a project requires EIA.

Table 2.3 Projects requiring EIA under EC Directive 85/337 (*as amended*)

Annex I (mandatory)

1. Crude oil refineries, coal/shale gasification and liquefaction
2. Thermal power stations and other combustion installations; nuclear power stations and other nuclear reactors
3. Radioactive waste processing and/or storage installations
4. Cast-iron and steel smelting works
5. Asbestos extraction, processing or transformation
6. Integrated chemical installations
7. Construction of motorways, express roads, other large roads, railways, airports
8. Trading ports and inland waterways
9. Installations for incinerating, treating or disposing of toxic and dangerous wastes
10. *Large-scale installation for incinerating or treating non-hazardous waste*
11. *Large-scale groundwater abstraction or recharge schemes*
12. *Large-scale transfer of water resources*
13. *Large-scale waste water treatment plants*
14. *Large-scale extraction of petroleum and natural gas*
15. *Large dams and reservoirs*
16. *Long pipelines for gas, oil or chemicals*
17. *Large-scale poultry or pig-rearing installations*
18. *Pulp, timber or board manufacture*
19. *Large-scale quarries or open-cast mines*
20. *Long overhead electrical power lines*
21. *Large-scale installations for petroleum, petrochemical or chemical products*

Annex II (discretionary)

1. Agriculture, silviculture and aquaculture
2. Extractive industry
3. Energy industry
4. Production and processing of metals
5. *Minerals industry* (projects not included in Annex I)
6. Chemical industry
7. Food industry
8. Textile, leather, wood and paper industries
9. Rubber industry
10. *Infrastructure projects*
11. *Other projects*
12. *Tourism and leisure*
13. Modification, extension or temporary testing of Annex I projects

Note: Amendments are shown in italic.

Similarly, the information required in an EIA is listed in Annex III of the Directive, but must only be provided

> inasmuch as: (a) The Member States consider that the information is relevant to a given stage of the consent procedure and to the specific characteristics of a particular project ... and of the environmental features likely to be affected; (b) The Member States consider that a developer may reasonably be required to compile this information having regard *inter alia* to current knowledge and methods of assessment. (Article 5.1)

Table 2.4 Information required in an EIA under EC Directive 85/337 (*as amended*)

Annex III (IV)

1. Description of the project
2. Where appropriate (*an outline of main alternatives studied and an indication of the main reasons for the final choice*).
3. Aspects of the environment likely to be significantly affected by the proposed project, including population, fauna, flora, soil, water, air climatic factors, material assets, architectural and archaeological heritage, landscape, and the interrelationship between them
4. Likely significant effects of the proposed project on the environment
5. Measures to prevent, reduce and where possible offset any significant adverse environmental effects
6. Non-technical summary
7. Any difficulties encountered in compiling the required information

Note: Amendment is shown in italic.

Table 2.4 summarizes the information required by Annex III (Annex IV, post-amendments). A developer is thus required to prepare an EIS that includes the information specified by the relevant Member State's interpretation of Annex III (Annex IV, post-amendments) and to submit it to the "competent authority". This EIS is then circulated to other relevant public authorities and made publicly available: "Member States shall take the measures necessary to ensure that the authorities likely to be concerned by the project...are given an opportunity to express their opinion" (Article 6.1).

Member States shall ensure that:

*   any request for development consent and any information gathered pursuant to [the Directive's provisions] are made available to the public;
*   the public concerned is given the opportunity to express an opinion before the project is initiated.

The detailed arrangements for such information and consultation shall be determined by the Member States (Articles 6.2 and 6.3) (see Section 6.2 also).

The competent authority must consider the information presented in an EIS, the comments of relevant authorities and the public, and the comments of other Member States (where applicable) in its consent procedure (Article 8). (The CEC (1994) published a checklist to help competent authorities to review environmental information.) It must then inform the public of the decision and any conditions attached to it (Article 9).

2.6 EC Directive 85/337, as amended by Directive 97/11/EC

Directive 85/337 included a requirement for a five-year review, and a report was published in 1993 (CEC 1993). Whilst there was general satisfaction that the "basics of the EIA are mostly in place", there has been concern about the incomplete coverage of certain projects, insufficient consultation and public participation, the lack of information about alternatives, weak monitoring and the lack of consistency in Member States' implementation. The review process, as with the original Directive, generated considerable debate between the Commission and the Member States, and the amended Directive went through several versions, with some weakening of the proposed changes. The outcome, finalized in March 1997, and to be implemented within 2 years, included the following amendments:

- Annex I (mandatory) – the addition of 12 new classes of project (e.g. dams and reservoirs, pipelines, quarries and open-cast mining) (Table 2.3).
- Annex II (discretionary) – the addition of 8 new sub-classes of project (plus extension to 10 others), including shopping and car parks, and particularly tourism and leisure (e.g. caravan sites and theme parks) (Table 2.3).
- New Annex III lists matters which must be considered in EIA including:

  - Characteristics of projects: size, cumulative impacts, the use of natural resources, the production of waste, pollution and nuisance, the risk of accidents.
  - Location of projects: designated areas and their characteristics, existing and previous land uses.
  - Characteristics of the potential impacts: geographical extent, trans-frontier effects, the magnitude and complexity of impacts, the probability of impact, the duration, frequency and reversibility of impacts.

- Change of previous Annex III to Annex IV: small changes in content.
- Other changes:

  - Article 2(3): There is no exemption from consultation with other Member States on transboundary effects.
  - Article 4: When deciding which Annex II projects will require EIA, Member States can use thresholds, case by case or a combination of the two.
  - Article 5.3: The minimum information provided by the developer *must include* an outline of the main alternatives studied and an indication of the main reasons for the final choice between alternatives.
  - Article 5.2: A developer may request an opinion about the information to be supplied in an environmental statement (ES), and a competent authority must provide that information. Member States may require authorities to give an opinion irrespective of the request from the developer.
  - Article 7: This requires consultation with affected Member States, and other countries, about transboundary effects.[6]
  - Article 9: A competent authority must make public the *main* reasons and considerations on which decisions are based, together with a description of the *main* mitigation measures (CEC 1997a).

A consolidated version of the full Directive, as amended by these changes, is included in Appendix 1. There will be more projects subject to mandatory EIA (Annex I) and discretionary EIA (Annex II). Alternatives also become mandatory, and there is emphasis on consultation and participation. The likely implication is more EIA activity in the EU Member States over the next decade. Member States will also have to face up to some challenging issues when dealing with topics such as alternatives, risk assessment (RA) and cumulative impacts.

2.7 An overview of EIA systems in the EU – divergent practice in a converging system?

The EU has been active in the field of environmental policy, and the EIA Directive is widely regarded as one of its more significant environmental achievements (see CEC 2001). However, there has been, and continues to be, concern about the inconsistency of application across the (increasing number of) Member States (see CEC 1993, CEC 2003, Glasson & Bellanger 2003). This partly reflects the nature of EC/EU directives, which seek to establish a mandatory framework for European policies whilst leaving the "scope and method" of implementation to each Member State. In addition, whatever the degree of "legal harmonization" of Member State EIA policies, there is also the issue of "practical harmonization". Implementation depends on practitioners from public and private sectors, who invariably have their own national cultures and approaches.

An early inconsistency was in the timing of implementation of the original Directive. Some countries, including France, the Netherlands and the UK, implemented the Directive relatively on time; others (e.g. Belgium, Portugal) did not. Other differences, understandably, reflected variations in legal systems, governance and culture between the Member States, and several of these differences are outlined below:

- The legal implementation of the Directive by the Member States differed considerably. For some, the regulations come under the broad remit of nature conservation (e.g. France, Greece, the Netherlands, Portugal); for some they come under the planning system (e.g. Denmark, Ireland, Sweden, the UK); in others specific EIA legislation was enacted (e.g. Belgium, Italy). In addition, in Belgium, and to an extent in Germany and Spain, the responsibility for EIA was devolved to the regional level.
- In most Member States, EIAs are carried out and paid for by the developers or consultants commissioned by them. However, in Flanders (Belgium) EIAs are carried out by experts approved by the authority responsible for environmental matters, and in Spain the competent authority carries out an EIA based on studies carried out by the developer.
- In a few countries, or national regions, EIA commissions have been established. In the Netherlands the commission assists in the scoping process, reviews the adequacy of an EIS and receives monitoring information from the competent authority. In Flanders, it reviews the qualifications of the people carrying out an EIA, determines its scope and reviews an EIS for compliance with legal requirements. Italy also has an EIA commission.
- The decision to proceed with a project is, in the simplest case, the responsibility of the competent authority (e.g. in Flanders, Germany, the UK). However, in some cases the minister responsible for the environment must first decide whether a project is environmentally compatible (e.g. in Denmark, Italy, Portugal).

2.7.1 *Reviews of the original Directive*

The first five-year review of the original Directive (CEC 1993), as noted, expressed concern about a range of inconsistencies in the operational procedures across the Member States (project coverage, alternatives, public participation, etc). As a result, several Member States strengthened their regulations to achieve a fuller implementation. A second five-year review in 1997 (CEC 1997b) had the following key findings:

- EIA is a regular feature of project licensing/authorization systems, yet wide variation exists in relation to those procedures (e.g. different procedural steps, relationships with other relevant procedures);
- Whilst all Member States had made provision for the EIA of the projects listed in Annex I, there were different interpretations and procedures for Annex II projects;
- Quality control over the EIA process is deficient;
- Member States did not give enough attention to the consideration of alternatives;
- Improvements had been made on public participation and consultation; and
- Member States, themselves, complained about the ambiguity and lack of definitions of several key terms in the Directive.

The amendments of 1997 sought to reduce further several of the remaining differences. In addition to the substantial extensions and modification to the list of projects in Annex I and Annex II, the amended Directive (CEC 1997a) also strengthened the procedural base of the EIA Directive. This included a provision for new screening arrangements, including new screening criteria (in Annex III) for Annex II projects. It also introduced EIS content changes, including an obligation on developers to include an outline of the main alternatives studied, and an indication of the main reasons for their choices, taking into account environmental effects. The amended Directive also enables a developer, if it so wishes, to ask a competent authority for formal advice on the scope of the information that should be included in a particular EIS. Member States, if they so wish, can require competent authorities to give an opinion on the scope of any new proposed EIS, whether the developer has requested one or not. The amended Directive also strengthens consultation and publicity, obliging competent authorities to take into account the results of consultations with the public and the reasons and considerations on which the decision on a project proposal has been based.

2.7.2 *Review of the amended Directive*

A third review of the original Directive, as amended by Directive 97/11/EC (CEC 1997a), undertaken for the EC by the Impacts Assessment Unit at Oxford Brookes University (UK), provided a detailed overview of the implementation of the Directive (as amended) by Member States, and recommendations for further enhancement of application and effectiveness (CEC 2003). Some of the key implementation issues identified included:

- Further delays in the transposition of the Directive. Many Member States missed the 1999 deadline, and by the end of 2002 transposition was still incomplete for Austria, France and Walloon and Flanders regions of Belgium. There was a complete lack of transposition for Luxembourg.
- Variations in thresholds used to specify EIA for Annex II projects. In all Member States, EIA is mandatory for Annex I projects, and some countries have added in

additional Annex I categories. Until the amendments to the Directive, Member States differed considerably in their interpretation of which Annex II projects required EIA. In some (e.g. the Netherlands), a compiled list specified projects requiring EIA. Subsequent to the amendments, most of the Member States now appear to make use of a combination of both thresholds and a case-by-case approach for Annex II projects. However, the 2003 review revealed that there are still major variations in the nature of the thresholds used. For example, with afforestation projects the area of planting that triggers mandatory EIA ranges from 30 ha in Denmark to 350 ha in Portugal. Similarly, 3 turbines would trigger mandatory EIA for a wind farm in Sweden, compared with 50 in Spain. Considerable variations also continue to exist in the detailed specification of which projects are covered by some Annex II categories, with 10(b) (urban development) being particularly problematic.

- Considerable variation in the number of EIAs being carried out in Member States. Documentation is complicated by inadequate data in some countries, but Table 2.5 shows the continuing great variation in annual output from over 7,000 (in France, where a relatively low financial criterion is a key trigger) to fewer than 20 (in Austria). Whilst some of the variation may be explained by the relative economic conditions within countries, it also relates to the variations in levels at which thresholds have been set. The amendments to the Directive do seem to be bringing more projects into the EIA process in some Member States.

- Some improvements, but still issues in relation to the scoping stage, and consideration of alternatives. Until the amendments made it a more formal stage of the EIA process, scoping was carried out as a discrete and mandatory step in only a few countries. The amended Directive allows Member States to make this a mandatory procedure if they so wish, and seven of the Member States have such

Table 2.5 Change in the amount of EIA activity in EU Member States

County	Pre-1999 (average p.a.)	Post-1999 (average p.a.)
Austria	4	10–20
Belgium – Brussels	20	20
Belgium – Flanders	No data	20 per cent increase
Belgium – Walloon	No data	est. increase
Denmark	28	100
Finland	22	25
France	6,000–7,000	7,000+
Germany	1,000	est. increase
Greece	1,600	1,600
Ireland	140	178
Italy	37	No data
Luxembourg	20	20
Netherlands	70	70
Portugal	87	92
Spain	120	290
Sweden	1,000	1,000
UK	300	500

Source: CEC (2003).

procedures in place. Commitment to scoping in the other Member States is more variable. Similarly, the consideration of alternatives to a proposed project was mandatory in only a very few countries, including the Netherlands which also required an analysis of the most environmentally acceptable alternatives in each case. The amended Directive requires developers to include an outline of the main alternatives studied. The 2003 review shows that in some Member States the consideration of alternatives is a central focus of the EIA process; elsewhere the coverage is less adequate – although the majority of countries do now require assessment of the zero (do minimum) alternative.

- Variations in nature of public consultation required in the EIA process. The Directive requires an EIS to be made available after it is handed to the competent authority, and throughout the EU the public is given an opportunity to comment on the projects that are subject to EIA. However, the extent of public involvement and the interpretation of "the public concerned" varies from quite narrow to wide. In Denmark, the Netherlands and Wallonia, the public is consulted during the scoping process. In the Netherlands and Flanders, a public hearing must be held after the EIS is submitted. In Spain, the public must be consulted before the EIS is submitted. In Austria, the public can participate at several stages of an EIA, and citizens' groups and the Ombudsman for the Environment have special status. The transposition of the Aarhus Convention into EIA legislation may provide an opportunity for improvements in public participation in EIA (CEC 2001).

- Variations in some key elements of EIA/EIS content, relating in particular to biodiversity, human health, risk and cumulative impacts. Whilst the EIA Directive does not make explicit reference to biodiversity and to health impacts, both can be seen as of increasing importance for EIA. There are some examples of good practice, in the Netherlands and Finland for biodiversity, and in the Netherlands again for health impact assessment. On the other hand, the amended Directive (Annex III) now includes risk and cumulative impacts. The 2003 review shows that although RAs appear in many EISs, for most Member States risk is seen as separate from the EIA process and handled by other control regimes. The review also shows a growing awareness of cumulative impacts, with measures put in place in many Member States (e.g. France, Portugal, Finland, Germany, Sweden and Denmark) to address them. However, it would seem that Member States are still grappling with the nature and dimensions of cumulative impacts.

- Lack of systematic monitoring of a project's actual impacts by the competent authority. Despite widespread concern about this Achilles' heel in the EIA Directive, there was considerable resistance to the inclusion of a requirement for mandatory monitoring. As such, there are very few good examples (e.g. the Netherlands) of a mandatory and systematic approach. Dutch legislation requires the competent authority to draw up an evaluation programme, which compares actual outcomes with those predicted in the EIS. If the evaluation shows effects worse than predicted, the competent authority may order extra environmental measures. In Greece, legislation provides for a review of the EIA outcome as part of the renewal procedure for an environmental permit.

Overall, the 2003 review showed that there are both strengths and weaknesses in the operation of the Directive, as amended. There are many examples of good practice, and the amendments have provided a significant strengthening of the procedural base of EIA, and have brought more harmonization in some areas – for example on the projects subject to EIA. Yet, as noted here, there is still a wide disparity in both the approach and the application of EIA in the Member States, and significant weaknesses remain to be addressed. The review concluded with a number of recommendations. These included advice to Member States to, *inter alia*, better record on an annual basis the nature of EIA activity; check national legislation with regard to aspects such as thresholds, quality control, cumulative impacts; make more use of EC guidance (e.g. on screening, scoping and review); and improve training provision for EIA.

2.8 Summary

This chapter has reviewed the development of EIA worldwide, from its unexpectedly successful beginnings in the USA to recent developments in the EU. In practice, EIA ranges from the production of very simple *ad hoc* reports to the production of extremely bulky and complex documents, from wide-ranging to non-existent consultation with the public, from detailed quantitative predictions to broad statements about likely future trends. In the EU, reviews of EIA experience show that "overall, although practice is divergent, it may not be diverging, and recent actions such as the amended Directive appear to be 'hardening up' the regulatory framework and may encourage more convergence" (Glasson & Bellanger 2003). All these systems worldwide have the broad aim of improving decision-making by raising decision-makers' awareness of a proposed action's environmental consequences. Over the past 35 years, EIA has become an important tool in project planning, and its applications are likely to expand further. Chapter 10 provides further discussion of EIA systems internationally and Chapter 12 discusses the widening of scope to strategic environmental assessment of policies, plans and programmes. The next chapter focuses on EIA in the UK context.

Notes

1. For example, *Ely v. Velds*, 451 F.2d 1130, 4th Cir. 1971; *Carolina Action v. Simon*, 522 F.2d 295, 4th Cir. 1975.
2. *Calvert Cliff's Coordinating Committee, Inc. v. United States Atomic Energy Commission* 449 F.2d 1109, DC Cir. 1971.
3. *Natural Resources Defense Council, Inc. v. Morton*, 458 F.2d 827, DC Cir. 1972.
4. California, Connecticut, Georgia, Hawaii, Indiana, Maryland, Massachusetts, Minnesota, Montana, New York, North Carolina, South Dakota, Virginia, Washington and Wisconsin, plus the District of Columbia and Puerto Rica.
5. Arizona, Arkansas, Delaware, Florida, Louisiana, Michigan, New Jersey, North Dakota, Oregon, Pennsylvania, Rhode Island and Utah.
6. Amendments to Articles 7 and 9 were influenced by the requirements of the Espoo Convention on EIA in a Transboundary Context, signed by 29 countries and the EU in 1991. This widened and strengthened the requirements for consultation with Member States where a significant transboundary impact is identified. The Convention deals with both projects and impacts that cross boundaries and is not limited to a consideration of projects that are in close proximity to a boundary.

References

Anderson, F. R., D. R. Mandelker, A. D. Tarlock 1984. *Environmental protection: law and policy.* Boston: Little, Brown.

Bear, D. 1990. EIA in the USA after twenty years of NEPA. *EIA newsletter* **4**, EIA Centre, University of Manchester.

Canter, L. W. 1996. *Environmental impact assessment,* 2nd edn. London: McGraw-Hill.

Catlow, J. & C. G. Thirwall 1976. *Environmental impact analysis* (DoE Research Report 11). London: HMSO.

CEC (Commission of the European Communities) 1973. First action programme on the environment. *Official Journal* C112, 20 December.

CEC 1980. Draft directive concerning the assessment of the environmental effects of certain public and private projects. COM(80), 313 final. *Official Journal* C169, 9 July.

CEC 1982. Proposal to amend the proposal for a Council directive concerning the environmental effects of certain public and private projects. COM(82), 158 final. *Official Journal* C110, 1 May.

CEC 1985. On the assessment of the effects of certain public and private projects on the environment. *Official Journal* L175, 5 July.

CEC 1992. *Towards sustainability.* Brussels: CEC.

CEC 1993. *Report from the Commission of the Implementation of Directive 85/337/EEC on the assessment of the effects of certain public and private projects on the environment.* COM(93), 28 final. Brussels: CEC.

CEC 1994. *Environmental impact assessment review checklist.* Brussels: EC Directorate-General XI.

CEC 1995. *Environmental impact assessment: guidance on screening DG XI.* Brussels: CEC.

CEC 1997a. Council Directive 97/11/EC of 3 March 1997 amending Directive 85/337/EEC on the assessment of certain public and private projects on the environment. *Official Journal* L73/5, 3 March.

CEC 1997b. *Report from the Commission of the Implementation of Directive 85/337/EEC on the Assessment of the Effects of Certain Public and Private Projects on the Environment.* Brussels: CEC.

CEC 2001. *Proposal for a Directive of the European Parliament and of the Council providing for public participation in respect of the drawing up of certain plans and programmes relating to the environment and amending Council Directives 85/337/EEC and 96/61/EC.* COM(2000), 839 final, 18 January 2001. Brussels: DG Environment, EC.

CEC 2003. (Impacts Assessment Unit, Oxford Brookes University). *Five Years Report to the European Parliament and the Council on the Application and Effectiveness of the EIA Directive.* website of DG Environment, EC: http://europa.eu.int/comm/environment/eia/home.htm.

CEQ (Council on Environmental Quality) 1978. National Environmental Policy Act. Implementation of procedural provisions: final regulations. *Federal Register* **43**(230), 55977–6007, 29 November.

CEQ 1993. *Environmental quality: 23rd annual report.* Washington, DC: US Government Printing Office.

Clark, B. D. & R. G. H. Turnbull 1984. Proposals for environmental impact assessment procedures in the UK. In *Planning and ecology,* R. D. Roberts & T. M. Roberts (eds), 135–44. London: Chapman & Hall.

Clark, B. D., K. Chapman, R. Bisset, P. Wathern 1976. *Assessment of major industrial applications: a manual* (DoE Research Report 13). London: HMSO.

Clark, B. D., K. Chapman, R. Bisset, P. Wathern, M. Barrett 1981. *A manual for the assessment of major industrial proposals.* London: HMSO.

CPRE 1991. *The environmental assessment directive: five years on.* London: Council for the Protection of Rural England.

Dobry, G. 1975. *Review of the development control system: final report.* London: HMSO.

DoE (Department of the Environment) 1977. Press Notice 68. London: Department of the Environment.

DoE 1978. Press Notice 488. London: Department of the Environment.

EIA (Environmental impact assessment) Centre. *EIA Newsletter* (annual). EIA Centre, University of Manchester.

European Bank for Reconstruction and Development 1992. *Environmental procedures.* London: European Bank for Reconstruction and Development.

Glasson, J. & C. Bellanger 2003. Divergent practice in a converging system? The case of EIA in France and the UK. *Environmental Impact Assessment Review* **23**, 605–24.

Glasson, J. & N. N. B. Salvador 2000. EIA in Brazil: a procedures-practice gap. A comparative study with reference to the EU, and especially the UK. *Environmental Impact Assessment Review* **20**, 191–225.

Hall, E. 1994. *The Environment versus people? A study of the treatment of social effects in EIA* (MSc dissertation). Oxford: Oxford Brookes University.

HMSO (Her Majesty's Stationery Office) 1971. *Report of the Roskill Commission on the Third London Airport.* London: HMSO.

House of Lords 1981a. *Environmental assessment of projects.* Select Committee on the European Communities, 11th Report, Session 1980–81. London: HMSO.

House of Lords 1981b. *Parliamentary Debates (Hansard) Official Report, Session 1980–81,* 30 April, 1311–47. London: HMSO.

Lee, N. & C. M. Wood 1984. Environmental impact assessment procedures within the European Economic Community. In *Planning and ecology,* R. D. Roberts & T. M. Roberts (eds), 128–34. London: Chapman & Hall.

Legore, S. 1984. Experience with environmental impact assessment in the USA. In *Planning and ecology,* R. D. Roberts & T. M. Roberts (eds), 103–12. London: Chapman & Hall.

Mandelker, D. R. 2000. *NEPA law and litigation,* 2nd edn. St Paul, MN: West Publishing.

Moreira, I. V. 1988. EIA in Latin America. In *Environmental impact assessment: theory and practice,* P. Wathern (ed.), 239–53. London: Unwin Hyman.

NEPA (National Environmental Policy Act) 1970. 42 USC 4321–4347, 1 January, as amended.

O'Riordan, T. & W. R. D. Sewell (eds) 1981. *Project appraisal and policy review.* Chichester: Wiley.

Orloff, N. 1980. *The National Environmental Policy Act: cases and materials.* Washington, DC: Bureau of National Affairs.

Overseas Development Administration 1996. *Manual of environmental appraisal: revised and updated.* London: ODA.

Petts, J. & P. Hills 1982. *Environmental assessment in the UK.* Institute of Planning Studies, University of Nottingham.

Sadler, B. 1996. *Environmental assessment in a changing world: evaluating practice to improve performance,* International Study on the Effectiveness of Environmental Assessment. Ottawa: Canadian Environmental Assessment Agency.

SDD (Scottish Development Department) 1974. *Appraisal of the impact of oil-related development,* DP/TAN/16. Edinburgh: SDA.

State of California, Governor's Office of Planning and Research 1992. CEQA: *California Environmental Quality Act – statutes and guidelines.* Sacramento: State of California.

Stevens Committee 1976. *Planning control over mineral working.* London: HMSO.

Swaffield, S. 1981. Environmental assessment: by administration or design? *Planning Outlook* **24**(3), 102–09.

Tomlinson, P. 1986. Environmental assessment in the UK: implementation of the EEC Directive. *Town Planning Review* **57**(4), 458–86.

Turner, T. 1988. The legal eagles. *Amicus Journal* (winter), 25–37.

UNEP (United Nations Environment Programme) 1997. *Environmental impact assessment training resource manual.* Stevenage: SMI Distribution.

White House 1994. *Memorandum from President Clinton for all heads of all departments and agencies on an executive order on federal actions to address environmental injustice in minority populations and low income populations*. Washington, DC: White House.

Williams, R. H. 1988. The environmental assessment directive of the European Community. In *The rôle of environmental impact assessment in the planning process*, M. Clark & J. Herington (eds), 74–87. London: Mansell.

Wood, C. 1981. The impact of the European Commission's directive on environmental planning in the United Kingdom. *Planning Outlook* **24**(3), 92–98.

Wood, C. 1988. The genesis and implementation of environmental impact assessment in Europe. In *The rôle of environmental impact assessment in the planning process*, M. Clark & J. Herington (eds), 88–102. London: Mansell.

World Bank 1992. *Environmental assessment sourcebook*. Washington, DC: World Bank.

World Bank 1995. *Environmental assessment: challenges and good practice*. Washington, DC: World Bank.

World Bank 1999. *Operational Policy, Bank Procedures and Good Practice4.01: Environmental Assessment*. Washington, DC: World Bank.

3 UK agency and legislative context

3.1 Introduction

This chapter discusses the legislative framework within which EIA is carried out in the UK. It begins with an outline of the principal actors involved in EIA and in the associated planning and development process. It follows with an overview of relevant regulations and the types of project to which they apply, then of the EIA procedures required by the T&CP (Assessment of Environmental Effects (AEE)) Regulations 1988 and the 1999 amendments. These can be considered the "generic" EIA regulations, which apply to most projects and provide a model for the other EIA regulations. The latter are then summarized. Readers should refer to Chapter 8 for a discussion of the main effects and limitations of the application of these regulations.

3.2 The principal actors

3.2.1 An overview

Any proposed major development has an underlying configuration of interests, strategies and perspectives. But whatever the development, be it a motorway, a power station, a reservoir or a forest, it is possible to divide those involved in the planning and development process broadly into four main groups. These are:

1. the developers;
2. those directly or indirectly affected by or having an interest in the development;
3. the government and regulatory agencies;
4. various intermediaries (consultants, advocates, advisers) with an interest in the interaction between the developer, the affected parties and the regulators (Figure 3.1).

An introduction to the range of "actors" involved is an important first step in understanding the UK legislative framework for EIA.

3.2.2 Developers

In the UK, EIA applies to projects in both the public and private sectors, although there are notable exemptions, including Ministry of Defence developments and those of the Crown Commission. Public sector developments are sponsored by

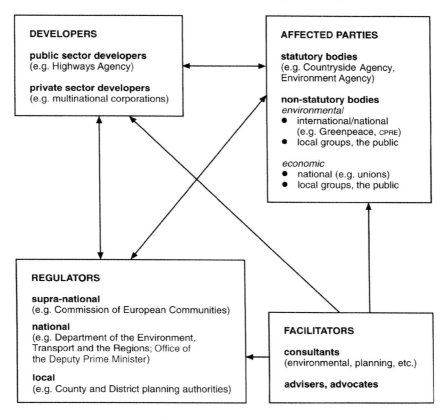

Figure 3.1 Principal actors in the EIA and planning and development processes.

central government departments (such as the Department for Transport (DfT)), by local authorities and by statutory bodies, such as the Environment Agency and the Highways Agency. Some were also sponsored by nationalized industries (such as the former British Rail and the nuclear industry), but the rapid privatization programme of the 1980s and 1990s has transferred many former nationalized industries to the private sector. Some, such as the major energy companies (Innogy, PowerGen, British Gas) and the regional water authorities, have major and continuing programmes of projects, where it may be possible to develop and refine EIA procedures, learning from experience. Many other private-sector companies, often of multinational form, may also produce a stream of projects. However, for many developers, a major project may be a one-off or "once in a lifetime" activity. For them, the EIA process, and the associated planning and development process, may be much less familiar, requiring quick learning and, it is to be hoped, the provision of some good advice.

3.2.3 Affected parties

Those parties directly or indirectly affected by such developments are many. In Figure 3.1 they have been broadly categorized, according to their role or degree of power (e.g. statutory, advisory), level of operation (e.g. international, national, local) or emphasis (e.g. environmental, economic). The growth in environmental groups, such as Greenpeace, Friends of the Earth, the Campaign to Protect Rural England (CPRE) and the Royal Society for the Protection of Birds (RSPB), is of particular note and is partly associated with the growing public interest in environmental issues. For instance, membership of the RSPB grew from 100,000 in 1970 to over a million in 1997. Membership of Sustrans, a charity which promotes car-free cycle routes, rose from 4,000 in 1993 to 20,000 in 1996. Such groups, although often limited in resources, may have considerable "moral weight". The accommodation of their interests by a developer is often viewed as an important step in the "legitimization" of a project. Like the developers, some environmental groups, especially at the national level, may have a long-term, continuing role. Some local amenity groups also may have a continuing role and an accumulation of valuable knowledge about the local environment. Others, usually at the local level, may have a short life, being associated with one particular project. In this latter category can be placed local pressure groups, which can spring up quickly to oppose developments. Such groups have sometimes been referred to as NIMBY ("not in my back yard"), and their aims often include the maintenance of property values and existing lifestyles, and the diversion of any necessary development elsewhere.

Statutory consultees are an important group in the EIA process. The planning authority must consult such bodies before making a decision on a major project requiring an EIA. Statutory consultees in England include the Countryside Agency, English Nature (EN), the Environment Agency (for certain developments) and the principal local council for the area in which the project is proposed. Other consultees often involved include the local highway authority and the county archaeologist. As noted above, non-statutory bodies, such as the RSPB and the general public, may provide additional valuable information on environmental issues.

3.2.4 *Regulators*

The government, at various levels, will normally have a significant role in regulating and managing the relationship between the groups previously outlined. As discussed in Chapter 2, the European Commission has adopted a Directive on EIA procedures (CEC 1985 and amendments). The UK government has subsequently implemented these through an array of regulations and guidance (see Section 3.3). The principal department involved currently is (2004) the Office of the Deputy Prime Minister (ODPM) (formerly DTLR, DETR and DoE!) through its London headquarters and regional offices. Notwithstanding the government scepticism noted in Chapter 2, William Waldegrave, UK Minister of State for the Environment, commented in 1987 that "...one of the most important tasks facing Government is to inspire a development process which takes into account not only the nature of any environmental risk but also the perceptions of the risk by the public who must suffer its consequences" (ESRC 1987).

Of particular importance in the EIA process is the local authority, and especially the relevant local planning authority (LPA). This may involve district, county and unitary authorities. Such authorities act as filters through which schemes proposed by developers usually have to pass. In addition, the LPA often opens the door for other agencies to become involved in the development process.

3.2.5 *Facilitators*

A final group, but one of particular significance in the EIA process, includes the various consultants, advocates and advisers who participate in the EIA and the planning and development processes. Such agents are often employed by developers; occasionally they may be employed by local groups, environmental groups and others to help to mount opposition to a proposal. They may also be employed by regulatory bodies to help them in their examination process.

A UK survey (Weston 1995) showed that environmental and planning consultancies carry out most of the EIA work, whereas consultancies specializing in such issues as archaeology or noise contribute less. There has been a massive growth in the number of environmental consultancies in the UK (Figure 3.2). The numbers have increased by over 400 per cent since the mid-1980s, and it has been estimated that clients in 2000 were spending approximately £1 billion on their services (ENDS 2001). Major factors underpinning the consultancy growth have been the advent of the UK Environmental Protection Act (EPA) in 1990, EIA regulations, the growing UK business interest in environmental management systems (e.g. BS 7750, ISO 14001), and the proposed EC regulations on eco-auditing and strategic environmental assessment (SEA).

Figure 3.3 provides a summary of the main work areas for environmental consultancies. Although the requirements of the EPA (with its "duty of care" regulations, which

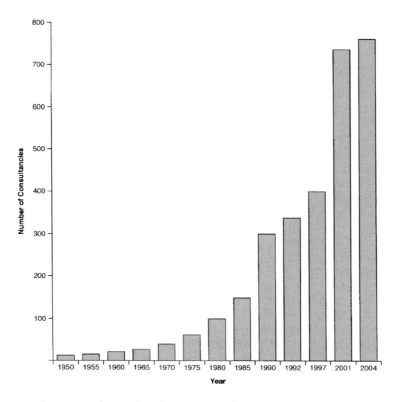

Figure 3.2 Increase in the number of environmental consultancies in the UK (1950–2004). (Based on ENDS 1993, 1997, 2001 and website.)

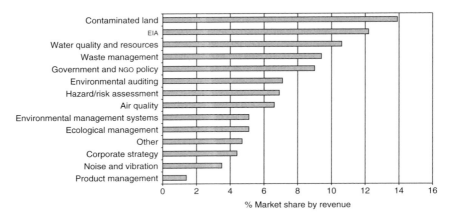

Figure 3.3 Main work areas for environmental consultancies (2002–03). (Adapted from ENDS 2003/2004 and website.)

came into force in April 1992) and the Water Resources Act of 1991 have concentrated the minds of developers and clients on water pollution and contaminated land in particular, there is no denying the significance of the EIA boom for consultants. Further characteristics of recent consultancy activity are discussed in Chapter 8.

3.2.6 Agency interaction

The various agencies outlined here represent a complex array of interests and aims, any combination of which may come into play for a particular development. This array has several dimensions, and within each there may be a range of often conflicting views. For example, there may be conflict between local and national views, between the interests of profit maximization and those of environmental conservation, between short-term and long-term perspectives and between corporate bodies and individuals. The agencies are also linked in various ways. Some links are statutory, others advisory. Some are contractual, others regulatory. The EIA regulations and guidance provide a set of procedures linking the various actors discussed, and these are now outlined.

3.3 EIA regulations: an overview

In the UK, EC Directive 85/337 was implemented through over 40 different secondary regulations under section 2(2) of the European Communities Act 1972. The large number of regulations was symptomatic of how EIA has been implemented in the UK. Different regulations apply to projects covered by the planning system, projects covered by other authorization systems and projects not covered by any authorization system but still requiring EIA. Different regulations apply to England and Wales, to Scotland and to Northern Ireland. The introduction of a new set of revised regulations from 1999 onwards, to implement the amended EC Directive, provided an opportunity for some tidying up of the list, but as Table 3.1 shows, there are still

Table 3.1 Key UK EIA regulations and dates of implementation

UK *regulations for projects subject to the Town and Country Planning system*

England and Wales

Town and Country Planning (Environmental Impact Assessment) Regulations 1999 (SI 293)

Town and Country Planning (General Permitted Development) Order 1995 (SI 418)

Scotland

Environmental Impact Assessment (Scotland) Regulations – Town and Country Planning, Roads and Bridges and Land Drainage 1999 (SSI 1)

Northern Ireland

Planning (Environmental Impact Assessment) Regulations (Northern Ireland) 1999 (SR 73)

UK EIA *regulations for projects subject to alternative consent systems*

Afforestation

Environmental Impact Assessment (Forestry) (England and Wales) Regulations 1999 (SI 2228)

Environmental Impact Assessment (Forestry) Regulations (Northern Ireland) 1999 (SR 84)

Environmental Impact Assessment (Forestry) Regulations (Scotland) 1999 (SSI 43)

Land drainage improvements

Environmental Impact Assessment (Land Drainage Improvement Works) Regulations 1999 (SI 1783)

Drainage (Environmental Assessment) Regulations (Northern Ireland) 1991 (SR 376)

Fish farming

Environmental Impact Assessment (Fish Farming in Marine Waters) Regulations 1999 (SI 367)

Environmental Impact Assessment (Fish Farming in Marine Waters) Regulations (Northern Ireland) 1999 (SI 415)

Trunk roads and motorways

Highways (Assessment of Environmental Effects) Regulations 1999 (SI 369)

Roads (Environmental Impact Assessment) Regulations (Northern Ireland) 1999 (SR 89)

Railways, tramways, inland waterways and works interfering with navigation rights

Transport and Works (Applications and Objections Procedure) (England and Wales) Rules 2000 (SI 2190)

Ports and harbours, and marine dredging

Environmental Impact Assessment and Habitats (Extraction of Minerals by Marine Dredging) Regulations 2002

Harbour Works (Environmental Impact Assessment) Regulations 1999 (SI 3445)

Harbour Works (Assessment of Environmental Effects) Regulations (Northern Ireland) 1990 (SR 181)

Power stations, overhead power lines and long-distance oil and gas pipeline

Electricity Works (Assessment of Environmental Effects) Regulations 2000 (SI 1927)

Electricity Works (Environmental Impact Assessment) (Scotland) Regulations 2000 (SSI 320)

Pipe-line Works (Environmental Impact Assessment) Regulations 2000 (SI 1928)

The Nuclear Reactors (Environmental Impact Assessment for Decommissioning) Regulations 1999 (SI 2892)

The Public Gas Transporter Pipe-line Works (Environmental Impact Assessment) Regulations 1999 (SI 1672)

Table 3.1 (*continued*)

Offshore Petroleum Production and Pipe-lines (Assessment of Environmental Effects) Regulations 1999 (SI 360)

Water Resources
The Water Resources (Environmental Impact Assessment) Regulations 2003 (SI 164)

Table 3.2 UK Government EIA guidance

DoE 1991. *Monitoring environmental assessment and planning*. London: HMSO

DoE 1994a. *Evaluation of environmental information for planning projects: a good practice guide*. London: HMSO

DoE 1994b. *Good practice on the evaluation of environmental information for planning projects: research report*. London: HMSO

DoE 1995. *Preparation of environmental statements for planning projects that require environmental assessment*. London: HMSO

DoE 1996. *Changes in the quality of environmental statements for planning projects*. London: HMSO

Environment Agency 1996. *A scoping handbook for projects*. London: HMSO

DETR 1997c. *Mitigation Measures in Environmental Statements*. London: HMSO

DETR 1999. *Circular 02/99 Environmental Impact Assessment*. London: HMSO

Planning Service (Northern Ireland) *Development Control Advice Note 10* 1999. Belfast: NI Planning Service

Scottish Executive Development Department *Circular 1999 15/99 The environmental impact assessment regulations* 1999. Edinburgh: SEDD

Scottish Executive Rural Affairs Department 1999. *Guide to the environmental impact assessment (Fish Farming in Marine Water) regulations* 1999. Edinburgh: SERAD

Scottish Executive Development Department 1999. *Planning Advice Note 58*. Edinburgh: SEDD

National Assembly for Wales 1999. *Circular 11/99 Environmental Impact Assessment*. Cardiff: National Assembly

DETR 2000. *Environmental impact assessment: a guide to the procedures*. Tonbridge: Thomas Telford

---

many regulations to ensure that all of the Directive's requirements are met. The regulations are supplemented by an array of EIA guidance from government and other bodies (Table 3.2). In addition, the Planning and Compensation Act 1991 allows the government to require EIA for other projects that fall outside the Directive.

In contrast to the US system of EIA, the EC EIA Directive applies to both public and private sector development. The developer carries out the EIA, and the resulting EIS must be handed in with the application for authorization. In England and Wales, most of the developments listed in Annexes I and II of the Directive fall under the remit of the planning system, and are thus covered by the Town and Country Planning (Environmental Impact Assessment) Regulations 1999 (the T&CP Regulations) (previously the Town and Country Planning (Assessment of Environmental Effects) Regulations 1988). Prior to the revised regulations in 1999 various additions and amendments were made to the 1988 T&CP Regulations to plug loopholes and extend the remit of the regulations, for instance:

- to expand and clarify the original list of projects for which EIA is required (e.g. to include motorway service areas and wind farms);
- to require EIA for projects that would otherwise be permitted (e.g. land reclamation, waste water treatment works, projects in Simplified Planning Zones);

- to require EIA for projects resulting from a successful appeal against a planning enforcement notice;
- to allow the then Secretary of State (SoS) for the Environment to direct that a particular development should be subject to EIA even if it is not listed in the regulations.

Other types of projects listed in the EC Directive require separate legislation, since they are not governed by the planning system. Of the various *transport* projects, local highway developments and airports are dealt with under the T&CP Regulations by the local planning (highways) authority, but motorways and trunk roads proposed and regulated by the Department for Transport (DfT) fall under the Highways (AEE) Regulations 1999. Applications for harbours are regulated by the DfT under the various Harbour Works (EIA) Regulations. New railways and tramways require EIA under the Transport and Works (Applications and Objections) Procedure 2000.

Energy projects producing less than 50 MW are regulated by the local authority under the T&CP Regulations. Those of 50 MW or over, most electricity power lines, and pipelines (in Scotland as well as in England and Wales) are controlled by the Department of Trade and Industry (DTI) under the various Electricity and Pipeline Works (EIA) Regulations 2000.

New *land drainage* works, including flood defence and coastal defence works, require planning permission and are thus covered by the T&CP Regulations. Improvements to drainage works carried out by the Environment Agency and other drainage bodies require EIA through the EIA (Land Drainage Improvement Works) Regulations, which are regulated by the Department for Environment, Food and Rural Affairs (DEFRA).

Forestry projects require EIA under the EIA (Forestry) Regulations 1999.

Marine fish farming within 2 km of the coast of England, Wales or Scotland requires a lease from the Crown Estates Commission, but not planning permission. For these developments, EIA is required under the EA (Fish Farming in Marine Waters) Regulations 1999.

Most other developments in Scotland are covered by the EIA (Scotland) Regulations 1999, including developments related to town and country planning, electricity, roads and bridges, development by planning authorities and land drainage. The British regulations apply to harbours, pipelines and forestry projects. Northern Ireland has separate legislation in parallel with that of England and Wales.

As will be discussed in Chapter 8, about 70 per cent of all the EIAs prepared in the UK fall under the T&CP Regulations, about 10 per cent fall under each of the EIA (Scotland) Regulations 1999 and the Highways (EIA) Regulations; almost all the rest involve land drainage, electricity and pipeline works, forestry projects in England and Wales and planning-related developments in Northern Ireland.

The enactment of this wide range of EIA regulations has made many of the early concerns regarding procedural loopholes (e.g. CPRE 1991, Fortlage 1990) obsolete. However several issues still remain. First is the ambiguity inherent in the term "project". An example of this is the EIA procedures for electricity generation and transmission, in which a power station and the transmission lines to and from it are seen as separate projects for the purposes of EIA, despite the fact that they are inextricably linked (Sheate 1995, and see Section 9.2). Another example is the division of road construction into several separate projects for planning and EIA purposes even though none of them would be independently viable. This is discussed further with regard to SEA in Chapter 12.

3.4 The Town and Country Planning (Environmental Impact Assessment) (England and Wales) Regulations 1999 (S1 293) (previously the Town and Country Planning (Assessment of Environmental Effects) Regulations 1988 (ESI 1199))

The T&CP Regulations implement the EC Directive for those projects that require planning permission in England and Wales. They are the central form in which the Directive is implemented in the UK; the other UK EIA regulations were established to cover projects that are not covered by the T&CP Regulations. As a result, the T&CP Regulations are the main focus of discussions on EIA procedures and effectiveness. This section presents the procedures of the T&CP Regulations. Figure 3.4 summarizes these procedures; the letters in the figure correspond to the letters in bold preceding the explanatory paragraphs below. Section 3.5 considers other main EIA regulations as variations of the T&CP Regulations and Section 3.6 comments on the changes following from the amended EC Directive.

The original T&CP Regulations were issued on 15 July 1988, 12 days after Directive 85/337 was to have been implemented. Guidance on the Regulations, aimed primarily at local planning authorities, was given in DoE Circular 15/88 (Welsh Office Circular 23/88). A guidebook entitled *Environmental assessment: a guide to the procedures* (DoE 1989), aimed primarily at developers and their advisers, was released in November 1989. Further DoE guidance on good practice in carrying out and reviewing EIAS was published in 1994 and 1995 (DoE 1994a,b, 1995), and in 1997 (DETR). The new 1999 T&CP Regulations were accompanied by new circulars on EIA (DETR 1999, SEDD 1999, and NAFW 1999), which give comprehensive guidance on the Regulations. A new guidebook, *Environmental Impact Assessment: a guide to the procedures* (DETR 2000) was also issued. This, the circulars, and other government guidance are strongly recommended reading. However, only the regulations are mandatory: the guidance interprets and advises, but cannot be enforced.

3.4.1 Which projects require EIA?

The T&CP Regulations require EIAs to be carried out for two broad categories of project, given in Schedules 1 and 2. These broadly corresponded[1] to Annexes I and II of Directive 85/337 before it was amended, excluding those projects that do not require planning permission. The amended schedules correspond very closely to Annexes I and II in the amended Directive, as outlined in Table 2.3 and detailed in Appendix 1. Schedule 1 has very minor wording changes from Annex I, plus the switch of Annex I, 1.20, long overhead electrical power lines, to the Electricity Works (AEE) Regulations 2000 (S1 1927). Schedule 2 has only very minor modifications from Annex II; primarily in 2.10 where (b) also includes sports stadiums, leisure centres and multiplex cinemas. Also, there is a separate category (p) for motorway service areas, and a few other categories are split or relocated. Schedule 2.12 also includes an additional category (f) for golf courses and associated developments. For Schedule 1 projects, EIA is required in every case. A Schedule 2 project requires EIA if it is deemed "likely to give rise to significant environmental effects".

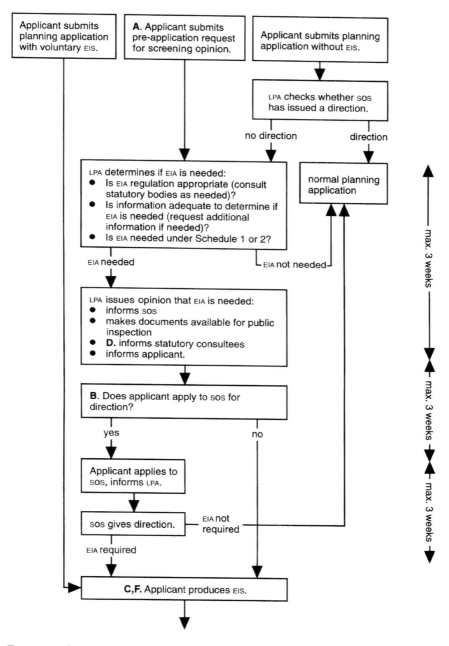

Figure 3.4 Summary of T&CP Regulations in EIA procedure. (Based on DETR 2000.)

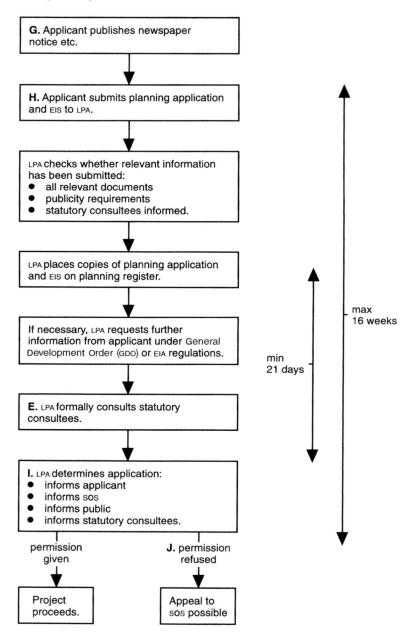

Figure 3.4 (*continued*)

The "significance" of a project's environmental effects is determined on the basis of three criteria (DETR 2000):

- whether it is a development of more than local importance [for example, in terms of physical scale];

- whether the development is proposed for a particularly environmentally sensitive or vulnerable location [for example, a national park or a site of special scientific interest];
- whether the development is likely to have unusually complex and potentially hazardous environmental effects [for example, in terms of the discharge of pollutants].

The guidebook (DETR 2000) includes indicative criteria and thresholds for a range of Schedule 2 projects which "should provide a starting point for consideration by the developer and the planning authority of the need for EIA". For instance, pig-rearing installations for more than 750 sows, industrial estate developments of more than 20 ha and new roads of over 2 km not located in a designated area may require EIA, according to the guidebook. The UK Government was one of the first to revise its thresholds in the context of the amended Directive. This followed extensive research commissioned from the Oxford Brookes' Impacts Assessment Unit (IAU), leading to a Government Consultation Paper (DETR 1997b). Table 3.3 lists the indicative thresholds and criteria.

Table 3.3 Indicative thresholds and criteria for identification of Schedule 2 development requiring EIA (1999)

The criteria and thresholds are only indicative. In determining whether significant effects are likely, the location of a development is of crucial importance. The more environmentally sensitive the location, the lower will be the threshold at which significant effects will be likely.

1. Agriculture and aquaculture
    In general, agricultural operations fall outside the scope of the Town and Country Planning system and, where relevant, will be regulated under other consent procedures. The descriptions below apply only to projects that are considered to be "development" for the purposes of the T&CP Act 1990.

    (a) *Use of uncultivated or semi-natural land for intensive agricultural purposes.* Development (such as greenhouses, farm buildings, etc.) on previously uncultivated land is unlikely to require EIA unless it covers more than 5 ha. In considering whether a particular development is likely to have significant effects, consideration should be given to impacts on the surrounding ecology, hydrology and landscape.
    (b) *Water management for agriculture, including irrigation and land drainage works.* EIA is more likely to be required if the development would result in permanent changes to the character of more than 5 ha of land. In assessing the significance of any likely effects, particular regard should be had to whether the development would have damaging wider impacts on hydrology and surrounding ecosystems. It follows that EIA will not normally be required for routine water management projects undertaken by farmers.
    (c) *Intensive livestock installations.* The significance or otherwise of the impacts of intensive livestock installations will often depend upon the level of odours, increased traffic and the arrangements for waste handling. EIA is more likely to be required for intensive livestock installations if they are designed to house more than 750 sows, 2,000 fattening pigs, 60,000 broilers or 50,000 layers, turkeys or other poultry.
    (d) *Intensive fish farming.* Apart from the physical scale of any development, the likelihood of significant effects will generally depend on the extent of any likely wider impacts on the hydrology and ecology of the surrounding area. Developments designed to produce more than 100 tonnes (dead weight) of fish per year will be more likely to require EIA.

Table 3.3 (*continued*)

(e)	*Reclamation of land from the sea.* In assessing the significance of any development, regard should be had to the likely wider impacts on natural coastal processes beyond the site itself, as well as to the scale of reclamation works themselves. EIA is more likely to be required where work is proposed on a site which exceeds 1 ha.

2. Extractive Industry

(a + b)	*Surface and underground mineral working.* The likelihood of significant effects will tend to depend on the scale and duration of the works, and the likely consequent impact of noise, dust, discharges to water and visual intrusion. All new open-cast mines and underground mines will generally require EIA. For clay, sand and gravel workings, quarries and peat extraction sites, EIA is more likely to be required if they would cover more than 15 ha or involve the extraction of more than 30,000 tonnes of mineral per year.
(c)	*Extraction of minerals by dredging in fluvial waters.* Particular consideration should be given to noise, and any wider impacts on the surrounding hydrology and ecology. EIA is more likely to be required where it is expected that more than 100,000 tonnes of mineral will be extracted per year.
(d)	*Deep drilling.* EIA is more likely to be required where the scale of the drilling operations involves development of a surface site of more than 5 ha. Regard should be had to the likely wider impacts on surrounding hydrology and ecology. On its own, exploratory deep drilling is unlikely to require EIA. It would not be appropriate to require EIA for exploratory activity simply because it might eventually lead to some form of permanent activity.
(e)	*Surface industrial installations for the extraction of coal, petroleum, natural gas, ores or bituminous shale.* The main considerations are likely to be the scale of development, emissions to air, discharges to water, the risk of accident and the arrangements for transporting the fuel. EIA is more likely to be required if the development is on a major scale (site of 10 ha or more) or where production is expected to be substantial (e.g. more than 100,000 tonnes of petroleum per year).

3. Energy industry

(a + b)	*Power stations.* EIA will normally be required for power stations which require approval from the Secretary of State at the Department of Trade and Industry (i.e. those with a thermal output of more than 50 MW). EIA is unlikely to be required for smaller new conventional power stations. Small stations using novel forms of generation should be considered carefully in line with the guidance in PPG 22 (Renewable Energy). The main considerations are likely to be the level of emissions to air, arrangements for the transport of fuel and any visual impact.
(c + d + e + f)	*Surface storage of fossil fuel and natural gas, underground storage of combustible gases, storage facilities for petroleum, petrochemical and chemical products.* In addition to the scale of the development, significant effects are likely to depend on discharges to water, emissions to air and risk of accidents. EIA is more likely to be required where it is proposed to store more than 100,000 tonnes of fuel. Smaller installations are unlikely to require EIA unless hazardous chemicals are stored.
(g)	*Installations for the processing and storage of radioactive waste.* EIA will normally be required for new installations whose primary purpose is to

Table 3.3 (*continued*)

process and store radioactive waste, and which are located on sites not previously authorised for such use. In addition to the scale of any development, significant effects are likely to depend on the extent of routine discharges of radiation to the environment. In this context EIA is unlikely to be required for installations where the processing or storage of radioactive waste is incidental to the main purpose of the development (e.g. installations at hospitals or research facilities).

(h) *Installations for hydroelectric energy production.* In addition to the physical scale of the development, particular regard should be had to the potential wider impacts on hydrology and ecology. EIA is more likely to be required for new hydroelectric developments which have more than 5 MW of generating capacity.

(i) *Wind farms.* The likelihood of significant effects will generally depend upon the scale of the development, and its visual impact, as well as potential noise impacts. EIA is more likely to be required for commercial developments of five or more turbines, or more than 5 MW of new generating capacity.

4–9. Industrial and manufacturing development

New manufacturing or industrial plants of the types listed in the Regulations may well require EIA if the operational development covers a site of more than 10 ha. Smaller developments are more likely to require EIA if they are expected to give rise to significant discharges of waste, emission of pollutants or operational noise. Among the factors to be taken into account in assessing the significance of such effects are:

- whether the development involves a process designated as a "scheduled process" for the purpose of air pollution control;
- whether the process involves discharges to water which require the consent of the Environment Agency;
- whether the installation would give rise to the presence of environmentally significant quantities of potentially hazardous or polluting substances;
- whether the process would give rise to radioactive or other hazardous waste;
- whether the development would fall under Council Directive 96/82/EC on the control of major accident hazards (COMAH) involving dangerous substances.

10.   Infrastructure developments

(a) *Industrial estates.* EIA is more likely to be required if the site area of the new development is more than 20 ha. In determining whether significant effects are likely, particular consideration should be given to the potential increase in traffic, emissions and noise.

(b) *Urban development projects (including the construction of shopping centres and car parks, sports stadiums, leisure centres and multiplex cinemas).* In addition to the physical scale of such developments, particular consideration should be given to the potential increase in traffic, emissions and noise. EIA is unlikely to be required for the redevelopment of land unless the new development is on a significantly greater scale than the previous use, or the types of impact are of a markedly different nature or there is a high level of contamination.

Developments proposed for sites which have not previously been intensively developed are more likely to require EIA if:

- the site area of the scheme is more than 5 ha; or
- it would provide a total of more than 10,000 m^2 of new commercial floorspace; or
- the development would have significant urbanizing effects in a previously non-urbanized area (e.g. a new development of more than 1,000 dwellings).

Table 3.3 (*continued*)

(c)	*Intermodal transhipment facilities and intermodal terminals.* In addition to the physical scale of the development, particular impacts for consideration are increased traffic, noise, emissions to air and water. Developments of more than 5 ha are more likely to require EIA.
(d+f+j)	*Construction of roads, railways (including elevated and underground) and tramways (unless included in Schedule 1).* For linear transport schemes, the likelihood of significant effects will generally depend on the estimated emissions, traffic, noise and vibration and degree of visual intrusion and impact on the surrounding ecology. EIA is more likely to be required for new development over 2 km in length.
(e)	*Construction of airfields (projects not included in Schedule 1).* The main impacts to be considered in judging significance are noise, traffic generation and emissions. New permanent airfields will normally require EIA, as will major works (such as new runways or terminals with a site area of more than 10 ha) at existing airports. Smaller scale development at existing airports is unlikely to require EIA unless it would lead to significant increases in air or road traffic.
(g)	*Construction of harbours and port installations, including fishing harbours.* Primary impacts for consideration are those on hydrology, ecology, noise and increased traffic. EIA is more likely to be required if the development is on a major scale (e.g. would cover a site of more than 10 ha). Smaller developments may also have significant effects where they include a quay or pier which would extend beyond the high water mark or would affect wider coastal processes.
(h)	*Construction of inland waterways, canalization and flood relief works.* The likelihood of significant impacts is likely to depend primarily on the potential wider impacts on the surrounding hydrology and ecology. EIA is more likely to be required for development of over 2 km of canal. The impact of flood relief works is especially dependent upon the nature of the location and the potential effects on the surrounding ecology and hydrology. Schemes for which the area of the works would exceed 5 ha or which are more than 2 km in length would normally require EIA.
(i)	*Dams and other installations designed to hold water or store it on a long-term basis.* In considering such developments, particular regard should be had to the potential wider impacts on the hydrology and ecology, as well as to the physical scale of the development. EIA is likely to be required for any major new dam (e.g. where the construction site exceeds 20 ha).
(k+l)	*Installation of oil pipelines, gas pipelines and long-distance aqueducts (including water and sewerage pipelines).* For underground pipelines, the major impact to be considered will generally be the disruption to the surrounding ecosystems during construction, while for overground pipelines visual impact will be a key consideration. EIA is more likely to be required for any pipeline over 6 km long. EIA is unlikely to be required for pipelines laid underneath a road, or for those installed entirely by means of tunnelling.
(m)	*Coastal works to combat erosion and maritime works capable of altering the coast.* The impact of such works will depend largely on the nature of the particular site and the likely wider impacts on natural coastal processes outside the site. EIA will be more likely where the area of the works would exceed 1 ha.

Table 3.3 (*continued*)

(n + o) *Groundwater abstraction and artificial groundwater recharge schemes, works for the transfer of water resources between river basins.* Impacts likely to be significant are those on hydrology and ecology. Developments of this sort can have significant effects on environments some kilometres distant. This is particularly important for wetland and other sites where the habitat and species are particularly dependent on an aquatic environment. EIA is likely to be required for developments where the area of the works exceeds 1 ha.

(p) *Motorway service areas.* Impacts likely to be significant are traffic, noise, air quality, ecology and visual impact. EIA is more likely to be required for new motorway service areas which are proposed for previously undeveloped sites and if the proposed development would cover an area of more than 5 ha.

11. Other projects

(a) *Permanent racing and test tracks for motorized vehicles.* Particular consideration should be given to the size, noise impacts, emissions and the potential traffic generation. EIA is more likely to be required for developments with a site area of 20 ha or more.

(b) *Installations for the disposal of non-hazardous waste.* The likelihood of significant effects will depend on the scale of the development and the nature of the potential impact in terms of discharges, emissions or odour. For installations (including landfill sites) for the deposit, recovery and/or disposal of household, industrial and/or commercial wastes (as defined by the Controlled Waste Regulations 1992), EIA is more likely to be required where new capacity is created to hold more than 50,000 tonnes per year, or to hold waste on a site of 10 ha or more. Sites taking smaller quantities of these wastes, sites seeking only to accept inert wastes (demolition rubble, etc.) or Civic Amenity sites, are unlikely to require EIA.

(c) *Waste-water treatment plants.* Particular consideration should be given to the size, treatment process, pollution and nuisance potential, topography, proximity of dwellings and the potential impact of traffic movements. EIA is more likely to be required if the development would be on a substantial scale (e.g. site area of more than 10 ha) or if it would lead to significant discharges (e.g. capacity exceeding 100,000 population equivalent). EIA should not be required simply because a plant is on a scale which requires compliance with the Urban Waste Water Treatment Directive (91/271/EEC).

(d) *Sludge-deposition sites (sewage sludge lagoons).* Similar considerations will apply for sewage sludge lagoons as for waste disposal installations. EIA is more likely to be required where the site is intended to hold more than 5,000 m^3 of sewage sludge.

(e) *Storage of scrap iron, including scrap vehicles.* Major impacts are likely to be discharges to soil, site noise and traffic generation. EIA is more likely to be required where it is proposed to store scrap on an area of 10 ha or more.

12. Tourism and leisure

(a) *Ski runs, ski lifts and cable cars and associated developments.* EIA is more likely to be required if the development is over 500 m in length or if it requires a site of more than 5 ha. In addition to any visual or ecological impacts, particular regard should also be had to the potential traffic generation.

(b) *Marinas.* In assessing whether significant effects are likely, particular regard should be had to any wider impacts on natural coastal processes outside the site, as well as the potential noise and traffic generation. EIA is more likely to be required for large new marinas, for example where

Table 3.3 (*continued*)

	the proposal is for more than 300 berths (seawater site) or 100 berths (freshwater site). EIA is unlikely to be required where the development is located solely within an existing dock or basin.
(c + d + e)	*Holiday villages and hotel complexes outside urban areas and associated developments, permanent camp sites and caravan sites, and theme parks.* In assessing the significance of tourism development, visual impacts, impacts on ecosystems and traffic generation will be key considerations. The effects of new theme parks are more likely to be significant if it is expected that they will generate more than 250,000 visitors per year. EIA is likely to be required for major new tourism and leisure developments which require a site of more than 10 ha. In particular, EIA is more likely to be required for holiday villages or hotel complexes with more than 300 bed spaces, or for permanent camp sites or caravan sites with more than 200 pitches.
(f)	*Golf courses.* New 18-hole golf courses are likely to require EIA. The main impacts are likely to be those on the surrounding hydrology, ecosystems and landscape, as well as those from traffic generation. Developments at existing golf courses are unlikely to require EIA.

(*Source*: DETR 1999, 2000; ODPM 2003.)

A. A developer may decide that a project requires EIA under the T&CP Regulations, or may want to carry out an EIA even if it is not required. If the developer is uncertain, the LPA can be asked for an opinion ("screening opinion") on whether an EIA is needed. To do this the developer must provide the LPA with a plan showing the development site, a description of the proposed development and an indication of its possible environmental impacts. The LPA must then make a decision within three weeks. The LPA can ask for more information from the developer, but this does not extend the three week decision-making period.

If the LPA decides that no EIA is needed, the application is processed as a normal planning application. If instead the LPA decides that an EIA is needed, it must explain why, and make both the developer's information and the decision publicly available. If the LPA receives a planning application without an EIA when it feels that it is needed, the LPA must notify the developer within three weeks, explaining why an EIA is needed. The developer then has three weeks in which to notify the LPA of the intention either to prepare an EIS or to appeal to the SoS; if the developer does not do so, the planning application is refused.

B. If the LPA decides that an EIA is needed but the developer disagrees, the developer can refer the matter to the SoS for a ruling.[2] The SoS must give a decision within three weeks. If the SoS decides that an EIA is needed, an explanation is needed; it is published in the *Journal of Planning and Environment Law*. No explanation is needed if no EIA is required. The SoS may make a decision if a developer has not requested an opinion, and may rule, usually as a result of information made available by other bodies, that an EIA is needed where the LPA has decided that it is not needed.

3.4.2 *The contents of the EIA*

Schedule 4 of the T&CP Regulations, which is shown in Table 3.4, lists the information that should be included in an EIA. Schedule 4 interprets the requirements of the EIA

Directive Annex IV (Annex III pre-amendments) according to the criteria set out in Article 5 of the Directive, namely:

> Member States shall adopt the necessary measures to ensure that the developer supplies in an appropriate form the information specified in Annex III (Annex IV, post-amendments) inasmuch as:
>
> (a) the Member States consider that the information is relevant to a given stage of the consent procedure and to the specific characteristics of a particular project or type of project and of the environmental features likely to be affected;
>
> (b) the Member States consider that a developer may reasonably be required to compile this information having regard *inter alia* to current knowledge and methods of assessment.

Table 3.4 Content of EIS required by the T&CP Regulations (1999) – Schedule 4

Under the definition in Regulation 2(1), "environmental statement" means a statement:

    (a)   that includes such of the information referred to in Part I of Schedule 4 as is reasonably required to assess the environmental effects of the development and which the applicant can, having regard in particular to current knowledge and methods of assessment, reasonably be required to compile, but

    (b)   that includes at least the information referred to in Part II of Schedule 4.

Part I

1.   Description of the development, including in particular:

    (a)   a description of the physical characteristics of the whole development and the land-use requirements during the construction and operational phases;

    (b)   a description of the main characteristics of the production processes, for instance, nature and quantity of the materials used;

    (c)   an estimate, by type and quantity, of expected residues and emissions (water, air and soil pollution, noise, vibration, light, heat, radiation, etc.) resulting from the operation of the proposed development.

2.   An outline of the main alternatives studied by the applicant or appellant and an indication of the main reasons for his choice, taking into account the environmental effects.

3.   A description of the aspects of the environment likely to be significantly affected by the development, including, in particular, population, fauna, flora, soil, water, air, climatic factors, material assets, including the architectural and archaeological heritage, landscape and the inter-relationship between the above factors.

4.   A description of the likely significant effects of the development on the environment, which should cover the direct effects and any indirect, secondary, cumulative, short, medium and long-term, permanent and temporary, positive and negative effects of the development, resulting from:

    (a)   the existence of the development;

    (b)   the use of natural resources;

    (c)   the emission of pollutants, the creation of nuisances and the elimination of waste, and the description by the applicant of the forecasting methods used to assess the effects on the environment.

5.   A description of the measures envisaged to prevent, reduce and where possible offset any significant adverse effects on the environment.

Table 3.4   *(continued)*

6.	A non-technical summary of the information provided under paragraphs 1 to 5 of this Part.
7.	An indication of any difficulties (technical deficiencies or lack of know-how) encountered by the applicant in compiling the required information.

Part II

1.	A description of the development comprising information on the site, design and size of the development.
2.	A description of the measures envisaged in order to avoid, reduce and, if possible, remedy significant adverse effects.
3.	The data required to identify and assess the main effects which the development is likely to have on the environment.
4.	An outline of the main alternatives studied by the applicant or appellant and an indication of the main reasons for his choice, taking into account the environmental effects.
5.	A non-technical summary of the information provided under paragraphs 1 to 4 of this Part.

(*Source:* DETR 2000.)

In Schedule 4, the information required in Annex IV has been interpreted to fall into two parts. The EIS must contain the information specified in Part II, and such relevant information in Part I "as is reasonably required to assess the effects of the project and which the developer can reasonably be required to compile". This distinction is important: as will be seen in Chapter 8, the EISs prepared to date have generally been weaker on Part I information, although this includes such important matters as the alternatives that were considered and the expected wastes or emissions from the development. In addition, in Appendix 5 of the guidebook (DETR 2000), the DETR has given a longer checklist of matters which may be considered for inclusion in an EIA: this list is for guidance only, but it helps to ensure that all the possible significant effects of the development are considered (Table 3.5).

C. Until the implementation of the amended Directive in 1999, there was no mandatory requirement in the UK for a formal "scoping" stage at which the LPA, the developer and other interested parties could agree on what would be included in the EIA. Indeed, there was no requirement for any kind of consultation between the developer and other bodies before the submission of the formal EIA and planning application, although guidance (DoE 1989) did stress the benefits of early consultation and early agreement on the scope of the EIA. The 1999 Regulations now enable a developer to ask the LPA for a formal "scoping opinion" on the information to be included in an EIS – in advance of the actual planning application. This allows a developer to be clear on LPA views on the anticipated key significant effects. The request must be accompanied by the same information provided for a screening opinion, and may be made at the same time as for the screening opinion. The LPA must consult certain bodies (see **D**), and must produce the scoping opinion within five weeks. The time period may be extended if the developer agrees. There is no provision for appeal to the SoS if the LPA and developer disagree on the content of an EIS. But if the LPA fails to produce a scoping opinion within the required timescale, the developer may apply to the SoS (or Assembly) for a scoping direction, also to be produced within five weeks, and also to be subject to consultation with certain bodies. The checklist (DETR 2000) provides a useful aid to developer–LPA discussions (see Table 3.5).

Table 3.5 Checklist of matters to be considered for inclusion in an environmental statement

---

This checklist is intended as a guide to the subjects that need to be considered in the course of preparing an environmental statement. It is unlikely that all the items will be relevant to any one project.

The environmental effects of a development during its construction and commissioning phases should be considered separately from the effects arising whilst it is operational. Where the operational life of a development is expected to be limited, the effects of decommissioning or reinstating the land should also be considered separately.

Section 1
Information describing the project

1.1 Purpose and physical characteristics of the project, including details of proposed access and transport arrangements, and of numbers to be employed and where they will come from.
1.2 Land-use requirements and other physical features of the project:

- during construction;
- when operational;
- after use has ceased (where appropriate).

1.3 Production processes and operational features of the project:

- type and quantities of raw materials, energy and other resources consumed;
- residues and emissions by type, quantity, composition and strength including;

  - discharges to water;
  - emissions to air;
  - noise;
  - vibration;
  - light;
  - heat;
  - radiation;
  - deposits/residues to land and soil;
  - others.

1.4 Main alternative sites and processes considered, where appropriate, and reasons for final choice.

Section 2
Information describing the site and its environment
Physical features

2.1 Population – proximity and numbers.
2.2 Flora and fauna (including both habitats and species) – in particular, protected species and their habitats.
2.3 Soil: agricultural quality, geology and geomorphology.
2.4 Water: aquifers, watercourses, shoreline, including the type, quantity, composition and strength of any existing discharges.
2.5 Air: climatic factors, air quality, etc.
2.6 Architectural and historic heritage, archaeological sites and features, and other material assets.
2.7 Landscape and topography.
2.8 Recreational uses.
2.9 Any other relevant environmental features.

Table 3.5 (*continued*)

The policy framework

2.10 Where applicable, the information considered under this section should include all relevant statutory designations such as national nature reserves, sites of special scientific interest, national parks, areas of outstanding natural beauty, heritage coasts, regional parks, country parks and designated green belt, local nature reserves, areas affected by tree preservation orders, water protection zones, conservation areas, listed buildings, scheduled ancient monuments and designated areas of archaeological importance. It should also include references to relevant national policies (including Planning Policy Guidance (PPG) notes) and to regional and local plans and policies (including approved or emerging development plans).

2.11 Reference should also be made to international designations, e.g. those under the EC "Wild Birds" or "Habitats" Directives, the Biodiversity Convention and the Ramsar Convention.

Section 3
Assessment of effects

Including direct and indirect, secondary, cumulative, short-, medium- and long-term, permanent and temporary, positive and negative effects of the project.

Effects on human beings, buildings and man-made features

3.1 Change in population arising from the development, and consequential environment effects.
3.2 Visual effects of the development on the surrounding area and landscape.
3.3 Levels and effects of emissions from the development during normal operation.
3.4 Levels and effects of noise from the development.
3.5 Effects of the development on local roads and transport.
3.6 Effects of the development on buildings, the architectural and historic heritage, archaeological features and other human artefacts, e.g. through pollutants, visual intrusion, vibration.

Effects on flora, fauna and geology

3.7 Loss of, and damage to, habitats and plant and animal species.
3.8 Loss of, and damage to, geological, palaeontological and physiographic features.
3.9 Other ecological consequences.

Effects on land

3.10 Physical effects of the development, e.g. change in local topography, effect of earth-moving on stability, soil erosion, etc.
3.11 Effects of chemical emissions and deposits on soil of site and surrounding land.
3.12 Land-use/resource effects:

- quality and quantity of agricultural land to be taken;
- sterilization of mineral resources;
- other alternative uses of the site, including the "do-nothing" option;
- effect on surrounding land uses including agriculture;
- waste disposal.

Effects on water

3.13 Effects of development on drainage pattern in the area.
3.14 Changes to other hydrographic characteristics, e.g. groundwater level, watercourses, flow of underground water.
3.15 Effects on coastal or estuarine hydrology.
3.16 Effects of pollutants, waste, etc. on water quality.

Table 3.5 (*continued*)

Effects on air and climate

3.17 Level and concentration of chemical emissions and their environmental effects.
3.18 Particulate matter.
3.19 Offensive odours.
3.20 Any other climatic effects.

Other indirect and secondary effects associated with the project

3.21 Effects from traffic (road, rail, air, water) related to the development.
3.22 Effects arising from the extraction and consumption of materials, water, energy or other resources by the development.
3.23 Effects of other development associated with the project, e.g. new roads, sewers, housing, power lines, pipelines, telecommunications, etc.
3.24 Effects of association of the development with other existing or proposed development.
3.25 Secondary effects resulting from the interaction of separate direct effects listed above.

Section 4
Mitigating measures

4.1 Where significant adverse effects are identified, a description of the measures to be taken to avoid, reduce or remedy those effects, e.g.:

    (a) site planning;
    (b) technical measures, e.g.:

        • process selection;
        • recycling;
        • pollution control and treatment;
        • containment (e.g. bunding of storage vessels).

    (c) aesthetic and ecological measures, e.g.:

        • mounding;
        • design, colour, etc.;
        • landscaping;
        • tree plantings;
        • measures to preserve particular habitats or create alternative habitats;
        • recording of archaeological sites;
        • measures to safeguard historic buildings or sites.

4.2 Assessment of the likely effectiveness of mitigating measures.

Section 5
Risk of accidents and hazardous development

5.1 Risk of accidents as such is not covered in the EIA Directive or, consequently, in the implementing Regulations. However, when the proposed development involves materials that could be harmful to the environment (including people) in the event of an accident, the environmental statement should include an indication of the preventive measures that will be adopted so that such an occurrence is not likely to have a significant effect. This could, where appropriate, include reference to compliance with Health and Safety legislation.
5.2 There are separate arrangements in force relating to the keeping or use of hazardous substances and the HSE provides local planning authorities with expert advice about risk assessment on any planning application involving a hazardous installation.
5.3 Nevertheless, it is desirable that, wherever possible, the risk of accidents and the general environmental effects of developments should be considered together, and developers and planning authorities should bear this in mind.

(*Source:* DETR 2000, ODPM 2003.)

3.4.3 *Statutory and other consultees*

Under the T&CP Regulations, a number of statutory consultees are involved in the EIA process, as noted in Section 3.2. These bodies are involved at two stages of an EIA, in addition to possible involvement in the scoping stage.

D. First, when an LPA determines that an EIA is required, it must inform the statutory consultees of this. The consultees in turn must make available to the developer, if so requested and at a reasonable charge, any relevant environmental information in their possession. For example, English Nature might provide information about the ecology of the area. This does not include any confidential information or information that the consultees do not already have in their possession.

E. Second, once the EIS has been submitted, the LPA or developer must send a free copy to each of the statutory consultees. The consultees may make representations about the EIS to the LPA for at least two weeks after they receive the EIS. The LPA must take account of these representations when deciding whether to grant planning permission. The developer may also contact other consultees and the general public while preparing the EIS. The Government guidance explains that these bodies may have particular expertise in the subject or may highlight important environmental issues that could affect the project. The developer is under no obligation to contact any of these groups, but again the DETR guidance stresses the benefits of early and thorough consultation.

3.4.4 *Carrying out the EIA; preparing the EIS*

F. The DETR gives no formal guidance about what techniques and methodologies should be used in EIA, noting only that they will vary depending on the proposed development, the receiving environment and the information available, and that predictions of effects will often have some uncertainty attached to them.

3.4.5 *Submitting the EIS and planning application; public consultation*

G. When the EIS has been completed, the developer must publish a notice in a local newspaper and post notices at the site. These notices must fulfil the requirements of the Town and Country Planning Act (General Permitted Development) Order 1995 (GDPO), state that a copy of the EIS is available for public inspection, give a local address where copies may be obtained and state the cost of the EIS, if any. The public can make written representations to the LPA for at least 20 days after the publication of the notice, but within 21 days of the LPA's receipt of the planning application.

H. After the EIS has been publicly available for at least 21 days, the developer submits to the LPA the planning application, copies[3] of the EIS, and certification that the required public notices have been published and posted. The LPA must then send copies of the EIS to the statutory consultees, inviting written comments within a specified time (at least two weeks from receipt of the EIS), forward another copy to the SoS and place the EIS on the planning register. It must also decide whether any additional information about the project is needed before a decision can be made, and, if so, obtain it from the developer. The clock does not stop in this case: a decision must still be taken within the appropriate time.

3.4.6 *Planning decision*

I. Before making a decision about the planning application, the LPA must collect written representations from the public within three weeks of the receipt of the planning application, and from the statutory consultees at least two weeks from their receipt of the EIS. It must wait at least three weeks after receiving the planning application before making a decision. In contrast to normal planning applications, which must be decided within eight weeks, those accompanied by an EIS must be decided within 16 weeks. If the LPA has not made a decision after 16 weeks, the applicant can appeal to the SoS for a decision. The LPA cannot consider a planning application invalid because the accompanying EIS is felt to be inadequate: it can only ask for further information within the 16-week period.

In making its decision, the LPA must consider the EIS and any comments from the public and statutory consultees, as well as other material considerations. The environmental information is only part of the information that the LPA considers, along with other material considerations. The decision is essentially still a political one, but it comes with the assurance that the project's environmental implications are understood. The LPA may grant or refuse permission, with or without conditions. Further to the changes resulting from the amended Directive the LPA must, in addition to the normal requirements to notify the applicant, notify the SoS and publish a notice in the local press, giving the decision, the main reasons on which the decision was based, together with a description of the main mitigation measures.

J. If an LPA refuses planning permission, the developer may appeal to the SoS, as for a normal planning application. The SoS may request further information before making a decision.

3.5 Other EIA regulations

This section summarizes the procedures of the other EIA regulations under which a large number of EISs have been prepared to date. We discuss the regulations in approximate descending order of frequency of application to date:

1. Environmental Impact Assessment (Scotland) Regulations 1999
2. Highways (AEE) Regulations 1999
3. Environmental Impact Assessment (Land Drainage Improvement Works) Regulations 1999
4. Electricity Works (AEE) Regulations 2000
5. Pipe-line Works (Environmental Impact Assessment) Regulations 2000
6. Environmental Assessment (Forestry) Regulations 1999.

3.5.1 Environmental Impact Assessment (Scotland) Regulations 1999 (SSI 1)

The EIA (Scotland) Regulations are broadly similar to those for England and Wales. They implement the Directive for projects which are subject to planning permission, but also cover some land drainage and trunk road projects. There is separate guidance on the Scottish Regulations (see Table 3.2). For some projects, for example for the decommissioning of nuclear power stations, Scotland is included in regulations which also apply to other parts of the UK. In Northern Ireland, the Directive is implemented for projects subject to planning permission by the Planning (EIA) Regulations (Northern Ireland) 1999.

3.5.2 Highways (Assessment of Environmental Effects) Regulations 1999 (SI 369)

The Highways (AEE) Regulations apply to motorways and trunk roads proposed by the Department of Transport (DoT). The regulations are approved under procedures set out in the Highways Act 1980, which require the SoS for Transport to publish an EIS for the proposed route when draft orders for certain new highways, or major improvements to existing highways, are published. The SoS determines whether the proposed project comes under Annex I or Annex II of the Directive, and whether an EIA is needed. EIA is mandatory for projects to construct new motorways and certain other roads, including those with four or more lanes, and for certain road improvements. The regulations require an EIS to contain:

- a description of the published scheme and its site;
- a description of measures proposed to mitigate adverse environmental effects;
- sufficient data to identify and assess the main effects that the scheme is likely to have on the environment;
- a non-technical summary.

Before 1993, the requirements of the Highways (AEE) Regulations were further elaborated in DoT standard AD 18/88 (DoT 1989) and the *Manual of Environmental Appraisal* (DoT 1983). In response to strong criticism,[4] particularly by the SACTRA (1992), these were superseded in 1993 by the *Design manual for roads and bridges*, vol. II: *Environmental assessment* (DoT 1993). The manual proposed a three-stage EIA process and gave extensive, detailed advice on how these EIAs should be carried out. The transport analysis guidance (DfT 2003) provides the latest evolution of the guidance for road projects.

3.5.3 Environmental Impact Assessment (Land Drainage Improvement Works) Regulations 1999 (SI 1783)

The EIA (Land Drainage Improvement Works) Regulations apply to almost all watercourses in England and Wales except public-health sewers. If a drainage body (including a local authority acting as a drainage body) determines that its proposed improvement actions are likely to have a significant environmental effect, it must publish a description of the proposed actions in two local newspapers and indicate whether it intends to prepare an EIS. If it does not intend to prepare one, the public can make representations within 28 days concerning any possible environmental impacts of the proposal; if no representations are made, the drainage body can proceed without an EIS. If representations are made, but the drainage body still wants to proceed without an EIS, DEFRA (National Assembly in Wales) gives a decision on the issue at ministerial level.

The contents required of the EIS under these regulations are virtually identical to those under the T&CP Regulations. When the EIS is complete, the drainage body must publish a notice in two local newspapers, send copies to English Nature, the Countryside Agency and any other relevant bodies and make copies of the EIS available at a reasonable charge. Representations must be made within 28 days and are considered by the drainage body in making its decision. If all objections are then withdrawn, the works can proceed; otherwise the minister gives a decision. Overall, these regulations are considerably weaker than the T&CP Regulations because of their weighting in favour of consent, unless objections are raised, and their minimal requirements for consultation with environmental organizations.

3.5.4 Electricity Works (Assessment of Environmental Effects) Regulations 2000 (SI 1927) The Nuclear Reactors (Environmental Impact Assessment for Decommissioning) Regulations 1999 (SI 2892)

The construction or extension of power stations exceeding 50 MW, and the installation of overhead power lines, requires consent from the SoS for the DTI, under Sections 36 and 37 of the Electricity Act 1989. The Electricity Works (EIA) Regulations 2000 is part of the procedure for applications under these provisions. EIA is required for:

- all thermal and nuclear power stations which fall under Annex 1 of the Directive (i.e. thermal power stations of 300 MW or more, and nuclear power stations of at least 50 MW);
- construction of overhead power lines of 220 KV or more and over 15 km in length.

The regulations also require proposed power stations not covered by Annex I, and all overhead power lines of at least 132 KV, to be screened for EIA. Power stations of less than 50 MW are approved under the planning legislation, through the T&CP (EIA) Regulations. The Electricity Works Regulations allow a developer to make a written request to the SoS to decide whether an EIA is needed. The SoS must consult with the LPA before making a decision. When a developer gives notice that an EIS is being prepared, the SoS must notify the LPA or the principal council for the relevant area, the Countryside Agency, EN and the Environment Agency in the case of a power station, so that they can provide relevant information to the applicant. The contents required of the EIS are almost identical to those listed in the T&CP Regulations.

The regulations on decommissioning of nuclear power stations were added in 1999. Dismantling and decommissioning require the consent of the Health and Safety Executive (HSE). A licensee who applies for consent must provide the HSE with an EIS. The regulations apply also to changes to existing dismantling or decommissioning projects which may have significant effects on the environment.

3.5.5 The Pipe-line Works (Environmental Impact Assessment) Regulations 2000 (SI 1928) The Public Gas Transporter Pipe-line Works (Environmental Impact Assessment) Regulations 1999 (SI 1672) Offshore Petroleum Production and Pipe-lines (Assessment of Environmental Effects) Regulations (SI 360)

The evolving array of regulations relating to pipelines reflects not only the growing importance of such development, but also the continuation of the fragmented UK approach to EIA legislation. The on-shore Pipe-line Works, and the Public Gas Transporter Pipe-line Works, Regulations apply to England, Wales and Scotland; the Offshore Petroleum Production and Pipe-lines Regulations apply to the whole of the UK. Oil and gas pipelines with a diameter of more than 800 mm and longer than 40 km come within Annex I of the Directive; those that fall below either of these thresholds are in Annex II. For the latter, pipelines 10 miles long or less are approved under the planning legislation. The rest fall under the above pipeline regulations, normally with determination by the relevant SoS in relation to associated consent and authorisation procedures and to various criteria and thresholds. For example, on-shore gas pipe-line works in Annex II of the Directive may be subject to EIA if they have a design operating pressure exceeding 7 bar gauge or either they wholly or in part cross a sensitive area (e.g. national park).

3.5.6 Environmental Impact Assessment (forestry) (England and Wales) Regulations 1999 (SI 2228)

Under the original EIA Directive and associated UK regulations, forestry EIAs were limited to those projects where applicants wished to apply for a grant or loan, for afforestation purposes, from the Forestry Agency. The lack of EIA requirements for other forestry projects, the perceived vested interest of the Forestry Agency as a promoter of forestry and the lack of EIA requirements for the Agency's own projects have all been criticised (e.g. by the CPRE (CPRE 1991)). The amended Directive and associated UK legislation have subsequently brought about some changes.

Afforestation and deforestation come under Annex II of the Directive. Under the above Regulations, anyone who proposes to carry out a forestry project that is likely to have significant effects on the environment must apply for a consent from the Forestry Agency before starting work. Those who apply for consent will be required to produce an EIS. The Regulations include: afforestation (creating new woodlands), deforestation (conversion of woodland to another use), constructing forest roads and quarrying material to construct forest roads. Where projects are below 5 ha (afforestation) and 1 ha (others), they may be deemed unlikely to have significant effects on the environment, unless they are in sensitive areas. Given the variability of sites and projects, the Forestry Agency considers applications on a case-by-case basis. An applicant who disagrees with the Forestry Agency's opinion may apply to the relevant SoS for a direction. The contents required of an EIA under the Forestry Regulations are almost identical to those required under the T&CP Regulations.

3.6 Summary and conclusions on changing legislation

The original (and amended) EC EIA Directive has been implemented in the UK through an array of regulations that link those involved – developers, affected parties, regulators and facilitators – in a variety of ways. The T&CP (EIA) Regulations are central. Other regulations cover projects that do not fall under the English and Welsh planning systems, such as motorways and trunk roads, power stations, pipelines, land drainage works, forestry projects, and development projects in Scotland and Northern Ireland. In addition, other planning legislation, currently the Planning and Compensation Act 1991, allows other projects not listed in the Directive to be subject to EIA.

The original UK Regulations had a number of weaknesses, relating to the range of projects included in the ambit of the EIA procedures, approaches to screening, scoping, consideration of alternatives, mandatory and discretionary EIS content, public consultation and others. Directive 97/11/EC sought to address some of these issues which had arisen in the UK and other Member States. The UK Government response was generally positive to the implementation of the finally agreed, and compromised, amendments (DETR 1997a,b):

> The Government believes that the Directive represents a significant improvement on the original, particularly in clarifying a number of ambiguities. The new measures should improve the consistency with which the environment is taken into account in major development decisions throughout the Community. At the same time, the Directive offers Member States sufficient flexibility to achieve this without adding unnecessarily to bureaucracy of burdens on developers... wherever possible, new measures will be incorporated into existing procedures. (DETR 1997a)

The new UK Regulations replicate even more closely the now four annexes of the amended EC Directive. This has brought a wider array of projects into the UK EIA system, has increased the number of mandatory categories and has led to a growth in EIA activity and EIS output. Screening procedures have been developed, and the consideration of scoping and alternatives now have a higher profile. There has been some rationalization of legislation, but the array is still complex. EIA guidance is a particular strength of the UK system, and government publications (especially the relevant circulars and the Guide to the Procedures, DETR 1999, NAFW 1999, SEDD 1999 and DETR 2000) help to navigate the legislative array.

Notes

1. There are some discrepancies. For instance, power stations of 300 MW or more are included in Schedule 1, although they actually fall under the Electricity Works (AEE) Regulations, and all "special roads" are included, although the regulations should apply to special roads under local authority jurisdiction.
2. The decision is actually made by the relevant Government Office in the region concerned (or the Assembly). As will be discussed in Chapter 8, this has led to some discrepancies where two or more offices have made different decisions on very similar projects.
3. This includes enough copies for all the statutory consultees to whom the developer has not already sent copies, one copy for the LPA and several for the Secretary of State.
4. The criticism was well deserved. The circular's assertion that "... individual highway schemes do not have a significant effect on climatic factors and, in most cases, are unlikely to have significant effects on soil or water" is particularly interesting in view of the cumulative impact of private transport on air quality.

References

CEC (Commission of the European Communities) 1985. On the assessment of the effects of certain public and private projects on the environment. *Official Journal* L175, 5 July.

CPRE (Council for the Protection of Rural England) 1991. *The Environmental Assessment Directive: five years on.* London: CPRE.

DETR (Department of the Environment, Transport and the Regions) 1997a. *Consultation paper: implementation of EC Directive (97/11/EC) on environmental assessment.* London: DETR.

DETR 1997b. *Consultation paper: implementation of EC Directive (97/11/EC) – determining the need for environmental assessment.* London: DETR.

DETR 1997c. *Mitigation measures in environmental assessment.* London: HMSO.

DETR 1999. *Circular 02/99. Environmental impact assessment.* London: HMSO.

DETR 2000. *Environmental impact assessment: a guide to the procedures.* Tonbridge: Thomas Telford Publishing.

DfT (Department for Transport) 2003. *Transport analysis guidance.* www.webtag.org.uk.

DoE (Department of the Environment) 1989. *Environmental assessment: a guide to the procedures.* London: HMSO.

DoE 1991. *Monitoring environmental assessment and planning.* London: HMSO.

DoE 1994a. *Evaluation of environmental information for planning projects: a good practice guide.* London: HMSO.

DoE 1994b. *Good practice on the evaluation of environmental information for planning projects: research report.* London: HMSO.

DoE 1995. *Preparation of environmental statements for planning projects that require environmental assessment: a good practice guide.* London: HMSO.

DoE 1996. *Changes in the quality of environmental statements for planning projects.* London: HMSO.

DoT (Department of Transport) 1983. *Manual of environmental appraisal.* London: HMSO.

DoT 1989. *Departmental standard HD 18/88: environmental assessment under the EC Directive 85/337.* London: Department of Transport.

DoT 1993. *Design manual for roads and bridges,* vol. 11: *Environmental assessment.* London: HMSO.

ENDS (Environmental Data Services) 1992. *Directory of environmental consultants 1992/93.* London: Environmental Data Services.

ENDS 1993. *Directory of environmental consultants 1993/94.* London: Environmental Data Services.

ENDS 1997. *Directory of environmental consultants 1997/98.* London: Environmental Data Services.

ENDS 2001. *Environmental Consultancy Directory,* 8th edn. London: Environmental Data Services.

ENDS 2003. *Directory of environmental consultants 2003/2004.* London: Environmental Data Services.

Environment Agency 1996. *A Scoping Handbook for Projects.* London: HMSO.

ESRC (Economic and Social Research Council) 1987. *Newsletter. Environmental Issues.* London: Economic and Social Research Council, February.

Fortlage, C. 1990. *Environmental assessment: a practical guide.* Aldershot: Gower.

NAFW (National Assembly for Wales) 1999. *Circular 11/99 Environmental Impact Assessment.* Cardiff: National Assembly.

ODPM (Office of the Deputy Prime Minister) 2003. *Environmental impact assessment: guide to the procedures.* London: HMSO.

Planning Service (Northern Ireland) *Development Control Advice Note 10* 1999. Belfast: NI Planning Service.

SACTRA (Standing Advisory Committee on Trunk Road Assessment) 1992. *Assessing the environmental impact of road schemes.* London: HMSO.

Scottish Executive Development Department *Circular 1999 15/99 The environmental impact assessment regulations 1999.* Edinburgh: SEDD.

SEDD (Scottish Executive Development Department) 1999. *Planning Advice Note 58.* Edinburgh: SEDD.

Sheate, W. R. 1995. Electricity generation and transmission: a case study of problematic EIA implementation in the UK. *Environmental Policy and Practice* **5**(1), 17–25.

Weston, J. 1995. Consultants in the EIA process. *Environmental Policy and Practice* **5**(3), 131–34.

Part 2
Process

This illustration by Neil Bennett is reproduced from Bowers, J. (1990), *Economics of the environment: the conservationist's response to the Pearce Report*, British Association of Nature Conservationists, 69 Regent Street, Wellington, Telford, Shropshire, TE1 1PE.

4 Starting up; early stages

4.1 Introduction

This is the first of four chapters that discuss how an EIA is carried out. The focus throughout is on both the procedures required by UK legislation and the ideal of best practice. Although Chapters 4–7 seek to provide a logical step-by-step approach through the EIA process, there is no one exclusive approach. Every EIA process is set within an institutional context, and the context will vary from country to country (see Chapter 10). As already noted, even in one country, the UK, there may be a variety of regulations for different projects (see Chapter 9). The various steps in the process can be taken in different sequences. Some may be completely missing in certain cases. The process should also not just be linear but build in cycles, with feedback from later stages to the earlier ones.

Chapter 4 covers the early stages of the EIA process. These include setting up a management process for the EIA activity, clarifying whether an EIA is required at all ("screening") and an outline of the extent of the EIA ("scoping"), which may involve consultation between several of the key actors outlined in Chapter 3. Early stages of EIA should also include an exploration of possible alternative approaches for a project. Baseline studies, setting out the parameters of the development action (including associated policy positions) and the present and future state of the environment involved, are also included in Chapter 4. However, the main section in the chapter is devoted to impact identification. This is important in the early stages of the process, but, reflecting the cyclical, interactive nature of the process, some of the impact identification methods discussed here may also be used in the later stages. Conversely, some of the prediction, evaluation, communication and mitigation approaches discussed in Chapter 5 can be used in the early stages, as can the participation approaches outlined in Chapter 6. The discussion in this chapter starts, however, with a brief introduction to the management of the EIA process.

4.2 Managing the EIA process

Environmental impact assessment is a management-intensive process. EIAs often deal with major (and sometimes poorly defined) projects, with many wide-ranging and often controversial impacts. As we noted in Chapter 3, they can involve many participants with very different perspectives on the relative merits and impacts of projects. It is important that the EIA process is well managed. This section notes some of the elements involved in such management.

The EIA process invariably involves an *interdisciplinary* team approach. Early US legislation strongly advocated such an approach:

> Environmental impact statements shall be prepared using an interdisciplinary approach which will ensure the integrated use of the natural and social sciences and the environmental design arts. The disciplines of the preparers shall be appropriate to the scope and issues identified in the scoping approach. (CEQ 1978, par. 1502.6)

Such an interdisciplinary approach not only reflects the normal scope of EIA studies, from the biophysical to the socio-economic, but also brings to the process the advantages of multiple viewpoints and perspectives on the complex issues involved (Canter 1991).

The *team* producing the EIS may be one, or a combination, of proponent in-house, lead external consultant, external sub-consultants and individual specialists. The size of the team may vary from two (one person, although sometimes used, does not constitute a team) to more than a dozen for some projects; the average is three or four. Fortlage (1990) identified 17 relevant specialist types, including town planner, ecologist, chemist, archaeologist and lawyer. A team should cover the main issues involved. A small team of three could, as exemplified by Canter, cover the areas of physical/chemical, biological/ecological and cultural/socio-economic, with a membership that might include, for example, an environmental engineer, an ecologist and a planner, at least one member having training or experience in EIA and management. However, the finalization of a team's membership may be possible only after an initial scoping exercise has been undertaken.

Many EIA teams make a clear distinction between a "core/focal" management team and associated specialists, often reflecting the fact that no one organization can cover all the inputs needed in the production of an EIS for a major project. Some commentators (see Weaver et al. 1996) promote the virtues of this approach. On a study for a major open-cast mining project in South Africa, Weaver et al. had a core project team of five people: a project manager, two senior authors, an editorial consultant and a word processor. This team managed the inputs into the EIA process, co-ordinated over 60 scientific and non-scientific contributors, and organized various public participation and liaison programmes.

The team project manager obviously has a pivotal role. In addition to personnel and team management skills, the manager should have a broad appreciation of the project type under consideration, a knowledge of the relevant processes and impacts subject to EIA, the ability to identify important issues and preferably a substantial area of expertise. Petts & Eduljee (1994) identify the following core roles for a project manager:

- selecting an appropriate project team;
- managing specialist inputs;
- liaising with the people involved in the process;
- managing change in the internal and external environment of the project;
- co-ordinating the contributions of the team in the various documentary outputs.

The management team has to *co-ordinate resources* – information, people and equipment – to achieve an EIA study of quality, on time and within its budget.

Budgets vary, as we have noted elsewhere (see Chapters 1 and 8), but may involve major expenditure for large projects. The time available may also vary; the average is 4–6 months, but it could be much longer for complex projects. National and international quality assurance procedures (for example BS 5750/ ISO 9000) may also apply for the activities of many companies.

In interdisciplinary team work, *complementarity*, *comparability* and *co-ordination* are particularly important. Weaver et al. (1996) stress the importance of complementarity for the technical skills needed to compete the task, and of personal skills for those in the core management team. Fortlage and others stress that where there are various groups of consultants, it is important that findings and data are co-ordinated (e.g. that they should work to agreed map scales and to agreed chapter formats) and can be fed into a central source. "This is one of the weakest aspects of most assessment teams; all consultants must be aware, and stay aware of others' work in order to avoid lacunae, anomalies and contradictions which will be the delight of opposing counsel and the media" (Fortlage 1990).

Of course, basic management skills – including team management and time management – must not be overlooked. Cleland & Kerzner (1986) suggested the following factors were important in the successful management of an interdisciplinary team:

(a) a clear, concise statement of the mission or purpose of the team;
(b) a summary of the goals or milestones that the team is expected to accomplish in planning and conducting the EIA;
(c) a meaningful identification of the major tasks required to accomplish the team's purposes, with each task broken down by individual;
(d) a summary delineation of the strategy of the team relative to policies, programs, procedures, plans, budgets, and other resource allocation methods required in the conduct of the environmental impact study;
(e) a statement of the team's organizational design, with information included on the roles and authority and responsibility of all members of the team, including the team leader; and
(f) a clear delineation of the human and non-human resource support services available for usage by the interdisciplinary team.

This should all be documented in *a clear statement of the interdisciplinary team approach* in the EIS. This would indicate the specific roles of team members, and their titles, qualifications and experience. The nature of liaison with other parties in the process, including public and other meetings, should also be noted.

4.3 Project screening – is an EIA needed?

The number of projects that could be subject to EIA is potentially very large. Yet many projects have no substantial or significant environmental impact. A screening mechanism seeks to focus on those projects with potentially significant adverse environmental impacts or whose impacts are not fully known. Those with few or no impacts are "screened out" and allowed to proceed to the normal planning permission and administrative processes without any additional assessment or additional loss of time and expense.

Screening can be partly determined by the EIA regulations operating in a country at the time of an assessment. Chapter 3 indicated that in the EC, including the UK, there are some projects (Annex/Schedule 1) that will always be "screened in" for full assessment, by virtue of their scale and potential environmental impacts (for example a crude-oil refinery, a sizeable thermal power station, a special road). There are many other projects (Annex/Schedule 2) for which the screening decision is less clear. Here two examples of a particular project may be screened in different ways (one "in" for full assessment, one "out") by virtue of a combination of criteria, including project scale, the sensitivity of the proposed location and the expectation of adverse environmental impacts. In such cases it is important to have working guidelines, indicative criteria and thresholds on conditions considered likely to give rise to significant environmental impacts (see Section 3.4).

In California, the list of projects that must always have the full review is determined by project type, development and location. For example, type includes, *inter alia*, a proposed local general plan; development includes, *inter alia*, a residential development of more than 500 units, a hotel or motel of more than 500 rooms, a commercial office building of more than 250,000 square feet of floor space; location includes, *inter alia*, the Lake Tahoe Basin, the California Coastal Zone, an area within a quarter of a mile of a wild and scenic area (State of California 1992). This constitutes an "inclusion list" approach. In addition, there may be an "exclusion list", as used in California and Canada, identifying those categories of project for which an EIA is not required because experience has shown that the adverse effects are not significant.

Some EIA procedures include an initial outline EIA study to check on likely environmental impacts and on their significance. Under the California Environmental Quality Act a "negative declaration" can be produced by the project proponent, thereby claiming that the project has minimal significant effects and does not require a full EIA. The declaration must be substantiated by an initial study, which is usually a simple checklist against which environmental impacts must be ticked as *yes*, *maybe* or *no*. If the responses are primarily *no*, and most of the *yes* and *maybe* responses can be mitigated, then the project may be screened out from a full EIA. In Canada and Australia, the screening procedures are also well developed (see Chapter 10).

In general there are two main approaches to screening. The *use of thresholds* involves placing projects in categories and setting thresholds for each project type. These may relate, for example, to project characteristics (e.g. 20 ha and over), to anticipated project impacts (e.g. 50,000 tonnes or more of waste per annum to be taken from a site) or to project location (e.g. a designated landscape area). See Table 3.3 for UK indicative thresholds.

A *case-by-case* approach involves the appraisal of the characteristics of projects, as they are submitted for screening, against a checklist of guidelines and criteria. Some of the advantages and disadvantages of these two approaches are summarized in Table 4.1. The EC (2001a) has published guidance to help in such case-by-case screening processes. There are also many hybrid approaches with, for example, indicative thresholds used in combination with a flexible case-by-case approach. Figure 4.1 provides an illustrative guide to the threshold system adopted in the UK, with a range from mandatory Annex 1 thresholds, through indicative Annex II thresholds, to exclusive thresholds, where EIA is not usually required (outside "sensitive areas"). This is often referred to as the "traffic lights" approach – red (mandatory), orange (indicative) and green (exclusive).

Table 4.1 Thresholds versus case-by-case approach to screening: advantages and disadvantages

Advantages	Disadvantages
Thresholds	
Simple to use	Place arbitrary, inflexible rules on a variable environment (unless tiered)
Quick to use; more certainty	Less room for common sense or good judgement
Consistent between locations	May be or become inconsistent with relevant neighbours
Consistent between decisions within locations	Difficult to set and, once set, difficult to change
Consistent between project types	Lead to a proliferation of projects lying just below the thresholds
Case by Case	
Allows common sense and good judgement	Likely to be complex and ambiguous
Flexible – can incorporate variety in project and environment	Likely to be slow and costly
Can evolve (and improve) easily	Open to abuse by decision-makers because of political or financial interests
	Open to poor judgement of decision-makers
	Likely to be swayed by precedent and therefore lose flexibility

The DETR (2000), ODPM (2003b) give more detailed guidance on how screening is carried out for most English and Welsh development projects.

4.4 Scoping – which impacts and issues to consider?

The scope of an EIA is the impacts and issues it addresses. The process of scoping is that of deciding, from all of a project's possible impacts and from all the alternatives that could be addressed, which are the significant ones. An initial scoping of possible impacts may identify those impacts thought to be potentially significant, those thought to be not significant and those whose significance is unclear. Further study should examine impacts in the various categories. Those confirmed by such a study to be not significant are eliminated; those in the uncertain category are added to the initial category of other potentially significant impacts. This refining of focus onto the most significant impacts continues throughout the EIA process. Good scoping has been shown to be a key factor in good EIS (Mulvihill & Baker 2001, Wende 2002).

Scoping is generally carried out in discussions between the developer, the competent authority, other relevant agencies and, ideally, the public. It is often the first stage of negotiations and consultation between a developer and other interested parties. It is an important step in EIA because it enables the limited resources of the team carrying out an EIA to be allocated to best effect, and prevents misunderstanding between the parties concerned about the information required in an EIS. Scoping can also identify

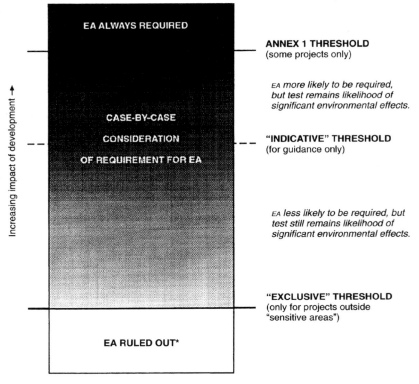

Figure 4.1 An illustrative guide to the threshold system. (*Source:* DETR 1997.)

issues that should later be monitored. Although it is an important step in the EIA process, it has not been a legally mandated step in the UK. Developer consultation with the competent authority and statutory consultees before preparing an EIS is very important but happened in only about half of all cases in early UK EIA practice (DoE 1996, Eastman 1997, Fuller 1992). This lack of early discussion was one of the principal limitations to effective EIA to date. The Government (DETR 2000, ODPM 2003a) strongly recommends such consultation and, as noted in Chapter 3, since the implementation of the amended Directive in 1999, the developer can now ask the LPA for a formal "scoping opinion" on the information to be included in an EIS.

Scoping should begin with the identification of individuals, communities, local authorities and statutory consultees likely to be affected by the project; good practice would be to bring them together in a working group and/or meetings with the developer. One or more of the impact identification techniques discussed in Section 4.8 can be used to structure a discussion and suggest important issues to consider. Other issues could include:

- particularly valued environmental attributes;
- those impacts considered of particular concern to the affected parties;

- the methodology that should be used to predict and evaluate different impacts;
- the scale at which those impacts should be considered;[1]
- broad alternatives that might be considered.

Reference should be made to relevant national, regional and local development plans, subject plans and government policies and guidelines, which we discuss in Section 4.7. Various alternatives should be considered, as discussed in Section 4.5. The result of this process of information collection and negotiation should be the identification of the chief issues and impacts, an explanation of why other issues are not considered significant, and, for each key impact, a defined temporal and spatial boundary within which it will be measured. Some developers, such as the Highways Agency for England, produce a scoping report as a matter of good practice. This indicates the proposed coverage of the EIA and the uncertainties that have been identified and can act as a basis for further studies and for public participation.

Other countries (e.g. Canada and The Netherlands) have a formal scoping stage, in which the developer agrees with the competent authority or an independent EIA commission, sometimes after public consultation, on the subjects the EIA will cover. The EC (2001b) has published a scoping checklist. In addition guidance on impacts normally associated with particular types of projects are being developed by various government and other regulatory agencies (e.g. (UK) Environment Agency (2002) and Government of New South Wales (1996)).

The importance of scoping and consultation early in the EIA process was highlighted in a research report for the UK Department of the Environment (DoE 1996). It identified early consultation and scoping as very important for the quality of the EIS, for all participants in the EIA process. Indeed it can be argued that one of the most valuable roles of the EIA process is to encourage such consultation.

4.5 The consideration of alternatives

4.5.1 *Regulatory requirements*

The US Council on Environmental Quality (CEQ 1978) calls the discussion of alternatives "the heart of the environmental impact statement": how an EIA addresses alternatives will determine its relation to the subsequent decision-making process. A discussion of alternatives ensures that the developer has considered both other approaches to the project and the means of preventing environmental damage. It encourages analysts to focus on the *differences* between real choices. It can allow people who were not directly involved in the decision-making process to evaluate various aspects of a proposed project and how decisions were arrived at. It also provides a framework for the competent authority's decision, rather than merely a justification for a particular action. Finally, if unforeseen difficulties arise during the construction or operation of a project, a re-examination of these alternatives may help to provide rapid and cost-effective solutions.

The original EC Directive 85/337 stated that alternative proposals should be considered in an EIA, subject to the requirements of Article 5 (if the information is relevant and if the developer may reasonably be required to compile this information). Annex III required "where appropriate, an outline of the main alternatives studied by the developer and an indication of the main reasons for this choice, taking into account the environmental effects". In the UK, this requirement was interpreted as being discretionary. This led to the consideration of alternatives being one of the weakest aspects of EIS quality (Barker & Wood 1999, Eastman 1997, Jones et al. 1991).

One of the main changes in the amendments of EC Directive 97/11 (CEC 1997) was to strengthen the requirements on alternatives: EISs are now required to include "an outline of the main alternatives studied by the developer and an indication of the main reasons for the developer's choice, taking into account the environmental effects". Current UK guidance (ODPM 2003a) is that:

> It is widely regarded as good practice to consider alternatives, as it results in a more robust application for planning permission. Also, the nature of certain developments and their location may make the consideration of alternatives a material consideration. Where alternatives are considered, the main ones must be outlined in the environmental statement.

The Department for Transport's Transport Analysis Guidance (DfT 2003) also encourages the consideration of alternatives.

4.5.2 Types of alternative

During the course of project planning, many decisions are made concerning the type and scale of the project proposed, its location and the processes involved. Most of the possible alternatives that arise will be rejected by the developer on economic, technical or regulatory grounds. The role of EIA is to ensure that environmental criteria are also considered at these early stages. A thorough consideration of alternatives would begin early in the planning process, before the type and scale of development and its location have been agreed on. A number of broad types of alternative can be considered: the "no action" option, alternative locations, alternative scales of the project, alternative processes or equipment, alternative site layouts, alternative operating conditions and alternative ways of dealing with environmental impacts. We shall discuss the last of these in Section 5.4.

The "no action" *option* refers to environmental conditions if a project were not to go ahead. In essence, consideration of the "no action" option is equivalent to a discussion of the need for the project: do the benefits of the project outweigh its costs? Consideration of this option is required in some countries, e.g. the USA,[2] but has been rarely discussed in UK EISs.

The consideration of alternative *locations* is an essential component of the project planning process. In some cases, a project's location is constrained in varying degrees: for instance, gravel extraction can take place only in areas with sufficient gravel deposits, and wind farms require locations with sufficient wind speed. In other cases, the best location can be chosen to maximize, for example, economic, planning and environmental considerations. For industrial projects, for instance, economic criteria such as land values, the availability of infrastructure, the distance from sources and markets, and the labour supply are likely to be important (Fortlage 1990). For road projects, engineering criteria strongly influence the alignment. In all these cases, however, siting the project in "environmentally robust" areas, or away from designated or environmentally sensitive areas, should be considered.

The consideration of different *scales* of development is also integral to project planning. In some cases, a project's scale will be flexible. For instance, the scale of a waste-disposal site can be changed, depending, for example, on the demand for landfill space, the availability of other sites and the presence of nearby residences or environmentally sensitive sites. The number of turbines on a wind farm could vary widely. In other cases, the developer will need to decide whether an entire unit should be built or not. For instance, the reactor building of a PWR nuclear power station is a large discrete structure that cannot easily be scaled down. Pipelines or bridges, to be functional, cannot be broken down into smaller sections.

Alternative *processes and equipment* involve the possibility of achieving the same objective by a different method. For instance, 1500 MW of electricity can be generated by one combined-cycle gas turbine power station, by a tidal barrage, by several waste-burning power stations or by hundreds of wind turbines. Gravel can be directly extracted or recycled, using wet or dry processes. Waste may be recycled, incinerated or put in a landfill.

Once the location, scale and processes of a development have been decided upon, different *site layouts* can still have different impacts. For instance, noisy plant can be sited near or away from residences. Power-station cooling towers can be few and tall (using less land) or many and short (causing less visual impact). Buildings can be sited either prominently or to minimize their visual impact. Similarly, *operating conditions* can be changed to minimize impacts. For instance, a level of noise at night is usually more annoying than the same level during the day, so night-time work could be avoided. Establishing designated routes for project-related traffic can help to minimize disturbance to local residents. Construction can take place at times of the year that minimize environmental impacts, for example on migratory and nesting birds. These kinds of "alternatives" act like mitigation measures.

Alternatives must be reasonable: they should not include ideas that are not technically possible, or illegal. The type of alternatives that can realistically be considered by a given developer will also vary: a mineral extraction company that has put a deposit on a parcel of land in the hope of extracting sand and gravel from it will not consider the option of using it for wind power generation: "reasonable" in such a case would be other sites for sand and gravel extraction, or other scales or processes. Essentially, alternatives should allow the competent authority to understand why this project, and not some other, is being proposed in this location and not some other.

On the other hand, from a US context (where EISs are prepared by government agencies) Steinemann (2001) argues that alternatives that do not meet a narrow definition of project objectives tend to be too easily rejected, and that alternatives should reflect social, not just agency, goals. She also suggests that

> the current sequence – propose action, define purpose and need, develop alternatives, then analyze alternatives – needs to be revised. Otherwise the proposed action can bias the set of alternatives for the analysis. Agencies should explore more environmentally sound approaches before proposing an action. Then, agencies should construct a purpose and need statement that would not summarily exclude less damaging alternatives, nor unduly favour the proposed action. Agencies should also be careful not to adhere to a single "problem" and "solution" early on.

Although private developers do not have the wider public remit of agencies, the basic approach of considering a broader range of alternatives and objectives than they would do without EIS can apply to private developers as well.

4.5.3 The presentation and comparison of alternatives

The costs of alternatives vary for different groups of people and for different environmental components. Discussions with local residents, statutory consultees and special
interest groups may rapidly eliminate some alternatives from consideration and suggest others. However, it is unlikely that one alternative will emerge as being most
acceptable to all the parties concerned. The EIS should distil information about a
reasonable number of realistic alternatives into a format that will facilitate public
discussion and, finally, decision-making. Methods for comparing and presenting
alternatives span the range from simple, non-quantitative descriptions, through
increasing levels of quantification, to a complete translation of all impacts into their
monetary values.

Many of the impact identification methods discussed later in this chapter can also
help to compare alternatives. Overlay maps compare the impacts of various locations
in a non-quantitative manner. Checklists or less complex matrices can also be
applied to various alternatives and compared; this may be the most effective way to
present the impacts of alternatives visually. Some of the other techniques used for
impact identification – the threshold-of-concern checklist, weighted matrix and EES –
allow alternatives to be implicitly compared. They do this by assigning quantitative
importance weightings to environmental components, rating each alternative
(quantitatively) according to its impact on each environmental component, multi-
plying the ratings by their weightings to obtain a weighted impact, and aggregating
these weighted impacts to obtain a total score for each alternative. These scores can
be compared with each other to identify preferable alternatives. With the exception
of the threshold-of-concern checklist, they do not lend themselves to the clear presentation of the alternatives in question, and none of them clearly states who will be
affected by the different alternatives.

4.6   Understanding the project/development action

4.6.1 Understanding the dimensions of the project

At first glance, this description of a proposed development would appear to be one of
the more straightforward steps in the EIA process. However, projects have many
dimensions, and relevant information may be limited. As a consequence, this first
step may pose some challenges. Crucial dimensions to be clarified include the
purpose of the project, its life cycle, physical presence, process(es), policy context
and associated policies.

Schedule 3 of the T&CP (EIA) (England and Wales) Regulations 1999 requires
"a description of the development proposed, comprising information about the site
and the design and scale or size of the development" and "the data necessary to identify
and assess the main effects which that development is likely to have on the environment".
It also requires:

- a description of the physical characteristics of the whole development and the
  land-use requirements during the construction and operational phases;
- a description of the main characteristics of the production processes, for
  instance, nature and quantity of the materials used;

- an estimate, by type and quantity, of expected residues and emissions (water, air and soil pollution, noise, vibration, light, heat, radiation, etc.) resulting from the operation of the proposed development;

where such information may be "reasonably required to assess the environmental effects of the development and which the applicant can, having regard in particular to current knowledge and methods of assessment, reasonably be required to compile". *Environmental impact assessment: guide to the procedures* (ODPM 2003a) provides a longer checklist of information that may be used to describe the project (see Table 3.5).

An outline of *the purpose and rationale* of a project provides a useful introduction to the project description. This may, for example, set the particular project in a wider context – the missing section of a major motorway, a power station in a programme of developments, a new settlement in an area of major population growth. A discussion of purpose may include the rationale for the particular type of project, for the choice of the project's location and for the timing of the development. It may also provide background information on planning and design activities to date.

As we noted in Section 1.5, all projects have *a life cycle of activities*, and a project description should clarify the various stages in the life cycle, and their relative duration, of the project under consideration. A minimum description would usually involve the identification of construction and operational stages and associated activities. Further refinement might include planning and design, project commissioning, expansion, close-down and site rehabilitation stages. The size of the development at various stages in its life cycle should also be specified. This can include reference to inputs, outputs, physical size and the number of people to be employed.

The *location and physical presence* of a project should also be clarified at an early stage. This should include its general location on a base map in relation to other activities and to administrative areas. A more detailed site layout of the proposed development, again on a large-scale base map, should illustrate the land area and the main disposition of the elements of the project (e.g. storage areas, main processing plant, waste-collection areas, transport connections to the site). Where the site layout may change substantially between different stages in the life cycle, it is valuable to have a sequence of anticipated layouts. Any associated projects and activities (e.g. transport connections to the site; pipes and transmission lines from the site) should also be identified and described, as should elements of a project that, although integral, may be detached from the main site (e.g. the construction of a barrage in one area may involve opening up a major quarry development in another area). A description of the physical presence of a project is invariably improved by a three-dimensional visual image, which may include a photo-montage of what the site layout may look like at, for example, full operation. A clear presentation of location and physical presence is important for an assessment of change in land uses, any physical disruption to other infrastructures, severance of activities (e.g. agricultural holdings, villages) and visual intrusion and landscape changes.

Understanding a project also involves an understanding of the *processes* integral to it. The nature of processes varies between industrial, service and infrastructure projects, but many can be described as a flow of inputs through a process and their transformation into outputs. The nature, origins and destinations of the inputs and outputs, and the timescale over which they are expected should be identified. This systematic identification should be undertaken for both physical and socio-economic characteristics, although the interaction should be clearly recognized, with many of the socio-economic characteristics following from the physical.

Physical characteristics may include:

- the land take and physical transformation of a site (e.g. clearing, grading), which may vary between different stages of a project's life cycle;
- the total operation of the process involved (usually illustrated with a process-flow diagram);
- the types and quantities of resources used (e.g. water abstraction, minerals, energy);
- transport requirements (of inputs and outputs);
- the generation of wastes, including estimates of types, quantity and strength of aqueous wastes, gaseous and particulate emissions, solid wastes, noise and vibration, heat and light, radiation, etc.;
- the potential for accidents, hazards and emergencies;
- processes for the containment, treatment and disposal of wastes and for the containment and handling of accidents; monitoring and surveillance systems.

Socio-economic characteristics may include:

- the labour requirements of a project – including size, duration, sources, particular skills categories and training;
- the provision or otherwise of housing, transport, health and other services for the workforce;
- the direct services required from local businesses or other commercial organizations;
- the flow of expenditure from the project into the wider community (from the employees and subcontracting);
- the flow of social activities (service demands, community participation, community conflict).

Figure 4.2 shows the interaction between the physical (ecological in this case) and socio-economic processes that may be associated with an industrial plant.

The projects may also have *associated policies*, not obvious from site layouts and process-flow diagrams, but are nevertheless significant for subsequent impacts. For example, shift-working will have implications for transport and noise that may be very significant for nearby residents. The use of a construction site hostel, camp or village can significantly internalize impacts on the local housing market and on the local community. The provision of on- or off-site training can greatly affect the mixture of local and non-local labour and the balance of socio-economic effects.

Projects should be seen in their *planning policy context*. In the UK, the main local policy context is outlined and detailed in Local Development Frameworks. The description of location must pay regard to land-use designations and development constraints that may be implicit in some of the designations. Of particular importance is a project's location in relation to various environmental designations (e.g. areas of outstanding natural beauty (AONBs), sites of special scientific interest (SSSIs), green belts and local and national nature reserves). Attention should also be given to Regional Spatial Strategies (RSSs) and to national planning guidance, provided in the UK by an important set of ODPM Planning Policy Statements (PPSs).

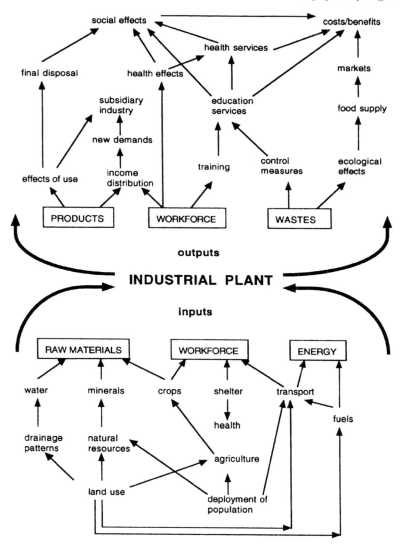

Figure 4.2 Interaction between an industrial plant and its socio-ecological environment. (*Source*: Marstrand 1976.)

4.6.2 *Sources and presentation of data*

The initial brief from the developer provides the starting point. Ideally, the developer may have detailed knowledge of the proposed project's characteristics, likely layout and production processes, drawing on previous experience. However, site layout diagrams and process-flow charts may be only in outline, provisional form at the initial design stage. Even in the ideal situation, there will need to be considerable interaction between the analyst and the developer to refine the project's characteristics. Unfortunately, the situation may often be far from ideal; Mills (1992) and Frost

(1994) provide interesting examples of major changes from the project description in EISs to the actual implemented action.

An analyst can supplement such information with reference to comparative studies, although the availability of such statements in the UK is still far from satisfactory, and their predictions are untested (see Chapters 7 and 8). The analyst may also draw on EIA literature (books and journals), guidelines, manuals and statistical sources, including Lee (1987), Wood & Lee (1987), CEC (1993), Morris & Therivel (2001) and Rodriguez-Bachiller with Glasson (2003). Site visits can be made to comparable projects, and advice can be gained from consultants with experience of the type of project under consideration.

As the project design and assessment process develop – in part in response to early EIA findings – so the developer will have to provide more detailed information on the characteristics specific to the project. The identification of sources of potential significant impacts may lead to changes in layout and process.

Data about the project can be presented in different ways. The life cycle of a project can be illustrated on a linear bar chart. Particular stages may be identified in more detail where the impacts are considered of particular significance; this is often the case for the construction stage of major projects. Location and physical presence are best illustrated on a map base, with varying scales to move from the broad location to the specific site layout. This may be supplemented by aerial photographs, photo-montages and visual mock-ups according to the resources and issues involved (Figures 4.3–4.6).

A process diagram for the different activities associated with a project should accompany the location and site-layout maps. This may be presented in the form of a simplified pictorial diagram or in a block flow chart. The latter can be presented simply to show the main interconnections between the elements of a project (see Figure 4.4 for socio-economic processes) or in sufficient detail to provide a comprehensive picture. Figure 4.5 shows a materials flow chart for a petroleum refinery; it outlines all the raw materials, additives, end products, by-products and atmospheric, liquid and solid wastes. A comprehensive flow chart of a production process should include the types, quantities and locations of resource inputs, intermediate and final product outputs and wastes generated by the total process.

The various information and illustrations should clearly identify the main variations between a project's stages. Figure 4.6 illustrates a labour-requirements diagram that identifies the widely differing requirements, in absolute numbers and in skill categories, of the construction and operational stages. In addition, more sophisticated flow diagrams could indicate the type, frequency (normal, batch, intermittent or emergency) and duration (minutes or hours per day or week) of each operation. Seasonal and material variations, including time periods of peak pollution loads, can also be documented.

4.7 Establishing the environmental baseline

4.7.1 *General considerations*

The establishment of an environmental baseline includes both the present and likely future state of the environment, assuming that a proposed project is not undertaken, taking into account changes resulting from natural events and from other human

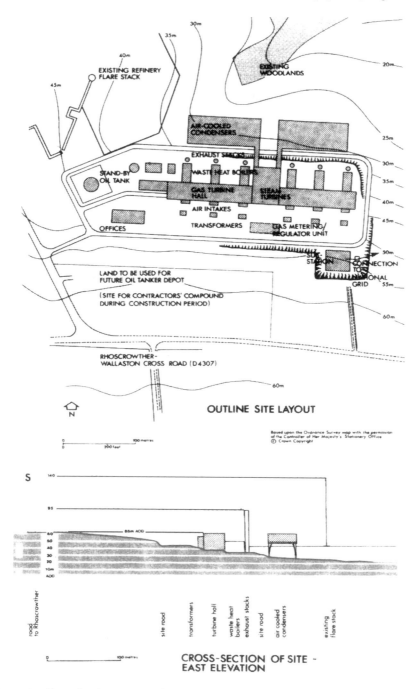

Figure 4.3 Example of a project site layout. (*Source*: Rendel Planning 1990, *Angle Bay Energy Project environmental statement*.)

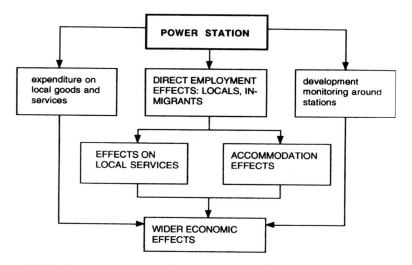

Figure 4.4 Socio-economic process diagram for a major project.

activities. For example, the population of a species of fish in a lake may already be declining before the proposed introduction of an industrial project on the lake shore. Figure 1.6 illustrated the various time, component and scale dimensions of the environment, and all these dimensions need to be considered in the establishment of the environmental baseline. The period for the prediction of the future state of the environment should be comparable with the life of the proposed development; this may mean predicting for several decades. Components include both the biophysical and socio-economic environment. Spatial coverage may focus on the local, but refer to the wider region and beyond for some environmental elements.

Initial baseline studies may be wide-ranging, but comprehensive overviews can be wasteful of resources. The studies should focus as quickly as possible on those aspects of the environment that may be significantly affected by the project, either directly or indirectly:

> environmental statements [need not] cover every conceivable aspect of a project's potential environmental effects at the same level of detail. They should be tailored to the nature of the project and its likely effects... In some cases, only a few [environmental aspects] will be significant in this sense and will need to be discussed in the statement at any great depth. Other issues may be of little or no significance for the particular project in question, and will need only very brief treatment, to indicate that their possible relevance has been considered (ODPM 2003a).

The rationale for the choice of focus should be explained as part of the documentation of the scoping process. Although the studies would normally consider the various environmental elements separately, it is also important to understand the interaction between them and the functional relationships involved; for instance, flora will be affected by air and water quality, and fauna will be affected by flora. This will facilitate prediction. As with most aspects of the EIA process, establishing the baseline is not a "one-off" activity. Studies will move from broad-brush to more detailed and

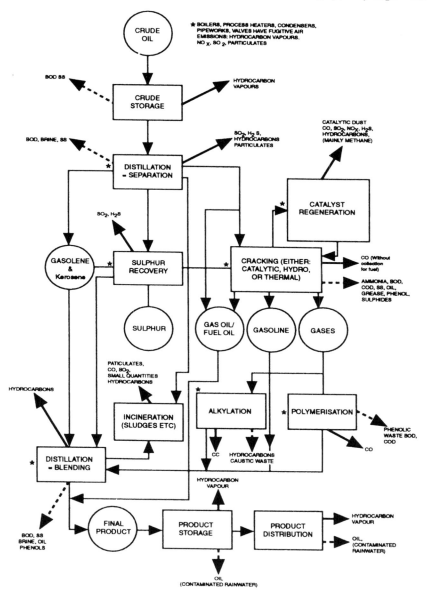

Figure 4.5 Materials flow chart for a petroleum refinery. (*Source:* UNEP 1981.)

focused approaches. The identification of new potential impacts may open up new elements of the environment for investigation; the identification of effective measures for mitigating impacts may curtail certain areas of investigation.

Environmental components or elements can be described simply in broad categories, as outlined in Table 1.3. *Environmental impact assessment: guide to the procedures* (ODPM 2003a) also provides information on what could be included in a project

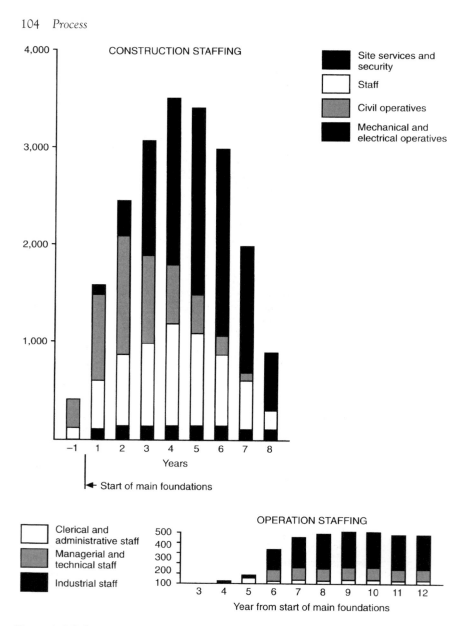

Figure 4.6 Labour requirements for a project over several stages of its life.

description (see Table 3.5), including an important distinction between physical features and policy framework. In contrast, Leopold has 88 components in his interactive matrix (see Figure 4.12), and each of these could be subdivided further. Several UN publications provide a more balanced listing of both the biophysical and socio-economic elements (see UNEP 1981). Table 4.2 provides an example of a framework for analysing each baseline sub-element.

Table 4.2 Framework for analysing baseline sub-elements: example of use

Sub-element	Objectives	Required information/ specialist(s)	Methodology	Findings/ measurements
Water quality	Protection of human health and aquatic life	Existing water quality; possible sources of pollution: run-off, leakage from waste treatment system, surface seepage of pollutants, intrusion of saline or polluted water; capacity of treatment system Water quality analyst; aquatic biologist; water pollution control engineer; sanitary and civil engineers	Laboratory analyses or field measurement of water quality; pollution indices	Potential for degradation of water quality; safety of potable water
Surface waters	Protection of: plant and animal life; water supply for domestic and industrial needs; natural water purification systems; groundwater recharge and discharge; recreation and aesthetic values	Location of surface waters streams, rivers, ponds, lakes, etc.; surface water volume, flow rates, frequency and duration of seasonal variations; 7-day, 10-year low flow; water uses; ecological characteristics; recreation and aesthetic uses Hydrologist; ecologist	Measurement of proximity of site to surface waters; field measurement of volume, rate and direction of water movement; categories of water usage; ecological assessment – see ecology element	Potential modification of volume, rate and direction of water movement; impact on ecological character; degree and type of water usage

(*Source*: UNEP (Industry and Environment Office) 1981.)

4.7.2 Sources and presentation of data

The quality and reliability of environmental data vary a great deal, and this can influence the use of such data in the assessment of impacts. Fortlage (1990) clarifies this in the following useful classification:

- "hard" data from reliable sources which can be verified and which are not subject to short-term change, such as geological records and physical surveys of topography and infrastructure;
- "intermediate" data which are reliable but not capable of absolute proof, such as water quality, land values, vegetation condition and traffic counts, which have variable values;

- "soft" data which are a matter of opinion or social values, such as opinion surveys, visual enjoyment of landscape and numbers of people using amenities, where the responses depend on human attitudes and the climate of public feeling.

Important UK data sources are the Regional Observatories and local authority monitoring units. These usually provide a range of very useful data on the physical, social and economic environment; they are reasonably up to date and are increasingly available on the Internet. Local data can be supplemented in the UK with published data from a wide range of national government sources – including the *Census of Population, Neighbourhood Statistics, Quality of Life Counts, Regional Trends, Digest of Environmental Statistics, Transport Statistics* – and increasingly from EU sources.

However, much useful information is unpublished or "semi-published" and internal to various organizations. In the UK, under the EIA regulations, statutory consultees (e.g. the Countryside Agency, EN, English Heritage and the Environment Agency) are put "under an obligation to provide the developer (on request) with any information in their possession which is likely to be relevant to the preparation of the environmental statement" (ODPM 2003a).

There are of course many other useful non-statutory consultees, at local and other levels, who may be able to provide valuable information. Local history, conservation and naturalist societies may have a wealth of information on, for example, local flora and fauna, rights of way and archaeological sites. National bodies, such as the RSPB and the Forestry Agency, may have particular knowledge and expertise to offer. Consultation with local amenity groups at an early stage in the EIA process can help not only with data but also with the identification of those key environmental issues for which data should be collected.

Every use should be made of data from existing sources, but there will invariably be gaps in the required environmental baseline data for the project under consideration. Environmental monitoring and surveys may be necessary. Surveys and monitoring raise a number of issues. They are inevitably constrained by budgets and time, and must be selective. However, such selectivity must ensure that the length of time over which monitoring and surveys are undertaken is appropriate to the task in hand. For example, for certain environmental features (e.g. many types of flora and fauna) a survey period of 12 months or more may be needed to take account of seasonal variations or migratory patterns. Sampling procedures will often be used for surveys; the extent and implications of the sampling error involved should be clearly established.

Baseline studies can be presented in the EIS in a variety of ways. These often involve either a brief overview of the biophysical and socio-economic environments for the area of study, following the project description, with the detailed focused studies in subsequent impact chapters (e.g. air quality, geology, employment), or a more comprehensive set of detailed studies at an early stage providing a point of reference for future and often briefer impact chapters.

A valuable innovation in the provision and presentation of environmental data is the increasing use of the Internet and geographical information systems (GIS). GIS are computer-based databases that include spatial references for the different variables stored, so that maps of such variables can be displayed, combined and analysed with speed and ease (Rodriguez-Bachiller 2000). The GIS market is developing rapidly, but initial setting-up costs are usually expensive, depending on the accessibility of relevant data. However, Rodriguez-Bachiller (2000) notes that, in practice:

a...paradox in the use of GIS [is that] their diffusion continues at a fast pace, while the realisation grows of the relative unsophistication of their functionality, illustrated in our review by the relatively narrow range of operations which they are called upon to perform in environmental matters: map display, map-overlay and intersection, buffering around given features, multi-factor map algebra, visibility analysis derived from terrain modelling. It is probably fair to say that, despite the technical power of GIS as databases, the purely "visual" appeal of their outputs (maps) has been and still is a major contributor to their success.

The analyst should also be wary of the seductive attraction of quantitative data at the expense of qualitative data; each type has a valuable role in establishing baseline conditions. Finally, it should be remembered that all data sources suffer from some uncertainty, and this needs to be explicitly recognized in the prediction of environmental effects (see Chapter 5).

4.8 Impact identification

4.8.1 *Aims and methods*

Impact identification brings together project characteristics and baseline environmental characteristics with the aim of ensuring that all potentially significant environmental impacts (adverse or favourable) are identified and taken into account in the EIA process. When choosing amongst the existing wide range of impact identification methods, the analyst needs to consider more specific aims, some of which conflict:

- to ensure compliance with regulations;
- to provide a comprehensive coverage of a full range of impacts, including social, economic and physical;
- to distinguish between positive and negative, large and small, long-term and short-term, reversible and irreversible impacts;
- to identify secondary, indirect and cumulative impacts as well as direct impacts;
- to distinguish between significant and insignificant impacts;
- to allow a comparison of alternative development proposals;
- to consider impacts within the constraints of an area's carrying capacity;
- to incorporate qualitative as well as quantitative information;
- to be easy and economical to use;
- to be unbiased and to give consistent results;
- to be of use in summarizing and presenting impacts in the EIS.

Many impact identification methods were developed in response to the NEPA and have since been expanded and refined. The simplest involve the use of lists of impacts to ensure that none has been forgotten. The most complex include the use of interactive computer programmes, networks showing energy flows and schemes to allocate significance weightings to various impacts. Many of the more complex methods were developed for (usually US) government agencies that deal with large numbers of fairly similar project types (e.g. the US Forest Service).

In the UK, the use of impact identification techniques is less well developed. Simple checklists or, at best, simple matrices are used to identify and summarize impacts. This may be attributable to the high degree of flexibility and discretion in the UK's implementation of the EIA Directive, to a general unwillingness in the UK to make the EIA process over-complex or to disillusionment with the more complex approaches that are available.

The aim of this section is to present a range of these methods, from the simplest checklists needed for compliance with regulations to complex approaches that developers, consultants and academics who aim to further "best practice" may wish to investigate further. The methods are divided into the following categories:

- checklists
- matrices
- quantitative methods
- networks
- overlay maps.

The discussion of the methods here relates primarily to impact identification, but most of the approaches are also of considerable (and sometimes more) use in other stages of the EIA process – in impact prediction, evaluation, communication, mitigation, presentation, monitoring and auditing. As such, there is considerable interaction between Chapters 4, 5, 6 and 7, paralleling the interaction in practice between these various stages.

For further information on the range of methods available we refer the reader to Rodriguez with Glasson (2003), Morris & Therivel (2001), Bregman & Mackenthun (1992), Wathern (1984), Sorensen & Moss (1973), Munn (1979), and Rau & Wooten (1980).

Checklists

Most checklists are based on a list of special biophysical, social and economic factors that may be affected by a development. The *simple checklist* can help only to identify impacts and ensure that impacts are not overlooked. Checklists do not usually include direct cause–effect links to project activities. Nevertheless, they have the advantage of being easy to use. Table 3.5 (ODPM 2003a) is an example of a simple checklist.

Questionnaire checklists are based on a set of questions to be answered. Some of the questions may concern indirect impacts and possible mitigation measures. They may also provide a scale for classifying estimated impacts, from highly adverse to highly beneficial. Figure 4.7 shows part of the EC's (2001b) questionnaire checklist.

Threshold-of-concern checklists consist of a list of environmental components and, for each component, a threshold at which those assessing a proposal should become concerned with an impact. The implications of alternative proposals can be seen by examining the number of times that an alternative exceeds the threshold of concern. For example, Figure 4.8 shows part of a checklist developed by the US Forest Service; it compares three alternative development proposals on the basis of various components. For the component of economic efficiency, a benefit:cost ratio of 1:1 is the threshold of concern; for spotted owls, 35 pairs is the threshold. In the example, alternative X causes two thresholds of concern to be exceeded, alternative Y one, alternative Z four; this would indicate that alternative Y is the least detrimental. Impacts are also rated according to their duration: A for 1 year or less, B for 1–10 years, C for 10–50 years and D for irreversible

No.	Questions to be considered in Scoping	Yes/No/?	Which Characteristics of the Project Environment could be affected and how?	Is the effect likely to be significant? Why?
7	Will the project lead to risks of contamination of land or water from releases of pollutants onto the ground or into sewers, surface waters, ground water, coastal waters or the sea?			
7.1	From handling, storage, use or spillage of hazardous or toxic materials?			
7.2	From discharge of sewage or other effluents (whether treated or untreated) to water or the land?			
7.3	By deposition of pollutants emitted to air, onto the land or into water?			
7.4	From any other sources?			
7.5	Is there a risk of long-term build-up of pollutants in the environment from these sources?			

Figure 4.7 Part of a questionnaire checklist. (*Source:* EC 2001b.)

impacts. Of the impacts listed, a reduction in the number of spotted owls would be irreversible, and the other impacts would last 10–50 years (Sassaman 1981).

Matrices

Matrices are the most commonly used method of impact identification in EIA. Simple matrices are merely two-dimensional charts showing environmental components on one axis and development actions on the other. They are, essentially, expansions of checklists that acknowledge the fact that various components of a development

Environmental component	Criterion	TOC	Alt X Imp	Alt X Imp > TOC?	Alt Y Imp	Alt Y Imp > TOC?	Alt Z Imp	Alt Z Imp > TOC?
Air quality	emission standards	1	2C	yes	1C	no	2C	yes
Economics	benefit:cost ratio	1:1	3:1	no	4:1	no	2:1	no
Endangered species	no. pairs of spotted owls	35	50D	no	35D	no	20D	yes
Water quality	water quality standards	1	1C	no	2C	yes	2C	yes
Recreation	no. camping sites	5000	2800C	yes	5000C	no	3500C	yes

Figure 4.8 Part of a threshold-of-concern (TOC) checklist. (Adapted from Sassaman 1981.)

Environmental component	Project action — Construction: Utilities	Construction: Residential and commercial buildings	Operation: Residential buildings	Operation: Commercial buildings	Operation: Parks and open spaces
Soil and geology	X	X			
Flora	X	X			X
Fauna	X	X			X
Air quality				X	
Water quality	X	X	X		
Population density			X	X	
Employment		X		X	
Traffic	X	X	X	X	
Housing			X		
Community structure		X	X		X

Figure 4.9 Part of a simple matrix.

project (e.g. construction, operation, decommissioning, buildings, access road) have different impacts. The action likely to have an impact on an environmental component is identified by placing a cross in the appropriate cell. Figure 4.9 shows an example of a *simple matrix*. Three-dimensional matrices have also been developed in which the third dimension refers to economic and social institutions: such an approach identifies the institutions from which data are needed for the EIA process, and highlights areas in which knowledge is lacking.

The *time-dependent matrix* (e.g. Parker & Howard 1977) includes a number sequence to represent the timescale of the impacts (e.g. one figure per year). Figure 4.10

Environmental component	Project action				
	Construction (3 years)		Operation (25 years, evens out after 4 years)		
	Utilities	Residential and commercial buildings	Residential buildings	Commercial buildings	Parks and open spaces
Soil and geology	211	321	0000	0000	0001
Flora	221	422	1223	1111	1123
Fauna	221	311	1100	1100	1122
Air quality	000	000	0123	0034	0011
Water quality	010	022	1223	0111	0000
Population density	011	112	2344	0222	0011
Employment	120	342	1111	1334	1111
Traffic	220	332	2333	2333	1111
Housing	010	121	2344	0000	0000
Community structure	010	232	2344	1111	1233

Figure 4.10 Part of a time-dependent matrix.

shows an example where magnitude is represented by numbers from 0 (none) to 4 (high), over a course of 7 years.

Magnitude matrices go beyond the mere identification of impacts by describing them according to their magnitude, importance and/or time frame (e.g. short-, medium- or long-term). Figure 4.11 is an example of a magnitude matrix.

The best known type of quantified matrix is the *Leopold matrix*, which was developed for the US Geological Survey by Leopold et al. (1971). It is based on a horizontal list of 100 project actions and a vertical list of 88 environmental components. Figure 4.12 shows a section of this matrix and lists all its elements. Of the 8,800 possible interactions between project action and environmental component, Leopold et al. estimate that an individual project is likely to result in 25–50. In each appropriate cell, two numbers are recorded. The number in the top left-hand corner represents the impact's magnitude, from +10 (very positive) to −10 (very negative). That in the bottom right-hand corner represents the impact's significance, from 10 (very significant) to 1 (insignificant); there is no negative significance. This distinction between magnitude and significance is important: an impact could be large but insignificant, or small but significant. For instance, in ecological terms, paving over a large field of intensively used farmland may be quite insignificant compared with the destruction of even a small area of an SSSI.

The Leopold matrix is easily understood, can be applied to a wide range of developments, and is reasonably comprehensive for first-order, direct impacts. However, it has disadvantages. The fact that it was designed for use on many different types of project makes it unwieldy for use on any one project. It cannot reveal indirect effects of developments: like checklists and most other matrices, it does not relate environmental components to one another, so the complex interactions between ecosystem components that lead to indirect impacts are not assessed. The inclusion

Environmental component	Project action				
	Construction		Operation		
	Utilities	Residential and commercial buildings	Residential buildings	Commercial buildings	Parks and open spaces
Soil and geology	•	•			
Flora	•	●			○
Fauna	•	•			o
Air quality				•	
Water quality	○	•	•		
Population density			o	o	
Employment		○		○	
Traffic	•	•	•	●	
Housing			○		
Community structure		•	○		o

• = small negative impact   o = small positive impact
● = large negative impact   ○ = large positive impact

Figure 4.11 Part of a magnitude matrix.

of magnitude/significance scores has additional drawbacks: it gives no indication whether the data on which these values are based are qualitative or quantitative; it does not specify the probability of an impact occurring; it excludes details of the techniques used to predict impacts; and the scoring system is inherently subject-ive and open to bias. People may also attempt to add the numerical values to produce a composite value for the development's impacts and compare this with that for other developments; this should not be done because the matrix does not assign weightings to different impacts to reflect their relative importance (Clark et al. 1979).

Weighted matrices were developed in an attempt to respond to some of the above problems. Importance weightings are assigned to environmental components, and sometimes to project components. The impact of the project (component) on the environmental component is then assessed and multiplied by the appropriate weight-ing(s), to obtain a total for the project. Figure 4.13 shows a small weighted matrix that compares three alternative project sites. Each environmental component is assigned an importance weighting (a), relative to other environmental components: in the example, air quality is weighted 21 per cent of the total environmental com-ponents. The magnitude (c) of the impact of each project on each environmental component is then assessed on a scale 0–10, and multiplied by (a) to obtain a weighted impact (a×c): for instance, site A has an impact of 3 out of 10 on air qual-ity, which is multiplied by 21 to give the weighted impact, 63. For each site, the weighted impacts can then be added up to give a project total. The site with the lowest total, in this case site B, is the least environmentally harmful. However, the evaluation procedure depends heavily on the weightings and impact scales assigned. The main problems implicit in such weighting approaches are considered further in Chapter 5. Also, the method does not consider indirect impacts.

1. Identify all actions (located across the top of the matrix) that are part of the proposed project

2. Under each of the proposed actions, place a slash at the inter section with each item on the side of the matrix if an impact is possible

3. Having completed the matrix, in the upper left-hand corner of each box with a slash, place a number from 1 to 10 which indicates the MAGNITUDE of the possible impact; 10 represents the greatest magnitude of impact and 1, the least (no zeroes). Before each number place a + (if the impact would be beneficial). In the lower right hand corner of the box place a number from 1 to 10 which indicates the IMPORTANCE of the possible impact (e.g. regional vs local); 10 represents the greatest importance and 1 the least (no zeroes)

4. The text which accompanies the matrix should be a discussion of the significant impacts, those columns and rows with large numbers of boxes marked and individual boxes with large numbers

	a	b	c	d	e
a	2				8
b	7	8	3	9	

Sample matrix

A. Modification of regime
a. Exotic flora or fauna introduction
b. Biological controls
c. Modification of habitat
d. Alteration of ground cover
e. Alteration of ground water hydrology
f. Alteration of drainage
g. River control and flow modification
h. Canalization
i. Irrigation
j. Weather modification
k. Burning
l. Surface or paving
m. Noise and vibration

B. Land transformation and construction
a. Urbanization
b. Industrial sites and buildings
c. Airports
d. Highways and bridges
e. Roads and trails
f. Railroads
g. Cables and lifts
h. Transmission lines, pipelines, corridors
i. Barriers including fencing
j. Channel dredging and straightening
k. Channel revetments
l. Canals
m. Dams and impoundments
n. Piers, seawall, marinas and sea terminals
o. Offshore structures
p. Recreational structures
q. Blasting and drilling
r. Cut and fill
s. Tunnels and underground structures

C. Resource extraction
a. Blasting and drilling
b. Surface excavation
c. Subsurface excavation and retorting
d. Well drilling and fluid removal
e. Dredging
f. Clear cutting and other lumbering
g. Commercial fishing and hunting

Proposed actions

CHEMICAL CHARACTERISTICS

1. Earth
a. Mineral resources
b. Construction material
c. Soils
d. Land form
e. Force fields and background radiation
f. Unique features

2. Water
a. Surface
b. Ocean
c. Underground
d. Quality
e. Temperature
f. Recharge
g. Snow, ice and permafrost

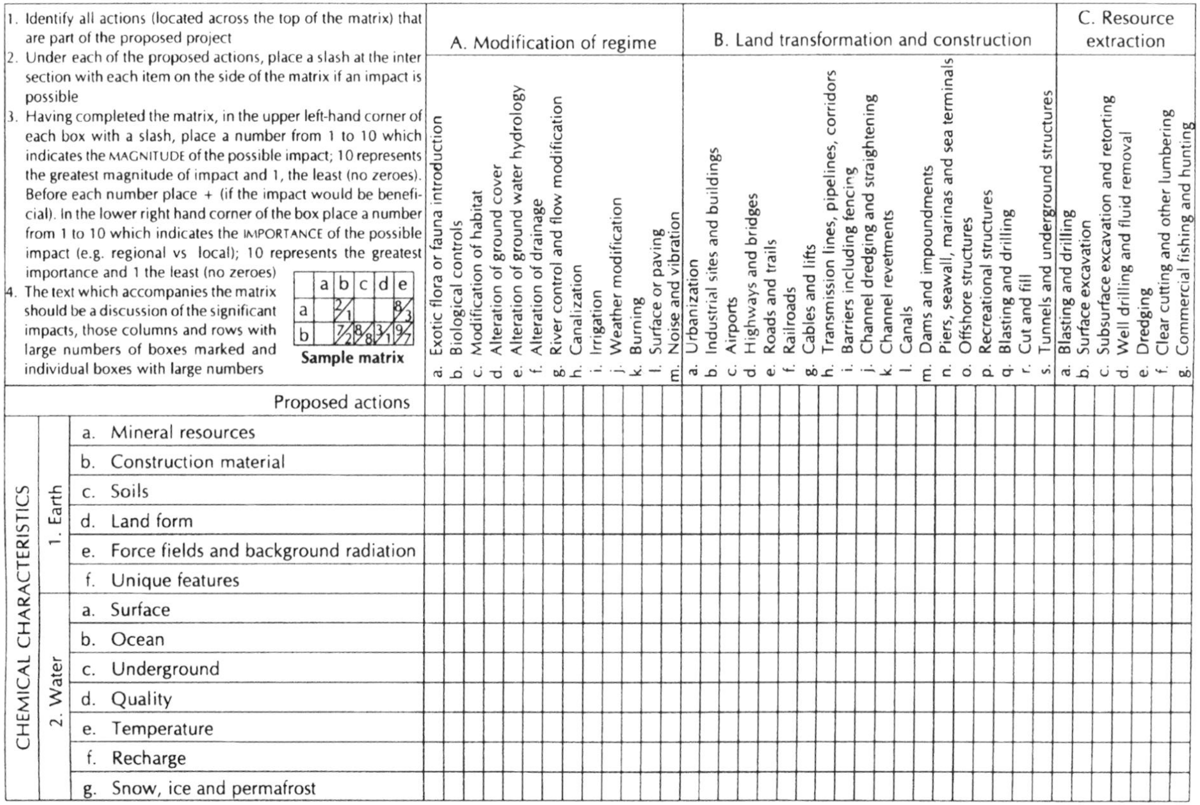

Figure 4.12 (a) Part of Leopold Matrix; (b) Leopold Matrix elements.

(b)

Part 1. Project actions

A. Modification of regime
a) exotic flora or fauna introduction
b) Biological controls
c) Modification of habitat
d) Alteration of ground cover
e) Alteration of groundwater hydrology
f) Alteration of drainage
g) River control and flow modification
h) Canalization
i) Irrigation
j) Weather modification
k) Burning
l) Surface or paving
m) Noise and vibration

B. Land transformation and construction
a) Urbanization
b) Industrial sites and buildings
c) Airports
d) Highways and bridges
e) Roads and trails
f) Railroads
g) Cables and lifts
h) Transmission lines, pipelines and corridors
i) Barriers, including fencing
j) Channel dredging and straightening
k) Channel revetments
l) Canals
m) Dams and impoundments
n) Piers, seawalls, marinas, and sea terminals
o) Offshore structures
p) Recreational structures
q) Blasting and drilling
r) Cut and fill
s) Tunnels and underground structures

C. Resource extraction
a) Blasting and drilling
b) Surface excavation
c) Subsurface excavation and retorting
d) Well drilling and fluid removal
e) Dredging
f) Clear cutting and other lumbering
g) Commercial fishing and hunting

D. Processing
a) Farming
b) Ranching and grazing
c) Feed lots
d) Dairying
e) Energy generation
f) Mineral processing
g) Metallurgical industry
h) Chemical industry
i) Textile industry
j) Automobile and aircraft
k) Oil refining
l) Food
m) Lumbering
n) Pulp and paper
o) Product storage

E. Land alteration
a) Erosion control and terracing
b) Mine sealing and waste control
c) Strip-mining rehabilitation
d) Landscaping
e) Harbour dredging
f) Marsh fill and drainage

F. Resource renewal
a) Reforestation
b) Wildlife stocking and management
c) Groundwater recharge
d) Fertilization application
e) Waste recycling

G. Changes in traffic
a) Railway
b) Automobile
c) Trucking
d) Shipping
e) Aircraft
f) River and canal traffic
g) Pleasure boating
h) Trails
i) Cables and lifts
j) Communication
k) Pipeline

H. Waste emplacement and treatment
a) Ocean dumping
b) Landfill
c) Emplacement of tailings, spoil and overburden
d) Underground storage
e) Junk disposal
f) Oil well flooding
g) Deep well emplacement
h) Cooling water discharge
i) Municipal waste discharge, including spray irrigation
j) Liquid effluent discharge
k) Stabilization and oxidation ponds
l) Septic tanks, commercial and domestic
m) Stack and exhaust emission
n) Spent lubricants

I. Chemical treatment
a) Fertilization
b) Chemical de-icing of highways, etc.
c) Chemical stabilization of soil
d) Weed control
e) Insect control (pesticides)

J. Accidents
a) Explosions
b) Spills and leaks
c) Operational failure

Others

Part 2. Natural and human environmental elements

A. Physical and chemical characteristics
1. Earth
a) Mineral resources
b) Construction material
c) Soils
d) Landform
e) Force fields and background radiation
f) Unique physical features
2. Water
a) Surface
b) Ocean
c) Underground
d) Quality
e) Temperature
f) Recharge
g) Snow, ice and permafrost
3. Atmosphere
a) Quality (gases, particulates)
b) Climate (micro, macro)
c) Temperature
4. Processes
a) Floods
b) Erosion
c) Deposition (sedimentation, precipitation)
d) Solution
e) Sorption (ion exchange, complexing)
f) Compaction and settling
g) Stability (slides, slumps)
h) Stress-strain (earthquakes)
i) Air movements

B. Biological conditions
1. Flora
a) Trees
b) Shrubs
c) Grass
d) Crops
e) Microflora
f) Aquatic plants
g) Endangered species
h) Barriers
i) Corridors
2. Fauna
a) Birds
b) Land animals, including reptiles
c) Fish and shellfish
d) Benthic organisms
e) Insects
f) Microfauna
g) Endangered species
h) Barriers
i) Corridors

C. Cultural factors
1. Land use
a) Wilderness and open spaces
b) Wetlands
c) Forestry
d) Grazing
e) Agriculture
f) Residential
g) Commercial
h) Industrial
i) Mining and quarrying
2. Recreation
a) Hunting
b) Fishing
c) Boating
d) Swimming
e) Camping and hiking
f) Picnicking
g) Resorts
3. Aesthetics and human interest
a) Scenic views and vistas
b) Wilderness qualities
c) Open space qualities
d) Landscape design
e) Unique physical features
f) Monuments
g) Parks and reserves
h) Rare and unique species or ecosystems
i) Historical or archaeological sites and objects
j) Presence of misfits
4. Cultural status
a) Cultural patterns, lifestyle
b) Health and safety
c) Employment
d) Population density
5. Man-made facilities and activities
a) Structures
b) Transportation network (movement access)
c) Utility networks
d) Waste disposal
e) Barriers
f) Corridors

D. Ecological relationships, such as
a) Salinization of water resources
b) Eutrophication
c) Disease – insect vectors
d) Food chains
e) Salinization of surficial material
f) Brush encroachment
g) Other

Others

Figure 4.12 (continued)

Environmental component	(a)	Alternative sites					
		Site A		Site B		Site C	
		(c)	(axc)	(c)	(axc)	(c)	(axc)
Air quality	21	3	63	5	105	3	63
Water quality	42	6	252	2	84	5	210
Noise	9	5	45	7	63	9	81
Ecosystem	28	5	140	4	112	3	84
Total	100		500		364		438

(a) = relative weighting of environmental component (total 100)
(c) = impact of project at particular site on environmental component (0–10)

Figure 4.13 A weighted matrix: alternative project sites.

Distributional impact matrices represent another possible development of the matrix approach. Such matrices can broadly identify who might lose and who might gain from the potential impacts of a development. This is useful information, which is rarely included in the matrix approach, and indeed is often missing from EISs. Impacts can have varying spatial impacts – varying, for example, between urban and rural areas. Spatial variations may be particularly marked for a linear project, such as a Light Rapid Transit system. A project can also have different impacts on different groups in society (for example the impacts of a proposed new settlement on old people, retired with their own houses, and young people, perhaps with children, seeking affordable housing and a way into the housing market) (see Figure 5.7, Chapter 5).

Quantitative methods

Quantitative methods attempt to compare the relative importance of all impacts by weighting, standardizing and aggregating them to produce a composite index. The best known of these methods is the environmental evaluation system (EES), devised by the Battelle Columbus Laboratories for the US Bureau of Land Reclamation to assess water resource developments, highways, nuclear power plants and other projects (Dee et al. 1973). It consists of a checklist of 74 environmental, social and economic parameters that may be affected by a proposal; these are shown in Figure 4.14. It assumes that these parameters can be expressed numerically and that they represent an aspect of environmental quality. For instance, the concentration of dissolved oxygen is a parameter that represents an aspect of the quality of an aquatic environment. For each parameter, functions were designed by experts to express environmental quality on a scale 0–1 (degraded–high quality). Two examples are shown in Figure 4.15. For instance, a stream with more than 10 mg/l of dissolved oxygen is felt to have a high level of environmental quality (1.0), whereas one with only 4 mg/l is felt to have an environmental quality of only about 0.35. Impacts are measured in terms of the likely change in environmental quality for each parameter. Two environmental quality scores are determined for each parameter, one for the current state of the environment and one for the state predicted once the project is in operation. If the post-development score is lower than the pre-development score, the impact is negative, and vice versa. To enable impacts to be compared directly, each parameter is

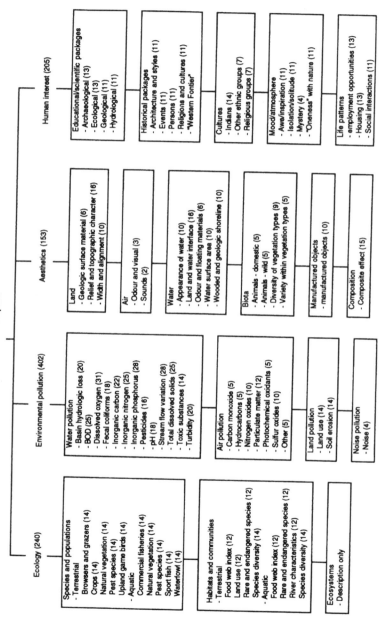

Figure 4.14 Framework for the Battelle Environmental Evaluation system. (*Source:* Dee et al. 1973.)

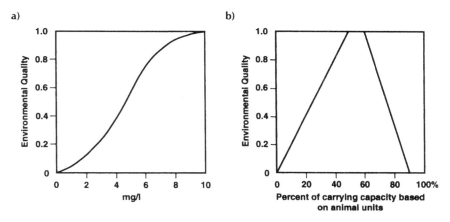

Figure 4.15 Environmental parameter functions for the Environmental Evaluation System: dissolved oxygen and deer: rangeland ratios: (a) dissolved oxygen; (b) browsers and grazers. (*Source*: Dee et al. 1973.)

given an importance weighting, which is then multiplied by the appropriate environmental quality score. The importance weightings (shown in parentheses in Figure 4.14) are determined by having a panel of experts distribute 1,000 points among the parameters. For instance, dissolved oxygen is considered quite important, at 31 points out of 1,000. A composite score for the beneficial and adverse effects of a single project, or for the net impact of alternative projects, can be obtained by adding up the weighted impact scores.

As an example of the full use of the EES, assume that the existing deer:rangeland ratio means that 40 per cent of the annual plant production is consumed (environmental quality score 0.8 in Figure 4.15). A project likely to halve the deer population would cause the score to drop to 0.4. The post-development score would be lower than the pre-development score, so the impact would be negative. This parameter's importance is 14 points out of 1,000, so the pre- and post-development scores would be multiplied by 14, and could then be compared with other parameters (Dee et al. 1973).

Another quantitative method developed to assess alternative highway proposals (Odum et al. 1975) considers impact duration: long-term irreversible impacts are considered to be more important than short-term reversible impacts and are given 10 times more weight. A sensitivity analysis showed that errors in impact estimation and weighting could significantly affect the rankings of alternative highway routes. Another method (Stover 1972) considers future impacts to be more important and gives them higher values than short-term impacts: it multiplies the numerical rating of each future impact by its duration in years.

The attraction of these quantitative methods lies in their ability to "substantiate" numerically that a particular course of action is better than others. This may save decision-makers considerable work, and it ensures consistency in assessment and results. However, these methods also have some fundamental weaknesses. They effectively take decisions away from decision-makers (Skutsch & Flowerdew 1976). The methods are difficult for lay people to understand, and their acceptability depends on the assumptions, especially the weighting schemes, built into them.[3] People carrying out assessments may manipulate results by changing assumptions

(Bisset 1978). Quantitative methods also treat the environment as if it consisted of discrete units. Impacts are related only to particular parameters, and much information is lost when impacts are reduced to numbers.

Networks

Network methods explicitly recognize that environmental systems consist of a complex web of relationships, and try to reproduce that web. Impact identification using networks involves following the effects of development through changes in the environmental parameters in the model. The *Sorensen network* was the first network method to be developed; it aimed to help planners reconcile conflicting land uses in California. Figure 4.16 shows a section of the network dealing with impacts on water quality. Water is one of the six environmental components, the others being climate, geophysical conditions, biota, access conditions and aesthetics.

The Sorensen method begins by identifying potential causes of environmental change associated with a proposed development action, using a matrix format; for instance, forestry potentially results in the clearing of vegetation and the use of herbicides and fertilizers. These environmental changes in turn result in specific environmental impacts; in the example, the clearing of vegetation could result in an increased flow of fresh water, which in turn could imperil cliff structures. The analyst stops following the network when an initial cause of change has been traced through all subsequent impacts and changes in environmental conditions, to its final impacts. Environmental impacts can result either directly from a development action or indirectly through induced changes in environmental conditions. A change in environmental conditions may result in several different types of impact. Sorensen argues that the method should lead to the identification of remedial measures and monitoring schemes (Sorensen 1971).

A simpler version of this technique is used in the development of many UK Local Transport Plans. *Causal chain analysis* (or cause–effect diagrams) are drawn by planners to identify how one action – say maintenance, renewal and improvements to carriageways and junctions (Figure 4.17) – leads to changes in social, economic and environmental conditions. They also identify what preconditions are needed to achieve a positive outcome, and problems to avoid.

Network methods do not establish the magnitude or significance of interrelationships between environmental components, or the extent of change. They can require considerable knowledge of the environment. Their main advantage is their ability to trace the higher-order impacts of proposed developments.

Overlay (or constraints) maps

Overlay maps have been used in environmental planning since the 1960s (McHarg 1968), before the NEPA was enacted. A series of transparencies is used to identify, predict, assign relative significance to and communicate impacts. A base map is prepared, showing the general area within which the project may be located. Successive transparent overlay maps are then prepared for the environmental components that, in the opinion of experts, are likely to be affected by the project (e.g. agriculture, woodland, noise). The project's degree of impact on the environmental feature is shown by the intensity of shading, darker shading representing a greater impact. The composite impact of the project is found by superimposing the overlay maps and noting the relative intensity of the total

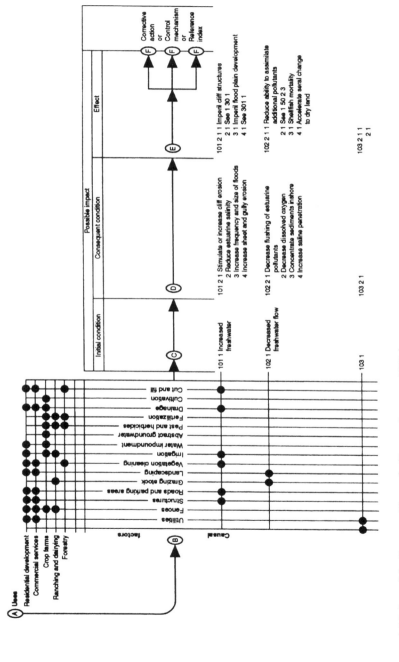

Figure 4.16 Part of the Sorensen Network. (*Source:* Sorensen 1971.)

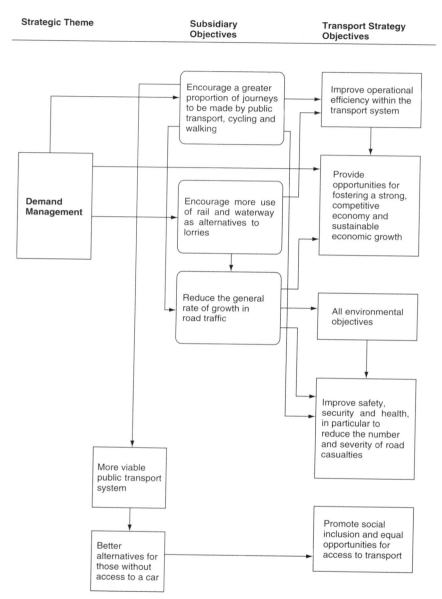

Figure 4.17 Causal chain analysis. (*Source*: West Yorkshire County Council 2004.)

shading. Unshaded areas are those where a development project would not have a significant impact. Figure 4.18 shows an example of this technique. Alternatively, the same process can be carried out using GIS and assigning different importance weightings to the impacts: this enables a sensitivity analysis to be carried out, to see whether changing assumptions about impact importance would alter the decision.

Overlay maps are easy to use and understand and are popular. They are an excellent way of showing the spatial distribution of impacts. They also lead intrinsically to a

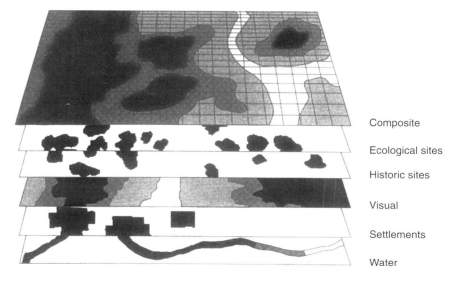

Composite

Ecological sites

Historic sites

Visual

Settlements

Water

Figure 4.18 An example of overlay maps.

low-impact decision. The overlay maps method is particularly useful for identifying optimum corridors for developments such as electricity lines and roads, for comparisons between alternatives, and for assessing large regional developments. However, the method is limited in that it does not consider factors such as the likelihood of an impact, secondary impacts or the difference between reversible and irreversible impacts. It requires the clear classification of often indeterminate boundaries (such as between forest and field), and so is not a true representation of conditions on the ground. It relies on the user to identify likely impacts before it can be used.

4.8.2 *Quality of life assessment*

The QOLA (or quality of life capital) approach was developed jointly by the Countryside Agency et al. (2001) as a way of integrating the different agencies' approaches to environmental management. QOLA focuses not on the *things* but on the *benefits* that would be affected by a development proposal. It starts with the assumption that things (e.g. woodlands, historical buildings) are important because of the benefits that they provide to people (e.g. visual amenity, recreation, CO_2 fixing), and conversely that management of those things should aim to optimize the benefits that they provide.

Quality of life assessment involves six steps (A–F). Having identified the purpose of the assessment (A) and described the proposed development site (B), the benefits/disbenefits that the site offers to sustainability, i.e. to present and future generations, are identified (C). The technique then asks the following questions (D):

- How important is each of these benefits or disbenefits, to whom, and why?
- On current trends, will there be enough of each of them?
- What (if anything) could substitute for the benefits?

The answers to these questions lead to a series of management implications (E) which allow a "shopping list" to be devised of things that any development/management on that site should achieve, how they could be achieved, and their relative importance. Finally, monitoring of these benefits is proposed (F). Thus the process concludes by clearly stipulating the benefits that the development would have to provide before it was considered acceptable and, as a corollary, indicates where development would not be appropriate. It can be used to set a management framework (e.g. for Section 106 obligations, planning conditions, etc.) for any development on a given site (and also for management of larger areas). The QOLA approach has been used to scope EIAs – notably that for the Bristol Arena – and can be used as a vehicle for public participation and/or the integration of different experts' analyses of a site.

4.8.3 Summary

Table 4.3 summarizes the respective advantages of the main impact identification methods discussed in this section.[4] Given the complexity of many impact identification techniques, it is understandable that many EIAs in the UK use checklists, simple matrices and simple networks, or some hybrid combination including elements from several of the methods discussed. Impact identification methods need to be chosen with care: they are not politically neutral, and the more sophisticated the method becomes, often the more difficult become clear communication and effective participation (see Chapter 6 for more discussion). The simpler methods are generally easier to use, more

Table 4.3 Comparison of impact identification methods

	Criterion										
	1	2	3	4	5	6	7	8	9	10	11
Checklists											
Simple/question	✓	✓						✓	✓	✓	✓
Threshold	✓	✓	✓		✓	✓	✓		✓	✓	✓
Matrices											
Simple	✓	✓						✓	✓	✓	✓
Magnitude/time-dependent	✓	✓	✓					✓	✓	✓	✓
Leopold	✓	✓	✓		✓			✓	✓		✓
Weighted	✓	✓			✓	✓		✓	✓		✓
Quantitative											
EES/WRAM	3		3		3	3	3				
Network											
Sorensen	✓			✓		✓		✓		✓	
Overlay maps		✓	✓		✓	✓		✓	✓	✓	✓

1. compliance with regulations;
2. comprehensive coverage (social, economic and physical impacts);
3. positive vs. negative, reversible vs. irreversible impacts, etc.;
4. secondary, indirect, cumulative impacts;
5. significant vs. insignificant impacts;
6. compare alternative options;
7. compare against carrying capacity;
8. uses qualitative and quantitative information;
9. easy to use;
10. unbiased, consistent;
11. summarizes impacts for use in EIS.

consistent and more effective in presenting information in the EIS, but their coverage of impact significance, indirect impacts or alternatives is either very limited or non-existent. The more complex models incorporate these aspects, but at the cost of immediacy.

4.9 Summary

The early stages of the EIA process are typified by several interacting steps. These include deciding whether an EIA is needed at all (screening), consulting with the various parties involved to produce an initial focus on some of the chief impacts (scoping), and an outline of possible alternative approaches to the project, including alternative locations, scales and processes. Scoping and the consideration of alternatives can greatly improve the quality of the process. Early in the process an analyst will also wish to understand the nature of the project concerned, and the environmental baseline conditions in the likely affected area. Projects have several dimensions (e.g. purpose, physical presence, processes and policies) over several stages in their life cycles; a consideration of the environmental baseline also involves several dimensions. For both projects and the affected environment, obtaining relevant data may present challenges.

Impact identification includes most of the activities already discussed. It usually involves the use of impact identification methods, ranging from simple checklists and matrices to complex computerized models and networks. In the UK, if any formal impact identification methods are used, they are normally of a simpler type. The methods discussed here have relevance also to the prediction, assessment, communication and mitigation of environmental impacts, which are discussed in the next chapter.

Notes

1. This refers both to the spatial extent that will be covered and to the scale at which it is covered. João (2002) suggests that the latter – which has been broadly ignored as an issue to date – could be crucial enough to lead to different decisions depending on the scale chosen.
2. In the US, "agencies should: consider the option of doing nothing; consider alternatives outside the remit of the agency; and consider achieving only a part of their objectives in order to reduce impact".
3. For instance, the EES's assumption that individual indicators of water quality (such as dissolved oxygen at 31 points) are more important than employment opportunities and housing put together (at 26 points) would certainly be challenged by large sectors of the public.
4. Another category of techniques, simulation models, was not discussed because they are still relatively undeveloped and have, to date, been applied only to problems involving a few environmental impacts.

References

Barker, A. & C. Wood 1999. An Evaluation of EIA System Performance in Eight EU Countries. *Environmental Impact Assessment Review* **19**, 387–404.
Bisset, R. 1978. Quantification, decision-making and environmental impact assessment in the United Kingdom. *Journal of Environmental Management* **7**(1), 43–58.
Bregman, J. I. & K. M. Mackenthun 1992. *Environmental impact statements*. Chelsea, Michigan: Lewis.

Canter, L. W. 1991. Interdisciplinary teams in environmental impact assessment. *Environmental Impact Assessment Review* **11**, 375–87.

CEC (Commission of the European Communities) 1993. *Environmental manual: user's guide; sectoral environment assessment sourcebook*. Brussels: CEC DG VIII.

CEC 1997. Council Directive 97/11/EC amending Directive 85/337/EEC on the assessment of certain public and private projects on the environment. *Official Journal* L73/5, 3 March.

CEQ (Council on Environmental Quality) 1978. National Environmental Policy Act. Implementation of procedural provision: final regulations. *Federal Register* **43**(230), 55977–6007, 29 November.

Clark, B. D., K. Chapman, R. Bisset, P. Wathern 1979. Environmental impact assessment. In *Land-use and landscape planning*, D. Lovejoy (ed.), 53–87. Glasgow: Leonard Hill.

Cleland, D. I. & H. Kerzner 1986. *Engineering team management*. New York: Van Nostrand Rheinhold.

Countryside Agency, English Nature, Environment Agency, English Heritage 2001. *Quality of Life Capital: What matters and why*, www.qualityoflifecapital.org.UK.

Dee, N., J. K. Baker, N. L. Drobny, K. M. Duke, I. Whitman, D. C. Fahringer 1973. An environmental evaluation system for water resources planning. *Water Resources Research* **3**, 523–35.

DETR (Department of the Environment, Transport and the Regions) 1997. *Consultation paper: Implementation of EC Directive 97/11/ec: determining the need for EIA*. London: HMSO.

DETR 2000. *Environmental Impact Assessment: a guide to the procedure*. Tonbridge: Thomas Telford Publishing.

DfT (Department for Transport) 2003. *Transport analysis guidance*, www.webtag.org.uk.

DoE (Department of Environment) 1996. *Changes in the quality of environmental statements for planning projects*. London: HMSO.

Eastman, C. 1997. *The treatment of alternatives in the environmental assessment process* (MSc dissertation). Oxford: Oxford Brookes University.

Environment Agency 2002. *Environmental impact assessment: scoping guidelines for the environmental impact assessment of projects*. Reading: Environment Agency.

EC (European Commission) 2001a. *Screening checklist*. Brussels: EC.

EC 2001b. *Scoping checklist*. Brussels: EC.

Fortlage, C. 1990. *Environmental assessment: a practical guide*. Aldershot: Gower.

Frost, R. 1994. *Planning beyond environmental statements* (MSc dissertation). Oxford: Oxford Brookes University, School of Planning.

Fuller, K. 1992. Working with assessment. In *Environmental assessment and audit, a user's guide 1992–1993* (special supplement to *Planning*), 14–15.

Government of New South Wales 1996. *EIS guidelines*. Sydney: Department of Urban Affairs and Planning.

João, E. 2002. How scale affects environmental impact assessment. *Environmental Impact Assessment Review* **22**, 289–310.

Jones, C. E., N. Lee, C. Wood 1991. *UK environmental statements 1988–1990: an analysis*. Occasional Paper 29, Department of Town and Country Planning, University of Manchester.

Lee, N. 1987. *Environmental impact assessment: a training guide*. Occasional Paper 18, Department of Town and Country Planning, University of Manchester.

Leopold, L. B., E. E. Clarke, B. B. Hanshaw, J. R. Balsley 1971. *A procedure for evaluating environmental impact*. Washington, DC: US Geological Survey Circular 645.

McHarg, I. 1968. *A comprehensive route selection method*. Highway Research Record 246. Washington, DC: Highway Research Board.

Marstrand, P. K. 1976. Ecological and social evaluation of industrial development. *Environmental Conservation* **3**(4), 303–08.

Mills, J. 1992. *Monitoring the visual impacts of major projects* (MSc dissertation). Oxford: Oxford Brookes University, School of Planning.

Morris, P. & R. Therivel (eds) 2001. *Methods of environmental impact assessment*. London: UCL Press.

Mulvihill, P. R. & D. C. Baker 2001. Ambitious and restrictive scoping: Case studies from Northern Canada. *Environmental Impact Assessment Review* **21**, 363–84.

Munn, R. E. 1979. *Environmental impact assessment: principles and procedures*, 2nd edn. New York: Wiley.

ODPM (Office of the Deputy Prime Minister) 2003a. *Environmental impact assessment: guide to the procedures*. London: HMSO.

ODPM 2003b. *Note on environmental impact assessment for local planning authorities*. London: ODPM.

Odum, E. P., J. C. Zieman, H. H. Shugart, A. Ike, J. R. Champlin 1975. In *Environmental impact assessment*, M. Blisset (ed.). Austin: University of Texas Press.

Parker, B. C. & R. V. Howard 1977. The first environmental monitoring and assessment in Antarctica: the Dry Valley drilling project. *Biological Conservation* 12(2), 163–77.

Petts, J. & G. Eduljee 1994. Integration of monitoring, auditing and environmental assessment: waste facility issues. *Project Appraisal* 9(4), 231–41.

Rau, J. G. & D. C. Wooten 1980. *Environmental impact analysis handbook*. New York: McGraw-Hill.

Rendel Planning 1990. *Angle Bay energy project environmental statement*. London: Rendel Planning.

Rodriguez-Bachiller, A. 2000. Geographical Information Systems and Expert Systems for Impact Assessment, Parts I and II. *Journal of Environmental Assessment Policy and Management* 2(3), 369–448.

Rodriguez-Bachiller, A. with J. Glasson 2003. *Expert Systems and Geographical Information Systems for Impact Assessment*. London: Taylor and Francis.

Sassaman, R. W. 1981. Threshold of concern a technique for evaluating environmental impacts and amenity values. *Journal of Forestry* **79**, 84–86.

Skutsch, M. M. & R. T. N. Flowerdew 1976. Measurement techniques in environmental impact assessment. *Environmental Conservation* 3(3), 209–17.

Sorensen, J. C. 1971. *A framework for the identification and control of resource degradation and conflict in multiple use of the coastal zone*. Berkeley: Department of Landscape Architecture, University of California.

Sorensen, J. C. & M. L. Moss 1973. *Procedures and programmes to assist in the environmental impact statement process*. Berkeley: Institute of Urban and Regional Development, University of California.

State of California, Governor's Office of Planning and Research 1992. *California Environmental Quality Act: statutes and guidelines*. Sacramento: State of California.

Steinemann, A. 2001. Improving alternatives for environmental impact assessment. *Environmental Impact Assessment Review* **21**, 3–21.

Stover, L. V. 1972. *Environmental impact assessment: a procedure*. Pottstown, Pennsylvania: Sanders & Thomas.

UNEP (United Nations Environment Programme) 1981. *Guidelines for assessing industrial environmental impact and environmental criteria for the siting of industry*. New York: United Nations.

Wathern, P. 1984. Ecological modeling in impact analysis. In *Planning and ecology*, R. D. Roberts & T. M. Roberts (eds), 80–93. London: Chapman & Hall.

Weaver, A. B., T. Greyling, B. W. Van Wilyer, F. J. Kruger 1996. Managing the ETA process. logistics and team management of a large environmental impact assessment – proposed dune mine at St. Lucia, South Africa. *Environmental Impact Assessment Review* **16**, 103–13.

Wende, W. 2002. Evaluation of the effectiveness and quality of environmental impact assessment in the Federal Republic of Germany. *Impact Assessment and Project Appraisal* 20(2), 93–99.

West Yorkshire 2004. *Local Transport Plan 2001–2006*. West Yorkshire County Council.

Wood, C. & N. Lee (eds) 1987. *Environmental impact assessment: five training case studies*. Occasional Paper 19, Department of Town and Country Planning, University of Manchester.

5 Impact prediction, evaluation and mitigation

5.1 Introduction

The focus of this chapter is the central steps of impact prediction, evaluation and mitigation. This is the heart of the EIA process, although, as we have already noted, the process is not linear. Indeed the whole EIA exercise is about prediction. It is needed at the earliest stages, when a project, including its alternatives, is being planned and designed, and it continues through to mitigation, monitoring and auditing. Yet, despite the centrality of prediction in EIA, there is a tendency for many studies to underemphasize it at the expense of more descriptive studies. Prediction is often not treated as an explicit stage in the process; clearly defined models are often missing from studies. Even when used, models are not detailed, and there is little discussion of limitations. Section 5.2 examines the dimensions of prediction (what to predict), the methods and models used in prediction (how to predict) and the limitations implicit in such exercises (living with uncertainty).

Evaluation follows from prediction and involves an assessment of the relative significance of the impacts. Methods range from the simple to the complex, from the intuitive to the analytical, from qualitative to quantitative, from formal to informal. CBA, monetary valuation techniques and multi-criteria/multi-attribute methods, with their scoring and weighting systems, provide a number of ways into the evaluation issue. The chapter concludes with a discussion of approaches to the mitigation of significant adverse effects. This may involve measures to avoid, reduce, remedy or compensate for the various impacts associated with projects.

5.2 Prediction

5.2.1 Dimensions of prediction (what to predict)

The object of prediction is to identify the magnitude and other dimensions of identified change in the environment *with* a project or action, in comparison with the situation *without* that project or action. Predictions also provide the basis for the assessment of significance, which we discuss in Section 5.3.

One starting point to identify the dimensions of prediction in the UK is the *legislative requirements* (see Table 3.4, Parts I and II). These basic specifications are amplified in guidance given in *Environmental assessment: a guide to the procedures* (DETR 2000, ODPM 2003) as outlined in Table 5.1. As already noted, this listing is limited on the assessment

Table 5.1 Assessment of effects, as outlined in UK regulations

---

Assessment of effects (including direct and indirect, secondary, cumulative, short-, medium- and long-term, permanent and temporary, positive and negative effects of project)

Effects on human beings, buildings and man-made features

1. Change in population arising from the development, and consequential environment effects.
2. Visual effects of the development on the surrounding area and landscape.
3. Levels and effects of emissions from the development during normal operation.
4. Levels and effects of noise from the development.
5. Effects of the development on local roads and transport.
6. Effects of the development on buildings, the architectural and historic heritage, archaeological features, and other human artefacts, e.g. through pollutants, visual intrusion, vibration.

Effects on flora, fauna and geology

7. Loss of, and damage to, habitats and plant and animal species.
8. Loss of, and damage to, geological, palaeotological and physiographic features.
9. Other ecological consequences.

Effects on land

10. Physical effects of the development, e.g. change in local topography, effect of earth-moving on stability, soil erosion, etc.
11. Effects of chemical emissions and deposits on soil of site and surrounding land.
12. Land-use/resource effects:

    (a) quality and quantity of agricultural land to be taken;
    (b) sterilization of mineral *resources*;
    (c) other alternative uses of the site, including the "do-nothing" option;
    (d) effect on surrounding land uses including agriculture;
    (e) waste disposal.

Effects on water

13. Effects of development on drainage pattern in the area.
14. Changes to other hydrographic characteristics, e.g. groundwater level, watercourses, flow of underground water.
15. Effects on coastal or estuarine hydrology.
16. Effects of pollutants, waste, etc. on water quality.

Effects on air and climate

17. Level and concentration of chemical emissions and their environmental effects.
18. Particulate matter.
19. Offensive odours.
20. Any other climatic effects.

Other indirect and secondary effects associated with the project

21. Effects from traffic (road, rail, air, water) related to the development.
22. Effects arising from the extraction and consumption of material, water, energy or other, resources by the development.
23. Effects of other development associated with the project, e.g. new roads, sewers, housing power lines, pipelines, telecommunications, etc.
24. Effects of association of the development with other existing or proposed development.
25. Secondary effects resulting from the interaction of separate direct effects listed above.

(*Sources*: DETR 2000, ODPM 2003.)

of socio-economic impacts. Table 1.3 provides a broader view of the scope of the environment, and of the environmental receptors that may be affected by a project.

Prediction involves the identification of potential change in indicators of such environment receptors. Scoping will have identified the broad categories of impact in relation to the project under consideration. If a particular environmental indicator (e.g. SO_2 levels in the air) revealed an increasing problem in an area, irrespective of the project or action (e.g. a power station), this should be predicted forwards as the baseline for this particular indicator. These indicators need to be disaggregated and specified to provide variables that are measurable and relevant. For example, an economic impact could be progressively specified as

direct employment → local employment → local skilled employment

In this way, a list of significant impact indicators of policy relevance can be developed.

An important distinction is often made between the prediction of the likely *magnitude* (i.e. size) and the *significance* (i.e. the importance for decision-making) of the impacts. Magnitude does not always equate with significance. For example, a large increase in one pollutant may still result in an outcome within generally accepted standards in a "robust environment", whereas a small increase in another may take it above the applicable standards in a "sensitive environment" (Figure 5.1). In terms of the Sassaman checklist (see Figure 4.8), the latter is crossing the threshold of concern and the former is not. This also highlights the distinction between *objective* and *subjective* approaches. The prediction of the magnitude of an impact should be an objective exercise, although it is not always easy. The determination of significance is often a more subjective exercise, as it normally involves value judgements.

As Table 1.4 showed, prediction should also identify *direct* and *indirect* impacts (simple cause–effect diagrams may be useful here), the *geographical extent* of impacts (e.g. local, regional, national), whether the impacts are *beneficial* or *adverse*, and the *duration* of the impacts. In addition to prediction over the life of a project (including, for example, its construction, operational and other stages), the analyst should also

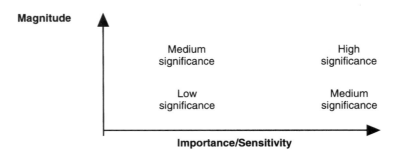

Figure 5.1 Significance expressed as a function of impact magnitude and the importance/ sensitivity of the resources or receptors. (Adapted from English Nature 1994, and Institute of Environmental Assessment (IEA) and Landscape Institute 1995.)

be alert to the *rate of change* of impacts. A slow build-up in an impact may be more acceptable than a rapid change; the development of tourism projects in formerly remote or undeveloped areas provides a topical example of the damaging impacts of rapid change. Projects may be characterized by non-linear processes, by delays between cause and effect, and the intermittent nature of some impacts should be anticipated. The *reversibility* or otherwise of impacts, their permanency, and their *cumulative* and synergistic impacts should also be predicted. Cumulative (or additive) impacts are the collective effects of impacts that may be individually minor but in combination, often over time, major. Such cumulative impacts are difficult to predict, and are often poorly covered or are missing altogether from EIA studies (see Chapter 11).

Another dimension is the unit of measurement, and the distinction between *quantitative* and *qualitative* impacts. Some indicators are more readily quantifiable than others (e.g. a change in the quality of drinking water, in comparison, for example, with changes in community stress associated with a project). Where possible, predictions should present impacts in explicit units, which can provide a basis for evaluation and trade-off. Quantification can allow predicted impacts to be assessed against various local, national and international standards. Predictions should also include estimates of the *probability* that an impact will occur, which raises the important issue of uncertainty.

5.2.2 Methods and models for prediction (how to predict)

There are many possible methods to predict impacts; a study undertaken by Environmental Resources Ltd for the Dutch government in the early 1980s identified 150 different prediction methods used in just 140 EIA studies from The Netherlands and North America (VROM 1984). None provides a magic solution to the prediction problem.

> All predictions are based on conceptual models of how the universe functions; they range in complexity from those that are totally intuitive to those based on explicit assumptions concerning the nature of environmental processes...the environment is never as well behaved as assumed in models, and the assessor is to be discouraged from accepting off-the-shelf formulae (Munn 1979).

Predictive methods can be classified in many ways; they are not mutually exclusive. In terms of *scope*, all methods are *partial* in their coverage of impacts, but some seek to be more *holistic* than others. Partial methods may be classified according to type of project (e.g. retail impact assessment) and type of impact (e.g. wider economic impacts). Some may be *extrapolative*, others may be more *normative*. For extrapolative methods, predictions are made that are consistent with past and present data. Extrapolative methods include, for example, trend analysis (extrapolating present trends, modified to take account of changes caused by the project), scenarios (common-sense forecasts of future state based on a variety of assumptions), analogies (transferring experience from elsewhere to the study in hand) and intuitive forecasting (e.g. the use of the Delphi technique to achieve group consensus on the impacts of a project) (Green et al. 1989). Normative approaches work backwards from desired outcomes to assess whether a project, in its environmental context, is adequate to achieve them. For example, a desired socio-economic outcome from the construction stage of a major project may be 50 per cent local employment. The achievement of this outcome may necessitate modifications to the project and/or to associated employment policies (e.g. on training). Various scenarios may be tested to determine the one most likely to achieve the desired outcomes.

Methods can also be classified according to their *form*, as the following types of model illustrate.

Mathematical and computer-based models

Mathematical models seek to represent the behaviour of aspects of the environment through the use of mathematical functions. They are usually based upon scientific laws, statistical analysis or some combination of the two, and are often computer based. The underpinning functions can range from simple direct input–output relationships to more complex dynamic mathematical models with a wide array of interrelationships. Mathematical models can be spatially aggregated (e.g. a model to predict the survival rate of a cohort population, or an economic multiplier for a particular area), or more locationally based, predicting net changes in detailed locations throughout a study area. Of the latter, retail impact models, which predict the distribution of retail expenditure using gravity model principles, provide a simple example; the comprehensive land-use locational models of Harris, Lowry, Cripps et al., provide more holistic examples (*Journal of American Institute of Planners* 1965). Mathematical models can also be divided into deterministic and stochastic models. Deterministic models, like the gravity model, depend on fixed relationships. In contrast, a stochastic model is probabilistic, and indicates "the degree of probability of the occurrence of a certain event by specifying the statistical probability that a certain number of events will take place in a given area and/or time interval" (Loewenstein 1966).

There are many mathematical models available for particular impacts. Reference to various EISs, especially from the USA, and to the literature (e.g. Bregman & Mackenthun 1992, Hansen & Jorgensen 1991, Rau & Wooten 1980, Suter 1993, US Environmental Protection Agency 1993, Westman 1985) reveals the availability of a rich array. For instance, Kristensen et al. (1990) list 21 mathematical models for phosphorus retention in lakes alone. Figure 5.2 provides a simple flow diagram for the prediction of the local socio-economic impacts of a power station development. Key determinants in the model are the details of the labour requirements for the project, the conditions in the local economy, and the policies of the relevant local authority and developer on topics such as training, local recruitment and travel allowances. The local recruitment ratio is a crucial factor in the determination of subsequent impacts.

An example of a deterministic mathematical model, often used in socio-economic impact predictions, is the multiplier (Lewis 1988), an example of which is shown in Figure 5.3. The injection of money into an economy – local, regional or national – will increase income in the economy by some multiple of the original injection. Modification of the basic model allows it to be used to predict income and employment impacts for various groups over the stages of the life of a project (Glasson et al. 1988). The more disaggregated (by industry type) input–output member of the multiplier family provides a particularly sophisticated method for predicting economic impacts, but with major data requirements.

Statistical models use statistical techniques such as regression or principal components analysis to describe the relationship between data, to test hypotheses or to extrapolate data. For instance, they can be used in a pollution-monitoring study to

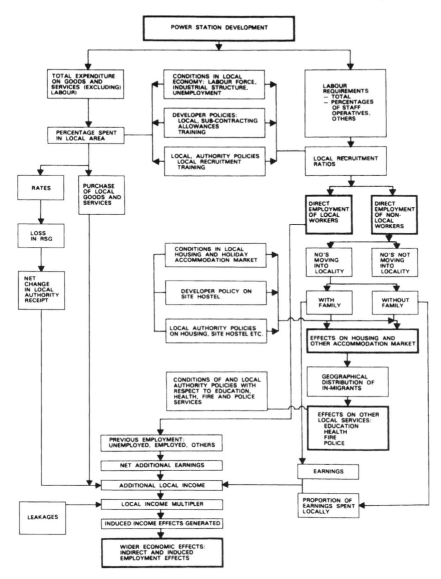

Figure 5.2 A cause–effect flow diagram for the local socio-economic impacts of a power station proposal. (*Source*: Glasson et al. 1987.)

describe the concentration of a pollutant as a function of the stream-flow rates and the distance downstream. They can compare conditions at a contaminated site and a control site to determine the significance of any differences in monitoring data. They can extrapolate a model to conditions outside the data range used to derive the model – e.g. from toxicity at high doses of a pollutant to toxicity at low doses – or from data that are available to data that are unavailable – e.g. from toxicity in rats to toxicity in humans.

$$Y_r = \frac{1}{1-(1-s)(1-t-u)(1-m)}\,J$$

where

Y_r = change in level of income (Y) in region (r), in £

J = initial income injection (or multiplicand)

t = proportion of additional income paid in direct taxation and National Insurance contributions

s = proportion of income saved (and therefore not spent locally)

u = decline in transfer payments (e.g. unemployment benefits) which result from the rise in local income and employment

m = proportion of additional income spent on imported consumer goods

Figure 5.3 A simple multiplier model for the prediction of local economic impacts.

Physical/architectural models and experimental methods

Physical, image or architectural models are illustrative or scale models that replicate some element of the project–environment interaction. For example, a scale model (or computer graphics) could be used to predict the impacts of a development on the landscape or built environment. Photo-montages can be used to show the views of the project site from the "receptor" areas, with images of the project superimposed to give an impression of visual impact. The image could be a photograph of a model of the project, or a simple "wire-line" profile of the project as it will appear to the viewer, showing just its skyline or a more sophisticated 3D impression.

Field and laboratory experimental methods use existing data inventories, often supplemented by special surveys, to predict impacts on receptors. Field tests are carried out in unconfined conditions, usually at approximately the same scale as the predicted impact; an example would be the testing of a pesticide in an outdoor pond. Laboratory tests, such as the testing of a pollutant on seedlings raised in a hydroponic solution, are usually cheaper to run but may not extrapolate well to conditions in natural systems.

Expert judgements and analogue models

All predictive methods in EIA make some use of expert judgement. Such judgement can make use of some of the other predictive methods, such as mathematical models and cause–effect networks or flow charts, as in Figure 5.2. Expert judgement can also draw on analogue models – making predictions based on analogous situations. They include comparing the impacts of a proposed development with a similar existing development; comparing the environmental conditions at one site with those at similar sites elsewhere; comparing an unknown environmental impact (e.g. of wind turbines on radio reception) with a known environmental impact (e.g. of other forms of development on radio reception). Analogue models can be developed from site visits, literature searches or the monitoring of similar projects.

Other methods for prediction

The various impact identification methods discussed in Chapter 4 may also be of value in impact prediction. The Sassaman threshold-of-concern checklist has already been noted; the Leopold matrix also includes magnitude predictions, although the objectivity of a system where each analyst is allowed to develop a ranking system on a scale of 1–10 is somewhat doubtful. Overlays can be used to predict spatial impacts, and the Sorensen network is useful in tracing through indirect impacts.

Choice of prediction methods

The nature and choice of prediction methods do vary according to the impacts under consideration, and Rodriguez-Bachiller with Glasson (2003) have identified the following types:

- Hard-modelled impacts: areas of impact prediction where mathematical simulation models play a central role. These include, for example, air and noise impacts. Air pollution impact prediction has been dominated by approaches based on the so-called "Gaussian dispersion model" which simulates the shape of the pollution plume from the development under concern (Elsom 2001).
- Soft-modelled impacts: areas of impact prediction where the use of mathematical simulation modelling is virtually non-existent. Examples here include terrestrial ecology and landscape. Terrestrial ecology depends very much on field sample survey for plant and animal species, where the expert's perception of what requires sampling plays an important role (Morris & Thurling 2001). Perception is also important in landscape assessment, but simple photo-montages, and the use of GIS, can help in the prediction of impacts (Therivel 2001, Wood 2000). Figure 5.4 provides an outline of key steps in landscape assessment.
- Mixed-modelled impacts: areas of impact prediction where simulation modelling is complemented (and sometimes replaced) by more technically lower-level approaches. Traffic impacts make considerable use of modelling, but often with some sample survey input. Socio-economic impacts may use simple flow diagrams, and mathematical models (as in Figures 5.2 and 5.3) particularly for economic impacts, but they tend to build a great deal on survey methods and expert judgement. This is particularly so with regard to social impacts.

When choosing prediction methods, an assessor should be concerned about their appropriateness for the task involved, in the context of the resources available (Lee 1987). Will the methods produce what is wanted (e.g. a range of impacts, for the appropriate geographical area, over various stages), from the resources available (including time, data, range of expertise)? In addition, the criteria of replicability (method is free from analyst bias), consistency (method can be applied to different projects to allow predictions to be compared) and adaptability should also be considered in the choice of methods. In many cases, more than one method may be appropriate. For instance, the range of methods available for predicting impacts on air quality is apparent from the 165 closely typed pages on the subject by Rau & Wooten (1980). Table 5.2 provides an overview of some of the methods of predicting the initial emissions of pollutants, which, with atmospheric interaction, may degrade air quality, which may then have adverse effects, for example on humans.

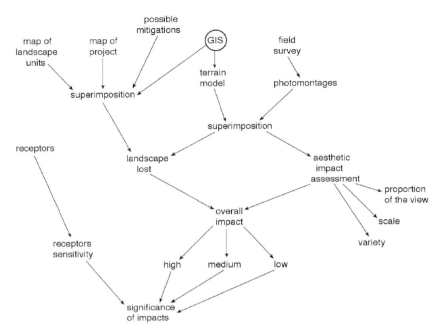

Figure 5.4 Key steps in landscape impact assessment. (*Source*: Rodriguez-Bachiller with Glasson 2003.)

Table 5.2 Examples of methods used in predicting air quality impacts

Sources	•	Original project design data on activity and emissions
↓		
POLLUTANT EMISSIONS	•	Published emission data for similar projects
↓	•	Emission factor models
↓	•	Emission standards
Atmospheric Interactions	•	Gaussian dispersion models (interactive programmes)
↓		
DEGRADED AIR QUALITY	•	Wind tunnel models
↓	•	Water analogue simulation models
↓	•	Expert opinion
↓	•	Mathematical deposition models
EFFECTS ON RECEPTORS	•	Laboratory or field experimental methods
e.g. humans	•	Inventories/surveys
	•	Dose–response factors

(*Sources*: VROM 1984, Rau & Wooten 1980.)

In practice, there has been a tendency to use the less formal predictive methods, and especially expert opinion (VROM 1984). Even where more formal methods have been used, they have tended to be simple, for example the use of photo-montages for visual impacts, or of simple dilution and steady-state dispersion models for water quality. However, simple methods need not be inappropriate, especially for early stages in the EIA process, nor need they be applied uncritically or in a simplistic way. Lee (1987) provides the following illustration:

(a) a single expert may be asked for a brief, qualitative opinion; or
(b) the expert may also be asked to justify that opinion (i) by verbal or mathematical description of the relationships he has taken into account and/or (ii) by indicating the empirical evidence which supports that opinion; or
(c) as in (b), except that opinions are also sought from other experts; or
(d) as in (c), except that the experts are also required to reach a common opinion, with supporting reasons, qualifications, etc.; or
(e) as in (d), except that the experts are expected to reach a common opinion using an agreed process of consensus building (e.g. based on "Delphi" techniques (Golden et al. 1979)).

The development of more complex methods can be very time-consuming and expensive, especially since many of these models are limited to specific environmental components and physical processes, and may only be justified when a number of relatively similar projects are proposed. However, notwithstanding the emphasis on the simple informal methods, there is scope for mathematical simulation models in the prediction stage. Munn (1979) identifies a number of criteria for situations in which computer-based simulation or mathematical models would be useful. The following are some of the most relevant:

• the assessment requires the handling of large numbers of simple calculations;
• there are many complex links between the elements of the EIA;
• the affected processes are time-dependent;
• increased definitions of assumptions and elements will be valuable in drawing together the many disciplines involved in the assessment;
• some or all of the relationships of the assessment can only be defined in terms of statistical probabilities.

5.2.3 *Living with uncertainty*

Environmental impact statements often appear more certain in their predictions than they should. This may reflect a concern not to undermine credibility and/or an unwillingness to attempt to allow for uncertainty. All predictions have an element of uncertainty, but it is only in recent years that such uncertainty has begun to be acknowledged in the EIA process (Beattie 1995, De Jongh 1988). The amended EIA Directive (CEC 1997) and subsequent UK Regulations (DETR 1999) include "the probability of the impact" in the characteristics of the potential impact of a project which must be considered. There are many sources of uncertainty relevant to the EIA process as a whole. In their classic works on strategic choice, Friend & Jessop (1977) and Friend & Hickling (1987) identified three broad classes of uncertainty: uncertainties about the physical, social and economic environment (UE), uncertainties about guiding values (UV) and uncertainties about related decisions (UR) (Figure 5.5). All three classes of uncertainty may affect the accuracy of predictions, but the focus in an EIA study is usually on

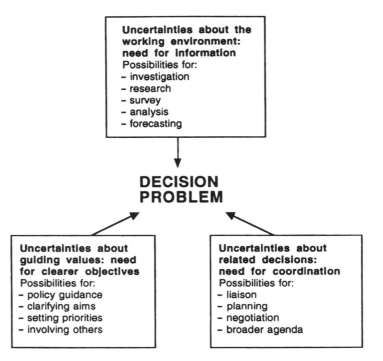

Figure 5.5 The types of uncertainty in decision-making. (*Source*: Friend & Hickling 1987.)

uncertainty about the environment. This may include the use of inaccurate and/or partial information on the project and on baseline-environmental conditions, unanticipated changes in the project during one or more of the stages of the life cycle, and oversimplification and errors in the application of methods and models. Socio-economic conditions may be particularly difficult to predict, as underlying societal values may change quite dramatically over the life, say 30–40 years, of a project.

Uncertainty in EIA predictive exercises can be handled in several ways. The assumptions underpinning predictions should be clearly stated (Voogd 1983). Issues of probability and confidence in predictions should be addressed, and ranges may be attached to predictions within which the analyst is n per cent confident that the actual outcome will lie. For example, scientific research may conclude that the 95 per cent confidence interval for the noise associated with a new industrial project is 65–70 dBA, which means that only 5 times out of 100 would the dBA be expected to be outside this range. Tomlinson (1989) draws attention to the twin issues of probability and confidence involved in predictions.

These twin factors are generally expressed through the same word. For example, in the prediction "a major oil spill would have major ecological consequences", a high degree of both probability and confidence exists. Situations may arise, however, where a low probability event based upon a low level of confidence is predicted. This is potentially more serious than a higher probability event with high confidence, since low levels of confidence may preclude expenditure on mitigating measures, ignoring issues of significance. Monitoring measures may be an appropriate response in such situations.

It may also be useful to show impacts under "peak" as well as "average" conditions for a particular stage of a project; this may be very relevant in the construction stage of major projects.

Sensitivity analysis may be used to assess the consistency of relationships between variables. If the relationship between input A and output B is such that whatever the changes in A there is little change in B, then no further information may be needed. However, where the effect is much more variable, there may be a need for further information. Of course, the best check on the accuracy of predictions is to check on the outcomes of the implementation of a project after the decision. This is too late for the project under consideration, but could be useful for future projects. Conversely, the monitoring of outcomes of similar projects may provide useful information for the project in hand. Holling (1978), who believes that the "core issue of EIA is how to cope with decision-making under uncertainty", recommends a policy of adaptive EIA, with periodic reviews of the EIA through a project's life cycle. Another procedural approach would be to require an *uncertainty report* as one step in the process; such a report would bring together the various sources of uncertainty associated with a project and the means by which they might be reduced (uncertainties are rarely eliminated).

5.3 Evaluation

5.3.1 Evaluation in the EIA process

Once impacts have been predicted, there is a need to assess their relative significance to inform decision-makers whether the impacts may be considered acceptable. Criteria for significance include the magnitude and likelihood of the impact and its spatial and temporal extent, the likely degree of the affected environment's recovery, the value of the affected environment, the level of public concern, and political repercussions. As with prediction, the choice of evaluation method should be related to the task in hand and to the resources available. Evaluation should feed into most stages of the EIA process, but the nature of the methods used may vary, for example, according to the number of alternatives under consideration, according to the level of aggregation of information and according to the number and type of parties involved (e.g. "in-house" and/or "external" consultation).

Evaluation methods can be of various types, including simple or complex, formal or informal, quantitative or qualitative, aggregated or disaggregated (see Maclaren & Whitney 1985, Voogd 1983). Much, if not most, current evaluation of significance in EIA is simple and often pragmatic, drawing on experience and expert opinion rather than on complex and sophisticated analysis. Table 5.3 provides an example of key factors used in Western Australia, where there is a particularly well-developed

Table 5.3 Determinants of environmental significance

Environmental significance is a judgement made by the Authority (West Australian Environmental Protection Authority) and takes into consideration the following factors:

(i) the extent and consequences of biophysical impacts;

(ii) the environmental value of the area affected;

(iii) the extent of emissions and their potential to unreasonably interfere with the health, welfare, convenience, comfort or amenity of people;

(iv) the potential for biophysical impacts of the proposal to significantly and adversely change people's social surroundings;

(v) the extent to and the rigour with which potential impacts have been investigated and described in the referral, and the confidence in the reliability of predicted impacts;

(vi) the extent to which the proposal implements the principle of sustainability;

(vii) the ability of decision-making authorities to place conditions on the proposals to ensure required environmental outcomes are achieved; and

(viii) the likely level of public interest, and the extent to which the proponent has consulted with interested and affected parties and responded to issues raised.

(*Source:* West Australian Environmental Protection Authority 2002.)

EIA system (see Chapter 10 also). To the factors in Table 5.3 could also be added scope for reversibility. The factor of public interest or perception ((viii) in Table 5.3) is an important consideration, and past and current perceptions of the significance of particular issues and impacts can raise their profile in the evaluation.

The most formal evaluation method is the *comparison of likely impacts against legal requirements and standards* (e.g. air quality standards, building regulations). Table 5.4 illustrates some of the standards which may be used to evaluate the traffic noise impacts of projects in Britain. Table 5.5 provides an example of more general guidance on standards and on environmental priorities and preferences, from the European Commission, for tourism developments. Of course, for some type of impacts, including socio-economic, there are no clear-cut standards. Socio-economic impacts provide a good example of "fuzziness" in assessment, where the line between being significant or not significant extends over a range of values which build on perceptions as much as facts. Socio-economic impacts can raise in particular the distributional dimension to evaluation, "who wins and who loses" (Glasson 2001, Vanclay 1999). Beyond the use of standards and legal requirements, all assessments of significance either implicitly or explicitly apply weights to the various impacts (i.e. some are assessed as more important than others). This involves interpretation and the application of judgement. Such judgement can be rationalized in various ways and a range of methods are available, but all involve values and all are subjective. Parkin (1992) sees judgements as being on a continuum between an analytical mode and an intuitive mode. In practice, many are at the intuitive end of the continuum, but such judgements, made without the benefit of analysis, are likely to be flawed, inconsistent and biased. The "social effects of resource allocation decisions are too extensive to allow the decision to 'emerge' from some opaque procedure free of overt political scrutiny" (Parkin 1992). Analytical methods seek to introduce a rational approach to evaluation.

Table 5.4 Examples of standards in relation to impacts of projects on traffic noise in Britain

---

- BS 7445 is the Standard for description and measurement of environmental noise. It is in three parts: Part 1: Guide to quantities and procedures, Part 2: Guide to acquisition of data, and Part 3: Guide to application of noise limits.
- Noise is measured in decibels (dB) at a given frequency. This is an objective measure of sound pressure. Measurements are made using a calibrated sound meter.
- Human hearing is approximately in the range 0–140 dBA.
  dB Example of noise
  <40   quiet bedroom
  60   busy office
  72   car at 60 km/h at a distance of 7 m
  85   Heavy goods vehicle (HGV) at 40 km/h at a distance of 7 m
  90   hazardous to hearing from continuous exposure
  105   jet flying overhead at 250 m
  120   threshold of pain.
- Traffic noise is perceived as a nuisance even at low dB levels. Noise comes from tyres on the road, engines, exhausts, brakes and HGV bodies. Poor maintenance of roads and vehicles and poor driving also increase road noise. Higher volumes of traffic and higher proportions of HGVs increase the noise levels. In general, annoyance is proportional to traffic flow for noise levels above 55 dB(A). People are sensitive to a change in noise levels of 1 dB (about 25 per cent change in flow).
- Assessment of traffic noise is assessed in terms of impacts within 300 m of the road. The EIA will estimate the number of properties and relevant locations (e.g. footpaths and sports fields) in bands of distance from the route: 0–50 m, 50–100 m, 100–200 m, 200–300 m, and then classify each group according to the baseline ambient noise levels (in bands of <50, 50–60, 60–70, >70 dB(A)) and the increase in noise (1–3, 3–5, 5–10, 10–15 and >15 dB(A)).
- Façade noise levels are measured at 1.7 m above ground, 1 m from façade or 10 m from kerb, and are usually predicted using the Department of Transport's Calculation of Road Traffic Noise (CRTN) which measures dB(A) $L_{A10,18h}$. This is the noise level exceeded 10 per cent of the time between 6:00 and 24:00. Noise levels at the façade are approximately 2 dB higher than 10 m from the building. PPG 13 uses dB(A) $L_{Aeq,16h}$. This is between 7:00 and 23:00. Most traffic noise meters use dB(A) L_{A10}, and an approximate conversion is:

$$L_{Aeq,16h} = L_{A10,18h} - 2 \text{ dB}.$$

- The DTP recommends an absolute upper limit for noise of 72 dB(A) $L_{eq,18h}$ (=70 dB(A) $L_{A10,18h}$) for residential properties. Compensation is payable to properties within 300 m of a road development for increases greater than 1 dB(A) which result in $L_{A10,18h}$ above 67.5.
- The DTP considers a change of 30 per cent slight, 60 per cent moderate and 90 per cent substantial. PPG 13 considers 5 per cent to be significant.

  There are four categories of noise in residential areas

day (16 h)	*night* (8 h)	
A <55 L_{Aeq}	<42 L_{Aeq}	Not determining the application
B 55–63 L_{Aeq}	<42–57 L_{Aeq}	Noice control measures are required
C 63–72 L_{Aeg}	57–66 L_{Aeq}	Strong presumption against developer
D >72 L_{Aeq}	>66 L_{Aeg}	Normally refuse the application

  For night-time noise, unless the noise is already in category D, a single event occurring regularly (e.g. HGV movements) where $L_{Aeg} > 82$ dB puts the noise in category C.

(*Source*: Bourdillon 1996.)

Table 5.5 Example of EC guidance on assessing significance of impacts for tourism projects for Asian, Caribbean and Pacific countries

---

The significance of certain environmental impacts can be assessed by contrasting the predicted magnitude of impact against a relevant environmental standard or value. For tourism projects in particular, impact significance should also be assessed by taking due regard of those environmental priorities and preferences held by society but for which there are no quantifiable objectives. Particular attention needs to be focused upon the environmental preferences and concerns of those likely to be directly affected by the project.

Environmental Standards

- Water quality standards
    - potable water supplies (*apply country standards; see also Section 1.3.2,* WHO *(1982) Guidelines for Drinking Water Quality Directives 80/778/EEC and 75/440/EEC*)
    - wastewater discharge (*apply country standards for wastewaters and fisheries; see also 76/160/EEC and 78/659/EEC*).
- National and local planning regulations
    - legislation concerning change in land use
    - regional/local land-use plans (particularly management plans for protected areas and coastal zones).
- National legislation to protect certain areas
    - national parks
    - forest reserves
    - nature reserves
    - natural, historical or cultural sites of importance.
- International agreements to protect certain areas
    - World Heritage Convention
    - Ramsar Convention on wetlands.
- Conservation/preservation of species likely to be sold to tourists or harmed by their activities
    - national legislation
    - international conventions
    - CITES Convention on trade in endangered species.

Environmental Priorities and Preferences

- Participation of affected people in project planning to determine priorities for environmental protection, including:
    - public health
    - revered areas, flora and fauna (e.g. cultural/medicinal value, visual landscape)
    - skills training to undertake local environmental mitigation measures
    - protection of potable water supply
    - conservation of wetland/tropical forest services and products, e.g. hunted wildlife, fish stocks
    - issues of sustainable income generation and employment (including significance of gender – *see* WID *manual*).
- Government policies for environmental protection (including, where appropriate, incorporation of objectives from Country Environmental Studies/Environmental Action Plans, etc.)
- Environmental priorities of tourism boards and trade associations representing tour operators.

(*Source:* CEC 1993.)

Two sets of methods are distinguished: those that assume a common utilitarian ethic with a single evaluation criterion (money), and those based on the measurement of personal utilities, including multiple criteria. The CBA approach, which seeks to express impacts in monetary units, falls into the former category. A variety of methods, including *multi-criteria analysis, decision analysis* and *goals achievement*, fall into the latter. The very growth of EIA is partly a response to the limitations of CBA and to the problems of the monetary valuation of environmental impacts. Yet, after two decades of limited concern, there is renewed interest in the mon-etizing of environmental costs and benefits (DoE 1991). The multi-criteria/ multi-attribute methods involve scoring and weighting systems that are also not problem-free. The various approaches are now outlined. In practice, there are many hybrid variations between these two main categories, and these are referred to in both categories.

5.3.2 Cost–benefit analysis and monetary valuation techniques

Cost–benefit analysis itself lies in a range of project and plan appraisal methods that seek to apply monetary values to costs and benefits (Lichfield et al. 1975). At one extreme are *partial* approaches, such as financial-appraisal, cost-minimization and cost-effectiveness methods, which consider only a subsection of the relevant population or only a subsection of the full range of consequences of a plan or project. *Financial appraisal* is limited to a narrow concern, usually of the developer, with the stream of financial costs and returns associated with an investment. *Cost effectiveness* involves selecting an option that achieves a goal at least cost (for example, devising a least-cost approach to produce coastal bathing waters that meet the CEC Blue Flag criteria). The cost-effectiveness approach is more problematic where there are a number of goals and where some actions achieve certain goals more fully than others (Winpenny 1991).

Cost–benefit analysis is more *comprehensive* in scope. It takes a long view of projects (farther as well as nearer future) and a wide view (in the sense of allowing for side effects). It is based in welfare economics and seeks to include all the relev-ant costs and benefits to evaluate the net social benefit of a project. It was used extensively in the UK in the 1960s and early 1970s for public sector projects, the most famous being the third London Airport (HMSO 1971). The methodology of CBA has several stages: project definition, the identification and enumeration of costs and benefits, the evaluation of costs and benefits, and the discounting and presentation of results. Several of the stages are similar to those in EIA. The basic evaluation principle is to measure in monetary terms where possible – as money is the common measure of value and monetary values are best understood by the community and decision-makers – and then reduce all costs and benefits to the same capital or annual basis. Future annual flows of costs and benefits are usually discounted to a net present value (Table 5.6). A range of interest rates may be used to show the sensitivity of the analysis to changes. If the net social benefit minus cost is positive, then there may be a presumption in favour of a project. However, the final outcome may not always be that clear. The presentation of results should distinguish between tangible and intangible costs and benefits, as relevant, allowing the decision-maker to consider the trade-offs involved in the choice of an option.

Table 5.6 Cost–benefit analysis: presentation of results: tangibles and intangibles

Category	Alternative 1	Alternative 2
Tangibles		
Annual benefits	£Bl	£bl
	£B2	£b2
	£B3	£b3
Total annual benefits	£B1 + B2 + B3	£bl + b2 + b3
Annual costs	£Cl	£cl
	£C2	£c2
	£C3	£c3
Total annual costs	£C1 + C2 + C3	£cl + c2 + c3
Net discounted present value (NDPV) of benefits and costs over "m" years at X%*	£D	£E
Intangibles		
Intangibles are likely to include costs and benefits	I1	i1
	I2	i2
	I3	i3
	I4	i4
Intangibles summation (undiscounted)	I1 + I2 + I3 + I4	i1 + i2 + i3 + i4

* e.g. NPDV (Alt 1).

$$D = \sum \left[\frac{B1}{(1+X)^1} + \frac{B1}{(1+X)^2} + \cdots + \frac{B1}{(1+X)^n} + \frac{B2}{(1+X)^1} + \cdots + \frac{B2}{(1+X)^n} + \frac{B3}{(1+X)^1} + \cdots + \frac{B3}{(1+X)^n} \right]$$

$$- \sum \left[\frac{C1}{(1+X)^1} + \frac{C1}{(1+X)^2} + \cdots + \frac{C1}{(1+X)^n} + \frac{C2}{(1+X)^1} + \cdots + \frac{C2}{(1+X)^n} + \frac{C3}{(1+X)^1} + \cdots + \frac{C3}{(1+X)^n} \right]$$

Cost–benefit analysis has excited both advocates (e.g. Dasgupta & Pearce 1978, Pearce 1989, Pearce et al. 1989) and opponents (e.g. Bowers 1990). It does have many problems, including identifying, enumerating and monetizing intangibles. Many environmental impacts fall into the intangible category, for example the loss of a rare species, the urbanization of a rural landscape and the saving of a human life. The incompatibility of monetary and non-monetary units makes decision-making problematic (Bateman 1991). Another problem is the choice of discount rate: for example, should a very low rate be used to prevent the rapid erosion of future costs and benefits in the analysis? This choice of rate has profound implications for the evaluation of resources for future generations. There is also the underlying and fundamental problem of the use of the single evaluation criterion of money, and the assumption that £1 is worth the same to any person, whether a tramp or a millionaire, a resident of a rich commuter belt or of a poor and remote rural community. CBA also ignores distribution effects and aggregates costs and benefits to estimate the change in the welfare of society as a whole.

The *planning balance sheet* (PBS) is a variation on the theme of CBA, and it goes beyond CBA in its attempts to identify, enumerate and evaluate the distribution of costs and benefits between the affected parties. It also acknowledges the difficulty of attempts to monetize the more intangible impacts. It was developed by Lichfield et al. (1975) to compare alternative town plans. PBS is basically a set of social accounts structured into sets of "producers" and "consumers" engaged in various transactions. The transaction could, for example, be an adverse impact, such as noise from an airport (the producer) on the local community (the consumers), or a beneficial impact, such as the time savings resulting from a new motorway development (the producer) for users of the motorway (the consumers). For each producer and consumer group, costs and benefits are quantified per transaction, in monetary terms or otherwise, and weighted according to the numbers involved. The findings are presented in tabular form, leaving the decision-maker to consider the trade-offs, but this time with some guidance on the distributional impacts of the options under consideration (Figure 5.6). More recently, Lichfield (1996) has sought to integrate EIA and PBS further in an approach he calls *community impact evaluation* (CIE).

Partly in response to the "intangibles" problem in CBA, there has also been considerable interest in the development of *monetary valuation techniques* to improve the economic measurement of the more intangible environmental impacts (Barde & Pearce 1991, DoE 1991, Winpenny 1991). The techniques can be broadly classified into direct and indirect, and they are concerned with the measurement of preferences about the environment rather than with the intrinsic values of the environment. The direct approaches seek to measure directly the monetary value of environmental gains – for example, better air quality or an improved scenic view. Indirect approaches measure preferences for a particular effect via the establishment of a "dose–response"-type relationship. The various techniques found under the direct and indirect categories are summarized in Table 5.7. Such techniques can contribute to the assessment of the total economic value of an action or project, which should not only include user values (preferences people have for using an environmental asset, such as a river for fishing) but also non-user values (where people value an asset but do not use it, although some may wish to do so some day). Of course, such techniques have their problems, for example the potential bias in people's replies in

	Plan A				Plan B			
	Benefits		Costs		Benefits		Costs	
	Capital	Annual	Capital	Annual	Capital	Annual	Capital	Annual
Producers								
X	£a	£b	–	£d	–	–	£b	£c
Y	i_1	i_2	–	–	i_3	i_4	–	–
Z	M_1	–	M_2	–	M_3	–	M_4	–
Consumers								
X´	–	£e	–	£f	–	£g	–	£h
Y´	i_5	i_6	–	–	i_7	i_8	–	–
Z´	M_1	–	M_3	–	M_2	–	M_4	–

£ = benefits and costs that can be monetized
M = where only a ranking of monetary values can be estimated
i = intangibles

Figure 5.6 Example of structure of a planning balance sheet.

Table 5.7 Summary of environmental monetary valuation techniques

---

Direct household production function (HPF)
HPF methods seek to determine expenditure on commodities that are substitutes or complements for an environmental characteristic to value changes in that characteristic. Subtypes include the following:

- Avertive expenditures: expenditure on various substitutes for environmental change (e.g. noise insulation as an estimate of the value of peace and quiet).
- Travel cost method: expenditure, in terms of cost and time, incurred in travelling to a particular location (e.g. a recreation site) is taken as an estimate of the value placed on the environmental good at that location (e.g. benefit arising from use of the site).

Direct hedonic price methods (HPM)
HPM methods seek to estimate the implicit price for environmental attributes by examining the real markets in which those attributes are traded. Again, there are two main subtypes:

1. Hedonic house land prices: these prices are used to value characteristics such as "clean air" and "peace and quiet", through cross-sectional data analysis (e.g. on house price sales in different locations).
2. Wage risk premia: the extra payments associated with certain higher risk occupations are used to value changes in morbidity and mortality (and implicitly human life) associated with such occupations.

Direct experimental markets
Survey methods are used to elicit individual values for non-market goods. Experimental markets are created to discover how people would value certain environmental changes. Two kinds of questioning, of a sample of the population, may be used:

1. Contingent valuation method (CVM): people are asked what they are willing to pay (WTP) for keeping X (e.g. a good view, a historic building) or preventing Y, or what they are willing to accept (WTA) for losing A, or tolerating B.
2. Contingent ranking method (CRM or stated preference): people are asked to rank their preferences for various environmental goods, which may then be valued by linking the preferences to the real price of something traded in the market (e.g. house prices).

Indirect methods
Indirect methods seek to establish preferences through the estimation of relationships between a "dose" (e.g. reduction in air pollution) and an effect (e.g. health improvement). Approaches include the following:

- Indirect market price approach: the dose–response approach seeks to measure the effect (e.g. value of loss of fish stock) resulting from an environmental change (e.g. oil pollution of a fish farm), by using the market value of the output involved. The replacement-cost approach uses the cose of replacing or restoring a damaged asset as a measure of the benefit of restoration (e.g. of an old stone bridge eroded by pollution and wear and tear).
- Effect on production approach: where a market exists for the goods and services involved, the environmental impact can be represented by the value of the change in output that it causes. It is widely used in developing countries, and is a continuation of the dose–response approach.

(Adapted from DoE 1991, Winpenny 1991, Pearce & Markandya 1990, Barde & Pearce 1991.)

the contingent valuation method (CVM) approach (for a fascinating example of this, see Willis & Powe 1998). However, simply through the act of seeking a value for various environmental features, such techniques help to reinforce the understanding that such features are not "free" goods and should not be treated as such.

5.3.3 Scoring and weighting and multi-criteria methods

Multi-criteria and multi-attribute methods seek to overcome some of the deficiencies of CBA; in particular they seek to allow for a pluralist view of society, composed of diverse "stakeholders" with diverse goals and with differing values concerning environmental changes. Most of the methods use – and sometimes misuse – some kind of simple scoring and weighting system; such systems generate considerable debate. Here we discuss some key elements of good practice, and then offer a brief overview of the range of multi-criteria/multi-attribute methods available to the analyst.

Scoring may use quantitative or qualitative scales, according to the availability of information on the impact under consideration. Lee (1987) provides an example (Table 5.8) of how different levels of impact (in this example noise, whose measurement is in units of $L_{10}dB_A$) can be scored in different systems. These systems seek to standardize the impact scores for purposes of comparison. Where quantitative data are not available, ranking of alternatives may use other approaches, for example using letters (A, B, C, etc.) or words (not significant, significant, very significant).

Weighting seeks to identify the relative importance of the various impact types for which scores of some sort may be available (for example, the relative importance of a water pollution impact, the impact on a rare flower). Different impacts may be allocated weights (normally numbers) out of a total budget (e.g. 10 points to be allocated between 3 impacts). But by whom?

Multi-criteria/multi-attribute methods seek to recognize the plurality of views and weights in their methods; the Delphi approach also uses individuals' weights, from which group weights are then derived. In many studies, however, the weights are those produced by the technical team. Indeed the decision-makers may be unwilling to reveal all their personal preferences, for fear of undermining their negotiating

Table 5.8 A comparison of different scoring systems

Method	Alternatives				Basis of score
	A (no action)	B	C	D	
Ratio	65	62	71	75	Absolute $L_{10}dB_A$ measure
Interval	0		−3	+6	+10	Difference in $L_{10}dB_A$ using alternative A as base
Ordinal	B		A	C	D	Ranking according to ascending value of $L_{10}dB_A$
Binary	0		0	1	1	0 = less than $70L_{10}dB_A$ 1 = $70L_{10}dB_A$ or more

Based on Lee (1987).

Table 5.9 Weighting, scoring and trade-offs

Impact	Weight (w)	Scheme A		Scheme B	
		Score (a)	(aw)	Score (b)	(bw)
Noise	2	5	10	1	2
Loss of flora	5	1	5	4	20
Air pollution	3	2	6	2	6
Total			21		28

positions. This internalization of the weighting exercise does not destroy the use of weights, but it does emphasize the need for clarification of scoring and weighting systems and, in particular, for the identification of the origin of the weightings used in an EIA. Wherever possible, scoring and weighting should be used to reveal the trade-offs in impacts involved in particular projects or in alternatives. For example, Table 5.9 shows that the main issue is the trade-off between the impact on flora of one scheme and the impact on noise of the other scheme.

Several approaches to the scoring and weighting of impacts have already been introduced in the outline of impact identification methods in Chapter 4. The Leopold matrix includes measures of the significance of impacts (on a scale of 1–10) as well as of their magnitude. The matrix approach can also be usefully modified to identify the distribution of impacts among geographical areas and/or among various affected parties (Figure 5.7). The quantitative EES and Water Resources Assessment Methods (WRAM) generate weights for different environmental parameters, drawn up by panels of experts. Weightings can also be built into overlay maps to identify areas with the most development potential according to various combinations of

Group environmental component	Project Action							
	Construction stage actions				Operational stage actions			
	A	B	C	D	a	b	c	d
Group 1 (e.g. indigenous population ≥ 45 years old) various								
• Social								
• Physical								
• Economic components								
Group 2 (e.g. indigenous population < 45 years old) various								
• Social								
• Physical								
• Economic components								

Figure 5.7 Simple matrix identification of distribution of impacts.

weightings. Some of the limitations of such approaches have already been noted in Chapter 4.

Other methods in the multi-criteria/multi-attribute category include decision analysis, the goals achievement matrix (GAM), multi-attribute utility theory (MAUT) and judgement analysis. *Decision analysis* is the operational form of decision theory, a theory of how individuals make decisions in the face of uncertainty, which owes its modern origins to Von Neuman & Morgenstern (1953). Decision analysis usually involves the construction of a decision tree, an example of which is shown in Figure 5.8. Each branch represents a potential action, with a probability of achievement attached to it.

The GAM was developed as a planning tool by Hill (1968) to overcome the perceived weaknesses of the PBS approach. GAM makes the goals and objectives of a project/plan explicit, and the evaluation of alternatives is accomplished by measuring the extent to which they achieve the stated goals. The existence of many diverse goals leads to a system of weights. Since all interested parties are not politically equal, the identified groups should also be weighted. The end result is a matrix of weighted objectives and weighted interests/agencies (Figure 5.9). The use of goals and value weights to evaluate plans in the interests of the community, and not just for economic efficiency, has much to commend it. The approach also provides an opportunity for public participation. Unfortunately, the complexity of the approach has limited its use, and the weights and goals used may often reflect the views of the analyst more than those of the interests and agencies involved.

MAUT has gained a certain prominence in recent years as an evaluation method that can incorporate the values of the key interests involved (Bisset 1988, Edwards & Newman 1982, Parkin 1992). MAUT involves a number of steps, including the identification of the entities (alternatives, objects) to be evaluated, and the identification and structuring of environmental attributes (e.g. noise level) to be measured. The latter may include a "value tree" with general objectives (values) at the top and specific attributes at the bottom. The ranking of attributes is by the central stakeholder/expert group whose values are to be maximized. Attributes are scaled and formal value or utility models developed to quantify trade-offs among attribute scales and attributes. For further reference, see Parkin (1992) for an outline of the main steps and an application of a "relatively" simple and well-proven version of MAUT known as the SMART method.

Finally, brief reference is made to the *Delphi method*, which can be used to incorporate the views of various stakeholders into the evaluation process. The method is an established means of collecting expert opinion and of gaining consensus among experts on various issues under consideration. It has the advantage of obtaining expert opinion from the individual, with guaranteed anonymity, avoiding the potential distortion caused by peer pressure in group situations. Compared with other evaluation methods it can also be quicker and cheaper.

There have been a number of interesting applications of the Delphi method in EIA (Green et al. 1989, 1990, Richey et al. 1985). Green et al. used the approach to assess the environmental impacts of the redevelopment and reorientation of Bradford's famous Salt Mill. The method involves drawing up a Delphi panel. In the Salt Mill case, the initial panel of 40 included experts with a working knowledge of the project (e.g. planners, tourism officers), councillors, employees, academics, local residents and traders. This was designed to provide a balanced view of interests and expertise.

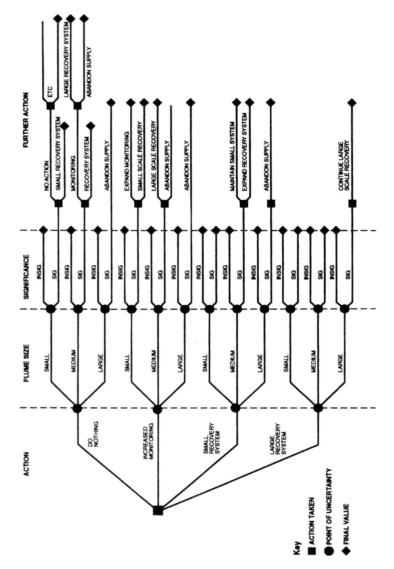

Figure 5.8 A decision tree: problem of groundwater contamination. (*Source*: De Jongh 1988.)

The Delphi exercise usually has a three-stage approach: (1) a general questionnaire asking panel members to identify important impacts (positive and negative); (2) a first-round questionnaire asking panel members to rate the importance of a list of impacts identified from the first stage; (3) a second-round questionnaire asking panel members to re-evaluate the importance of each impact in the light of the panel's response to the first round. However, the method is not without its limitations. The

Goal description:		α			β	
Relative weight:		2			3	
Incidence	Relative weight	Costs	Benefits	Relative weight	Costs	Benefits
Group a	1	A	D	5	E	1
Group b	3	H	J	4	M	2
Group c	1	L	J	3	M	3
Group d	2	–	J	2	V	4
Group e	1	–	K	1	T	5
		Σ	Σ	Σ	Σ	

Figure 5.9 Goals achievement matrix (section of). (Adapted from Hill 1968.)

potential user should be aware that it is difficult to draw up a "balanced" panel in the first place, and to avoid distorting the assessment by the varying drop-out rates of panel members between stages of the exercise, and by an overzealous structuring of the exercise by the organizers.

5.4 Mitigation

5.4.1 Types of mitigation measures

Mitigation is defined in EC Directive 97/11 as "measures envisaged in order to avoid, reduce and, if possible, remedy significant adverse effects" (CEC 1997). In similar vein, the US CEQ, in its regulations implementing the NEPA, defines mitigation as including: "not taking certain actions; limiting the proposed action and its implementation; repairing, rehabilitating, or restoring the affected environment; presentation and maintenance actions during the life of the action; and replacing or providing substitute resources or environments" (CEQ 1978). The guidance on mitigation measures provided by the UK government is set out in Table 5.10. It is not possible to specify here all the types of mitigation measures that could be used. Instead, the following text provides a few examples, relating to biophysical and socioeconomic impacts. The reader is also referred to Fortlage (1990), Morris & Therivel (2001) and Rodriguez-Bachiller with Glasson (2003) for useful coverage of mitigation measures for particular impact types. A review of EISs for developments similar to the development under consideration may also suggest useful mitigation measures.

At one extreme, the prediction and evaluation of impacts may reveal an array of impacts with such significant adverse effects that the only effective mitigation measure may be to abandon the proposal altogether. A less draconian, and more normal, situation would be to modify aspects of the development action to avoid various impacts. Examples of methods to *avoid* impacts include:

- the control of solid and liquid wastes by recycling on site or by removing them from the site for environmentally sensitive treatment elsewhere;

Table 5.10 Mitigation measures, as outlined in *UK guide to procedures*

Where significant adverse effects are identified, [describe] the measures to be taken to avoid, reduce or remedy those effects, e.g.:

(a) Site planning
(b) Technical measures, e.g.:

  (i) process selection
  (ii) recycling
  (iii) pollution control and treatment
  (iv) containment (e.g. bunding of storage vessels)

(c) Aesthetic and ecological measures, e.g.:

  (i) mounding
  (ii) design, colour, etc.
  (iii) landscaping
  (iv) tree plantings
  (v) measure to preserve particular habitats or create alternative habitats
  (vi) recording of archaeological sites
  (vii) measures to safeguard historic buildings or sites

[Assess] the likely effectiveness of mitigating measures.

(*Source*: ODPM 2003.)

- the use of a designated lorry route, and day-time working only, to avoid disturbance to village communities from construction lorry traffic and from night construction work;
- the establishment of buffer zones and the minimal use of toxic substances, to avoid impacts on local ecosystems.

Some adverse effects may be less easily avoided; there may also be less need to avoid them completely. Examples of methods to *reduce* adverse effects include:

- the sensitive design of structures, using simple profiles, local materials and muted colours, to reduce the visual impact of a development, and landscaping to hide it or blend it into the local environment;
- the use of construction-site hostels, and coaches for journeys to work to reduce the impact on the local housing market, and on the roads, of a project employing many workers during its construction stage;
- the use of silting basins or traps, the planting of temporary cover crops and the scheduling of activities during the dry months, to reduce erosion and sedimentation.

During one or more stages of the life of a project, certain environmental components may be temporarily lost or damaged. It may be possible to *repair*, *rehabilitate* or *restore* the affected component to varying degrees. For example:

- agricultural land used for the storage of materials during construction may be fully rehabilitated; land used for gravel extraction may be restored to agricultural use, but over a much longer period and with associated impacts according to the nature of the landfill material used;

- a river or stream diverted by a road project can be unconverted and re-established with similar flow patterns as far as is possible;
- a local community astride a route to a new tourism facility could be relieved of much of the adverse traffic effects by the construction of a bypass (which, of course, introduces a new flow of impacts).

There will invariably be some adverse effects that cannot be reduced. In such cases, it may be necessary to *compensate* people for adverse effects. For example:

- for the loss of public recreational space or a wildlife habitat, the provision of land with recreation facilities or the creation of a nature reserve elsewhere;
- for the loss of privacy, quietness and safety in houses next to a new road, the provision of sound insulation and/or the purchase by the developer of badly affected properties.

Mitigation measures can become linked with discussions between a developer and the local planning authority (LPA) on what is known in the UK as "planning gain". Fortlage (1990) talks of some of the potential complications associated with such discussions, and of the need to distinguish between mitigation measures and planning gain:

> Before any mitigating measures are put forward, the developer and the local planning authority must agree as to which effects are to be regarded as adverse, or sufficiently adverse to warrant the expense of remedial work, otherwise the whole exercise becomes a bargaining game which is likely to be unprofitable to both parties...
>
> Planning permission often includes conditions requiring the provision of planning gains by the developer to offset some deterioration of the area caused by the development, but it is essential to distinguish very clearly between those benefits offered by way of compensation for adverse environmental effects and those which are a formal part of planning consent. The local planning authority may decide to formulate the compensation proposals as a planning condition in order to ensure that they are carried out, so the developer should beware of putting forward proposals that he does not really intend to implement.

Mitigation measures must be planned in an integrated and coherent fashion to ensure that they are effective, that they do not conflict with each other and that they do not merely shift a problem from one medium to another. A *project may also benefit an area*, often socio-economically; where such benefits are identified, as a minimum there should be a concern to ensure that they do occur and do not become diluted, and that they may even be enhanced. For example, the potential local employment benefits of a project can be encouraged by the offer of appropriate skills training programmes to local people; various tenure arrangements can be used to make houses in new housing schemes available to local people in need.

The results of a more recent research project on the treatment of mitigation within EIA (DETR 1997) still found that UK practice varied considerably. For example, there was too much emphasis on physical measures, rather than on operational or management controls, and a lack of attention to the impacts of construction and to residual impacts after mitigation. Table 5.11 provides a wider

Table 5.11 A wider classification of mitigation

Levels of mitigation	Mitigation hierarchy	Project phase
• Alternatives (strategic, alternative locations and processes) • Physical design measures • Project management measures • Deferred mitigation	• Avoidance at source • Minimize at source • Abatement on site • Abatement at receptor • Repair • Compensation in kind • Other compensation and enhancement	• Construction • Commissioning • Operation • Decommissioning • Restoration, afteruse/aftercare

(*Source*: DETR 1997.)

classification of mitigation, adopted in the project, by levels of mitigation, mitigation hierarchy and project phase. The levels relate to broad decisions that are made during the design of a project, with the last two reflecting the fact that effective mitigation can be achieved through measures other than physical ones. The mitigation hierarchy focuses on the principle of prevention rather than cure where, in principle at least, the options higher in the list should be tried before those lower down the list. The project phases relate to the life cycle of the project first discussed in Chapter 1. Any particular mitigation measure can be classified in a combination of the three ways – for example, physical design measures can be used to minimize an impact at source, during the construction phase (DETR 1997).

5.4.2 Mitigation in the EIA process

Like many elements in the EIA process, and as noted in Table 5.11, mitigation is not limited to one point in the assessment. Although it may follow logically from the prediction and assessment of the relative significance of impacts, it is in fact inherent in all aspects of the process. An original project design may already have been modified, possibly in the light of mitigation changes made to earlier comparable projects or perhaps as a result of early consultation with the LPA or with the local community. The consideration of alternatives, initial scoping activities, baseline studies and impact identification studies may suggest further mitigation measures. Although more in-depth studies may identify new impacts, mitigation measures may alleviate others. The prediction and evaluation exercise can thus focus on a limited range of potential impacts.

Mitigation measures are normally discussed and documented in each topic section of the EIS (e.g. air quality, visual quality, transport, employment). Those discussions should clarify the extent to which the significance of each adverse impact has been offset by the mitigation measures proposed. A summary chart (Table 5.12) can provide a clear and very useful overview of the envisaged outcomes, and may be a useful basis for agreement on planning consents. Residual unmitigated or only partially mitigated impacts should be identified. These could be divided according to the degree of severity: for example, into "less than significant impacts" and "significant unavoidable impacts".

Table 5.12 Example of a section of a summary table for impacts and mitigation measures

Impact	Mitigation measure(s)	Level of significance after mitigation
1. 400 acres of prime agricultural land would be lost from the county to accommodate the petrochemical plant.	The only *full* mitigating measure for this impact would be to abandon the project.	SU
2. Additional lorry and car traffic on the adjacent hilly section of the motorway will increase traffic volumes by 10–20 per cent above those predicted on the basis of current trends.	A lorry crawler lane on the motorway, funded by the developer, will help to spread the volume, but effects may be partial and short-lived.	SU
3. The project would block the movement of most terrestrial species from the hilly areas to the east of the site to the wetlands to the west of the site.	A wildlife corridor should be developed and maintained along the entire length of the existing stream which runs through the site. The width of the corridor should be a minimum of 75 ft. The stream bed should be cleaned of silt and enhanced through the construction of occasional pools. The buffer zone should be planted with native riparian vegetation, including sycamore and willow.	LS

Note: SU = Significant unavoidable impact; LS = Less than significant impact.

Mitigation measures are of little or no value unless they are implemented. Hence there is a clear link between mitigation and the monitoring of outcomes, if and when a project is approved and moves to the construction and operational stages. Indeed, the incorporation of a clear monitoring programme can be one of the most important mitigation measures. Monitoring, which is discussed in Chapter 7, must include the effectiveness or otherwise of mitigation measures. The latter must therefore be devised with monitoring in mind; they must be clear enough to allow for the checking of effectiveness. The use of particular mitigation measures may also draw on previous experience of relative effectiveness, from previous monitoring activity in other relevant and comparable cases.

5.5 Summary

Impact prediction and the evaluation of the significance of impacts often constitute a "black box" in EIA studies. Intuition, often wrapped up as expert opinion, cannot provide a firm and defensible foundation for this important stage of the process. Various methods, ranging from simple to complex, are available to the analyst, and these can help to underpin analysis. Mitigation measures come into play particularly at this stage. However, the increasing sophistication of some methods does run the risk of cutting out key actors, and especially the public, from the EIA process. Chapter 6 discusses the important, but currently weak, role of public participation, the value of good presentation, and approaches to EIS review and decision-making.

References

Barde, J. P. & D. W. Pearce 1991. *Valuing the environment: six case studies.* London: Earthscan.

Bateman, I. 1991. Social discounting, monetary evaluation and practical sustainability. *Town and Country Planning* **60**(6), 174–76.

Beattie, R. 1995. Everything you already know about EIA, but don't often admit. *Environmental Impact Assessment Review* **15**.

Bisset, R. 1988. Developments in EIA methods. In *Environmental impact assessment: theory and practice*, P. Wathern (ed.), 47–60. London: Unwin Hyman.

Bourdillon, N. 1996. *Limits and standards in EIA.* Oxford: Oxford Brookes University, Impacts Assessment Unit, School of Planning.

Bowers, J. 1990. *Economics of the environment: the conservationists' response to the Pearce Report.* Wellington, Telford: British Association of Nature Conservationists.

Bregman, J. I. & K. M. Mackenthun 1992. *Environmental impact statements.* Boca Raton, Florida: Lewis.

CEC 1993. *Environmental manual: sectoral environmental assessment sourcebook.* Brussels: CEC, DG VIII.

CEC 1997. Council Directive 97/11/EC of 3 March 1997 amending Directive 85/337 EEC on the assessment of certain public and private projects on the environment. *Official Journal.* L73/5, 3 March.

CEQ (Council on Environmental Quality) 1978. *National Environmental Policy Act*, Code of Federal Regulations, Title 40, Section 1508.20.

Dasgupta, A. K. & D. W. Pearce 1978. *Cost–benefit analysis: theory and practice.* London: Macmillan.

De Jongh, P. E. 1988. Uncertainty in EIA. In *Environmental impact assessment: theory and practice*, P. Wathern (ed.), 62–83. London: Unwire Hyman.

DETR (Department of the Environment, Transport and the Regions) 1997. *Mitigation measures in environmental statements.* London: DETR.

DETR 1999. *Circular 02/99. Environmental impact assessment.* London: HMSO.

DETR 2000. *Environmental impact assessment: guide to the procedures.* Tonbridge: Thomas Telford.

DoE 1991. *Policy appraisal and the environment.* London: HMSO.

Edwards, W. & J. R. Newman 1982. *Multiattribute evaluation.* Los Angeles: Sage.

Elsom, D. 2001. Air quality and climate. In *Methods of Environmental Impact Assessment*, P. Morris & R. Therivel (eds), 2nd edn (Ch. 8). London: Spon Press.

English Nature 1994. *Nature conservation in environmental assessment.* Peterborough: English Nature.

Fortlage, C. 1990. *Environmental assessment: a practical guide.* Aldershot: Gower.

Friend, J. K. & A. Hickling 1987. *Planning under pressure: the strategic choice approach.* Oxford: Pergamon.

Friend, J. K. & W. N. Jessop 1977. *Local government and strategic choice: an operational research approach to the processes of public planning*, 2nd edn. Oxford: Pergamon.

Glasson, J. 2001. Socio-economic impacts 1: overview and economic impacts. In *Methods of Environmental Impact Assessment*, P. Morris & R. Therivel (eds), 2nd edn (Ch. 2). London: Spon Press.

Glasson, J., M. J. Elson, M. Van der Wee, B. Barrett 1987. *Socio-economic impact assessment of the proposed Hinkley Point C power station*. Oxford: Oxford Polytechnic, Impacts Assessment Unit.

Glasson, J., M. Van der Wee, B. Barrett 1988. A local income and employment multiplier analysis of a proposed nuclear power station development at Hinkley Point in Somerset. *Urban Studies* **25**, 248–61.

Golden, J., R. P. Duellette, S. Saari, P. N. Cheremisinoff 1979. *Environmental impact data book*. Ann Arbor: Ann Arbor Science Publishers.

Green, H., C. Hunter, B. Moore 1989. Assessing the environmental impact of tourism development – the use of the Delphi technique. *International Journal of Environmental Studies* **35**, 51–62.

Green, H., C. Hunter, B. Moore 1990. Assessing the environmental impact of tourism development. *Tourism Management*, June, 111–20.

Hansen, P. E. & S. E. Jorgensen (eds) 1991. *Introduction to environmental management*. New York: Elsevier.

Hill, M. 1968. A goals-achievement matrix for evaluating alternative plans. *Journal of the American Institute of Planners* **34**, 19.

HMSO 1971. *Report of the Roskill Commission on the Third London Airport*. London: HMSO.

Holling, C. S. (ed.) 1978. *Adaptive environmental assessment and management*. New York: Wiley.

Institute of Environmental Assessment and Landscape Institute 1995. *Guidelines for landscape and visual assessment*. Lincoln: IEA.

Journal of American Institute of Planners 1965. Theme issue on urban development models – new tools for planners, May.

Kristensen, P., J. P. Jensen, E. Jeppesen 1990. *Eutrophication models for lakes*. Research Report C9. Copenhagen: National Agency of Environmental Protection.

Lee, N. 1987. *Environmental impact assessment: a training guide*. Occasional Paper 18, Department of Town and Country Planning, University of Manchester.

Lewis, J. A. 1988. Economic impact analysis: a UK literature survey and bibliography. *Progress in Planning* **30**(3), 161–209.

Lichfield, N. 1996. *Community impact evaluation*. London: UCL Press.

Lichfield, N., P. Kettle, M. Whitbread 1975. *Evaluation in the planning process*. Oxford: Pergamon.

Loewenstein, L. K. 1966. On the nature of analytical models. *Urban Studies* **3**.

Maclaren, V. W. & J. B. Whitney (eds) 1985. *New directions in environmental impact assessment in Canada*. London: Methuen.

Morris, P. & R. Therivel (eds) 2001. *Methods of environmental impact assessment*, 2nd edn. London: Spon Press.

Morris, P. & D. Thurling 2001. Phase 2–3 ecological sampling methods. In *Methods of Environmental Impact Assessment*, P. Morris & R. Therivel (eds), 2nd edn (Appendix G). London: Spon Press.

Munn, R. E. 1979. *Environmental impact assessment: principles and procedures*. New York: Wiley.

ODPM (Office of the Deputy Prime Minister) 2003. *Environmental impact assessment: guide to the procedures*. London: HMSO.

Parkin, J. 1992. *Judging plans and projects*. Aldershot: Avebury.

Pearce, D. 1989. Keynote speech at the 10th International Seminar on Environmental Impact Assessment and Management, University of Aberdeen, 9–22 July.

Pearce, D. & A. Markandya 1990. *Environmental policy benefits: monetary valuation*. Paris: OECD.

Pearce, D., A. Markandya, E. B. Barbier 1989. *Blueprint for a green economy*. London: Earthscan.

Rau, J. G. & D. C. Wooten 1980. *Environmental impact analysis handbook*. New York: McGraw-Hill.

Richey, J. S., B. W. Mar, R. Homer 1985. The Delphi technique in environmental assessment. *Journal of Environmental Management* **21**(1), 135–46.

Rodriguez-Bachiller, A. with J. Glasson (2003). *Expert Systems and Geographical Information Systems*. London: Taylor and Francis.

Suter II, G. W. 1993. *Ecological risk assessment*. Chelsea, Michigan: Lewis.

Therivel, R. 2001. Landscape. In *Methods of Environmental Impact Assessment*, P. Morris & R. Therivel (eds), 2nd edn. London: Spon Press.

Tomlinson, P. 1989. Environmental statements: guidance for review and audit. *The Planner* **75**(28), 12–15.

US Environmental Protection Agency 1993. *Sourcebook for the environmental assessment (EA) process*.

Vanclay, F. 1999. Social Impact Assessment. In *Handbook of Environmental Impact Assessment*, J. Petts (ed.). Oxford: Blackwell Science Ltd (vol. 1, Ch. 14).

Von Neuman, J. & O. Morgenstern 1953. *Theory of games and economic behaviour*. Princeton: Princeton University Press.

Voogd, J. H. 1983. *Multicriteria evaluation for urban and regional planning*. London: Pion.

VROM 1984. *Prediction in environmental impact assessment*. The Hague: Netherlands Ministry of Public Housing, Physical Planning and Environmental Affairs.

West Australian Environmental Protection Authority 2002. *Environmental Impacts Assessment (Part IV Division 1) Administrative Procedures*. Perth: EPA.

Westman, W. E. 1985. *Ecology, impact assessment and environmental planning*. New York: Wiley.

Willis, K. G. & N. A. Powe 1998. Contingent valuation and real economic commitments: a private good experiment. *Journal of Environmental Planning and Management* **41**(5), 611–19.

Winpenny, J. T. 1991. *Values for the environment: a guide to economic appraisal* (Overseas Development Institute). London: HMSO.

Wood, G. 2000. Is what you see what you get? Post-development auditing of methods used for predicting the zone of visual influence in EIA. *Environmental Impact Assessment Review* **20/5**, 537–56.

6 Participation, presentation and review

6.1 Introduction

One of the key aims of the EIA process is to provide information about a proposal's likely environmental impacts to the developer, public and decision-makers, so that a better decision may be made. Consultation with the public and statutory consultees in the EIA process can help to ensure the quality, comprehensiveness and effectiveness of the EIA, as well as ensuring that the various groups' views are adequately taken into consideration in the decision-making process. Consultation and participation can be useful at most stages of the EIA process:

- in determining the scope of an EIA;
- in providing specialist knowledge about the site;
- in evaluating the relative significance of the likely impacts;
- in proposing mitigation measures;
- in ensuring that the EIS is objective, truthful and complete;
- in monitoring any conditions of the development agreement.

As such, how the information is presented, how the various interested parties use that information, and how the final decision incorporates the results of the EIA and the views of the various parties, are essential components in the EIA process.

Traditionally, the British system of decision-making has been characterized by administrative discretion and secrecy, with limited public input (McCormick 1991). However, there have been recent moves towards greater public participation in decision-making, and especially towards greater public access to information. In the environmental arena, the EPA of 1990 requires the Environment Agency and local authorities to establish public registers of information on potentially polluting processes; *Quality of Life Counts* (DETR 1999) and its annual updates compile environmental data in a publicly available form; the public participation requirements of EC Directives 85/337 and 2001/42 (on SEA) allow greater public access to information previously not compiled, or considered confidential; and EC Directive 90/313, which requires Member States to make provisions for freedom of access to information on the environment, has been implemented in the UK through the Environmental Information Regulations 1992.

However, despite the positive trends towards greater consultation and participation in the EIA process and the improved communication of EIA findings, both are still underdeveloped in the UK. Few developers make a real effort to gain a sense of the public's views before presenting their applications for authorization and EISs. Few competent authorities have the time or resources to gauge public opinion adequately before making their decisions. Few EISs are truly well presented.

This chapter discusses how consultation and participation of both the public (Section 6.2) and designated environmental consultees (Section 6.3) can be fostered, and how the results can be used to improve a proposed project and speed up its authorization process. The effective presentation of the EIS is then discussed in Section 6.4. The review of EISs and assessment of their accuracy and comprehensiveness are considered in Section 6.5. The chapter concludes with a discussion about decision-making and post-decision legal challenges.

6.2 Public consultation and participation

This section discusses how "best practice" public participation[1] can be encouraged. It begins by considering the advantages and disadvantages of public participation. It then establishes requirements for effective public participation and reviews methods of such participation. Finally, we discuss the UK approach to public participation. The reader is also referred to the Audit Commission (2000), Canter (1996), IEMA (2002) and Weston (1997) for further information on public participation.

6.2.1 Advantages and disadvantages of public participation

Developers do not usually favour public participation. It may upset a good relationship with the LPA. It carries the risk of giving a project a high profile, with attendant costs in time and money. It may not lead to a conclusive decision on a project, as diverse interest groups have different concerns and priorities; the decision may also represent the views of the most vocal interest groups rather than of the general public. Most developers' contact with the public comes only at the stage of planning appeals and inquiries; by this time, participation has often evolved into a systematic attempt to stop their projects. Thus, many developers never see the positive side of public participation because they do not give it a chance.

Historically, public participation has also had connotations of extremism, confrontation, delays and blocked development. In the USA, NEPA-related lawsuits have stopped major development projects, including oil and gas developments in Wyoming, a ski resort in California and clear-cut logging project in Alaska (Turner 1988). In Japan in the late 1960s and early 1970s, riots (so violent that six people died) delayed the construction of the Narita Airport near Tokyo by 5 years. In the UK, perhaps the most visible forms of public participation have been protesters wearing gas masks at nuclear power station sites, threatening to lie down in front of the bulldozers working on the M3 motorway at Twyford Down and being forcibly evicted from tunnels and tree houses on the Newbury bypass route, which cost more than £6 million for policing before construction even began. More typically, all planners are familiar with acrimonious public meetings and "ban the project" campaigns. Public participation may provide the legal means for intentionally obstructing development; the protracted delay of a project can be an effective method of defeating it.

On the other hand, public participation can be used positively to convey information about a development, clear up misunderstandings, allow a better understanding of relevant issues and how they will be dealt with, and identify and deal with areas of controversy while a project is still in its early planning phases. The process of considering and responding to the unique contributions of local people or special interest groups may suggest measures the developer could take to avoid local opposition and environmental problems. These measures are likely to be more innovative, viable and publicly acceptable than those proposed solely by the developer. Project modifications made early in the planning process, before plans have been fully developed, are more easily and cheaply accommodated than those made later. Projects that do not have to go to inquiry are considerably cheaper than those that do. Early public participation also prevents an escalation of frustration and anger, so it helps to avoid the possibility of more forceful "participation". The implementation of a project generally proceeds more cheaply and smoothly if local residents agree with the proposal, with fewer protests, a more willing labour force, and fewer complaints about impacts such as noise and traffic. Research by Morrison-Saunders et al. (2001) also suggests that public pressure is a key incentive to developers to prepare good EISs.

Past experience shows that the total benefits of openness can exceed its costs, despite the expenditure and delays associated with full-scale public participation in the project planning process. The case of British Gas has already been noted (House of Lords 1981). Similarly, the conservation manager of Europe's (then) largest zinc/lead mine noted that:

> properly defined and widely used, [EIA is] an advantage rather than a deterrent. It is a mechanism for ensuring the early and orderly consideration of all relevant issues and for the involvement of affected communities. It is in this last area that its true benefit lies. We have entered an era when the people decide. It is therefore in the interests of developers to ensure that they, the people, are equipped to do so with the confidence that their concern is recognized and their future life-style protected. (Dallas 1984)

Similarly, the developers of a motor-racing circuit noted:

> The [EIS] was the single most significant factor in convincing local members, residents and interested parties that measures designed to reduce existing environmental impacts of motor racing had been uppermost in the formulation of the new proposals. The extensive environmental studies which formed the basis of the statement proved to be a robust defence against the claims from objectors and provided reassurance to independent bodies such as the Countryside Commission and the Department of the Environment. Had this not been the case, the project would undoubtedly have needed to be considered at a public inquiry. (Hancock 1992)

6.2.2 Requirements for effective participation

The United Nations Environment Programme lists five interrelated components of effective public participation:

1. identification of the groups/individuals interested in or affected by the proposed development;
2. provision of accurate, understandable, pertinent and timely information;

3. dialogue between those responsible for the decisions and those affected by them;
4. assimilation of what the public say in the decision; and
5. feedback about actions taken and how the public influenced the decision (Clark 1994).

These points will be discussed in turn.

Although the *identification of relevant interest groups* seems superficially simple, it can be fraught with difficulty. The simple term "the public" actually refers to a complex amalgam of interest groups, which changes over time and from project to project. The public can be broadly classified into two main groups. The first consists of the voluntary groups, quasi-statutory bodies or issues-based pressure groups which are concerned with a specific aspect of the environment or with the environment as a whole. The second group consists of the people living near a proposed development who may be directly affected by it. These two groups can have very different interests and resources. The organized groups may have extensive financial and professional resources at their disposal, may concentrate on specific aspects of the development, and may see their participation as a way to gain political points or national publicity. People living locally may lack the technical, educational or financial resources, and familiarity with relevant procedures to put their points across effectively, yet they are the ones who will be the most directly affected by the development (Mollison 1992). The people in the two groups, in turn, come from a wide range of backgrounds and have a wide variety of opinions. A multiplicity of "publics" thus exits, each of which has specific views, which may well conflict with those of other groups and those of EIA "experts".

It is debatable whether all these publics should be involved in all decisions, for instance, whether "highly articulate members of the NGO, Greenpeace International, sitting in their office in Holland, also have a right to express their views on, and attempt to influence, a decision on a project which may be on the other side of the world" (Clark 1994). Participation may be rightly controlled by regulations specifying the groups and organizations that are eligible to participate or by criteria identifying those considered to be directly affected by a development (e.g. living within a certain distance of it).

Lack of information, or misinformation, about the nature of a proposed development prevents adequate public participation and causes resentment and criticism of the project. One objective of public participation is thus *to provide information* about the development and its likely impacts. Before an EIS is prepared, information may be provided at public meetings, exhibitions or telephone hotlines. This information should be as candid and truthful as possible: people will be on their guard against evasions or biased information, and will look for confirmation of their fears. A careful balance needs to be struck between consultation that is early enough to influence decisions and consultation that is so early that there is no real information on which to base any discussions. For instance, after several experiences of problematic pre-EIS consultation, one UK developer decided to conduct quite elaborate consultation exercises but only after the EISs were published (McNab 1997).

The way information is conveyed can influence public participation. Highly technical information can be understood by only a small proportion of the public. Information in different media (e.g. newspapers, radio) will reach different sectors of the public. Ensuring the participation of groups that generally do not take part in decision-making – notably minority and low-income groups – may be a special concern, especially in the light of the Brundtland Commission's emphasis on intra-generational equity and participation. Ross (2000) gives a compelling example of the difficulties of communicating technical information across a language barrier in Canada:

> At one of the hearings in an aboriginal community, there was a discussion of chlorinated organic emissions involving one of the elders, who was speaking in Cree through a translator. The translator needed to convey the discussions to the Elder. The difficult question was how to translate the phrase "chlorinated organic compounds" into Cree. Fellow panel member Jim Boucher...who spoke Cree, listened to the translator, who had solved the problem by using the translation "bad medicine".

Williams & Hill (1996) identified a number of disparities between traditional ways of communicating environmental information and the needs of minority and low-income groups in the US; for instance:

- agencies focus on desk studies rather than working actively with these groups;
- agencies often do not understand existing power structures, so do not involve community leaders such as preachers for low-income churches, or union leaders;
- agencies hold meetings where the target groups are not represented, for instance in city centres away from where the project will be located;
- agencies hold meetings in large "fancy" places which disenfranchised groups feel are "off-limits", rather than in local churches, schools or community centres;
- agencies use newspaper notices, publication in official journals and mass mailings instead of telephone trees or leaflets handed out in schools;
- agencies prepare thick reports which confuse and overwhelm;
- agencies use formal presentation techniques such as raised platforms and slide projections.

These points suggest that a wide variety of methods for conveying information should be used, with an emphasis on techniques that would be useful for traditionally less participative groups: EIS summaries with pictures and perhaps comics as well as technical reports, meetings in less formal venues, and contact through established community networks as well as through leaflets and newspaper notices.

Public participation in EIA also aims to *establish a dialogue* between the public and the decision-makers (both the project proponent and the authorizing body) and to ensure that decision-makers *assimilate the public's views* into their decisions. Public participation can help to identify issues that concern local residents. These issues are often not the same as those of concern to the developer or outside experts. Public participation exercises should thus achieve a two-way flow of information to allow residents to voice their views. The exercises may well identify conflicts between the needs of the developer and those of various sectors of the community; but this should ideally lead to solutions of these conflicts, and to agreement on future courses of action that reflect the joint objectives of all parties (Petts 1999, 2003).

Public participation is likely to be greatest where public comments are most likely to influence decisions. Arnstein (1971) identified "eight rungs on a ladder of citizen participation", ranging from non-participation (manipulation, therapy), through tokenism (informing, consultation, placation), to citizen power (partnership, delegated

Table 6.1 Advantages and disadvantages of levels of increasing public influence

Approaches	Extent of public power in decision-making	Advantages	Disadvantages
Information feedback Slide or film presentation, information kit, newspaper account, notices, etc.	Nil	Informative, quick	No feedback; presentation subject to bias
Consultation Public hearing, ombudsperson or representative, etc.	Low	Allows two-way information transfer; allows limited discussion	Does not permit ongoing communication; somewhat time-consuming
Joint planning Advisory committee, structured workshop, etc.	Moderate	Permits continuing input and feedback; increases education and involvement of citizens	Very time-consuming; dependent on what information is provided by planners
Delegated authority Citizens' review board, citizens' planning commission, etc.	High	Permits better access to relevant information; permits greater control over options and timing of decision	Long-term time commitment; difficult to include wide representation on small board

(*Source*: Westman 1985.)

power, citizen control). Similarly, Westman (1985) has identified four levels of increasing public power in participation methods: information-feedback approaches, consultation, joint planning and delegated authority. Table 6.1 lists advantages and disadvantages of these levels.

There are many different forms of public participation. A few are listed in Table 6.2, along with an indication of how well they provide information, cater for special interests, encourage dialogue and affect decision-making. Box 6.1 gives an example from Canada, where many of these techniques have been used in practice. The effectiveness of these techniques can vary widely. One UK local authority planner gives his views:

It is normal practice for controversial cases to be referred to a panel of committee members for a site visit. Frequently at these meetings the public attend and are invited to make comments. Often the applicant is encouraged to prepare a small exhibition of the proposals so that interested parties have the opportunity to examine the project in more detail and opinions can be exchanged. This practice gives the developer the chance to experience how local people feel about the proposal. How far this may cause a change in the details of the project

Table 6.2 Methods of public participation and their effectiveness

	Provide information	*Cater for special interests*	*Two-way communication*	*Impact decision-making*
Explanatory meeting, slide/ film presentation	✓	½	½	
Presentation to small groups	✓	✓	✓	½
Public display, exhibit, models	✓	–	–	–
Press release, legal notice	½	–	–	–
Written comment	–	½	½	½
Poll	½	–	✓	✓
Field office	✓	✓	½	–
Site visit	✓	✓	–	–
Advisory committee, task force, community representative	½	½	✓	✓
Working groups of key actors	✓	½	✓	✓
Citizen review board	½	½	✓	✓
Public inquiry	✓	½	½	✓/–
Litigation	½	–	½	✓/–
Demonstration protests, riots	–	–	½	✓/–

(Adapted from Westman 1985.)

Box 6.1 Grande-Baleine hydropower complex, Canada

Hydro-Québec has applied for permission to build a hydropower complex in north-ern Québec province, which could generate 16.2 TWh of energy annually. The complex would include three dammed-up reservoirs with a total area of 3,400 km^2, three generating stations, 136 dykes, a road system and three airfields. Likely envir-onmental impacts include impacts on flora, fauna and water quality (particularly methylmercury levels which would result in restrictions on fish consumption). The project would also affect about 500 Crees, 450 Inuit and 75 people of other origins, for most of whom hunting, fishing and trapping remains central to their identity as Native Peoples of northern Québec. As part of project planning, Hydro-Québec undertook extensive public consultation and description. The following description is verbatim from a leaflet summarising Hydro-Québec's communication activities (Hydro-Québec 1993):

Local populations were regularly consulted and kept informed from the start of phase I of the feasibility study for the Grande-Baleine complex, which ran from 1977 to 1981, when work was temporarily suspended. Hydro-Québec organized regular meetings with the Native communities directly affected by the project; their views were taken into consideration in conducting studies. These communities were subsequently informed of the study results and, once again, consulted about proposed mitigative and environmental enhancement measures. Based on these consultations – to take just one example – Hydro-Québec revised its scenario for the diversion of the Petite Rivière de la Baleine to eliminate environmental impacts on the drainable basin of the Rivière Nastapoka, further north.

Box 6.1 (*continued*)

In 1988, with the start of phase II of the Grande-Baleine feasibility study, Hydro-Québec relaunched its information and communication initiatives. At the local level, the communication program consisted of three phases: the general information phase, designed to provide information about various components of the project; the information-feedback phase, designed to gather reactions and data to guide Hydro-Québec in its decision making; and the information-consultation phase, in which Hydro-Québec presented options to the interested parties, analyzed the opinions expressed, and explained its decisions as they were made. Shortly after the phase II of the feasibility study was under way, the Crees informed Hydro-Québec that they no longer wished to maintain dialogue and that all communications should be addressed to their legal advisors. Starting in January 1989, Hydro-Québec's local information and consultation activities were directed mainly at the Inuit, who had formed a working group in 1988.

In the general information phase, Hydro-Québec held meetings with various organizations and clarified key aspects of the project, including the project rationale, environmental studies, employment opportunities, and the overall development calendar. A bulletin summarizing this information in French, English, Cree, and Inuktitut was sent to the persons, groups, and organizations that requested it, and all interested parties were able to express their concerns.

The information-feedback phase began with a helicopter tour over the affected area given to members of the Inuit working group. It continued with numerous meetings in which the Native peoples voiced their concerns in greater detail about the project and its diverse components. Thematic workshops focused on specific subjects such as employment and training for Native peoples.

In the information-consultation phase, members of the Inuit working groups flew over the sector chosen for the new Petite Riviére de la Baleine diversion option, which had been devised in response to concerns expressed in the earlier phases. The working group also flew over the La Grande complex, where members took a close look at a section of river where the flow of water had been reduced. During workshops, specific problems, such as impacts on the beluga whales and increased mercury levels were examined in greater depth. Hydro-Québec provided detailed data on all aspects of the project and gave updates on studies then in progress.

To keep the rest of the Québec population informed, Hydro-Québec held meetings with a cross section of groups and organizations, took part in public meetings, and distributed information bulletins. Once again, interested parties were given the opportunity to express their concerns about the project. These activities were part of a national communication campaign in the print media and on radio in which the public was invited to request more information by calling a toll-free number.

In addition, following Parliamentary Commission hearings in May 1990, Hydro-Québec worked with the Québec government to establish a framework for public consultation that was designed to integrate the expectations and concerns of Québec society into its proposed development plan. The opinions expressed in 47 meetings with 75 groups were made public in November 1992, when a new development plan proposal was tabled.

Internationally, Hydro-Québec undertook information campaigns in the northeastern United States and Europe after various groups took positions based on

Box 6.1 (*continued*)

erroneous data and the *New York Times* published a one-page advertisement in the fall of 1991 that was quite biased and took an unfavourable stance on the Grande-Baleine project. Hydro-Québec held conferences, organized visits to the La Grande complex, took part in college and university debates, set up a toll-free line in Vermont, and opened information offices in New York and Brussels. In February 1992, Hydro-Québec successfully defended the Grande-Baleine project and its assessment procedure before the International Water Tribunal in Amsterdam . . .

From the outset . . . Hydro-Québec has responded to thousands of questions from journalists and organized numerous news conferences and visits to James Bay. In its efforts to prevent the Grande-Baleine project from becoming a symbol of conflict between environmental protection and economic development Hydro-Québec has endeavoured to clarify the facts, set the record straight, and explain the complex issues involved. If the Grande-Baleine project is approved, Hydro-Québec will show the same commitment to maintaining open channels of communication and dialogue during the construction and operational phases. The utility will also remain in close contact with the communities concerned . . .

is another matter. Another area of publicity is the public meeting and it is probably the least productive . . . The public meeting appears not to be the right forum for the exchange of information or opinion. It might function well as a community safety valve . . . but as a contribution to environmental decision making it is often unhelpful . . .

It is becoming more usual with planning cases for them to be placed before community forums of local people and other interested parties [since] the earlier the community is involved in planning matters the better chance a project has of eventually being implemented . . . The resource implications of servicing forums are considerable and indeed risky, as the debate may go in unexpected directions. Also, importantly, such a process cannot be hurried. (Read 1997)

Finally, an essential part of effective public participation is *feedback about any decisions* and actions taken, and how the public's views affected those decisions. In the US, for instance, comments on a draft EIS are incorporated into the final EIS along with the agency's response to those comments. For example:

> *comment*: I am strongly opposed to the use of herbicides in the forest. I believe in a poison-free forest! *response*: Your opposition to use of herbicides was included in the content analysis of all comments received. However, evidence in the EIS indicates that low risk use of selected herbicides is assured when properly controlled – the evaluated herbicides pose minimal risk as long as mitigation measures are enforced.

Without such feedback, people are likely to question the use to which their input was put, and whether their participation had any effect at all; this could affect their approach to subsequent projects as well as their view of the one under consideration.

6.2.3 UK procedures

Article 6 of EC Directive 85/337 (as amended by Directive 97/11) requires Member States to ensure that:

- any request for development consent and any information gathered pursuant to Article 5 are made available to the public;
- the public concerned is given the opportunity to express an opinion before development consent is granted.

The detailed arrangements for such information and consultation are determined by the Member States which may in particular, depending on the particular characteristics of the projects or sites concerned:

- determine the public concerned;
- specify the places where the information can be consulted;
- specify the ways in which the public may be informed, for example by bill posting within a certain radius, publication in local newspapers, organization of exhibitions with plans, drawings, tables, graphs and models;
- determine the manner in which the public is to be consulted, for example by written submissions, by public enquiry;
- fix time limits for the various stages of the procedure in order to ensure that a decision is taken within a reasonable period.

In the UK, this has been translated by the various EIA regulations (with minor differences) into the following general requirements. Notices must be published in two local newspapers and posted at a proposed site at least seven days before the submission of the development application and EIS. These notices must describe the proposed development, state that a copy of the EIS is available for public inspection with other documents relating to the development application for at least 21 days, give an address where copies of the EIS may be obtained and the charge for the EIS, and state that written representations on the application may be made to the competent authority for at least 28 days after the notice is published. When a charge is made for an EIS, it must be reasonable, taking into account printing and distribution costs.

Environmental impact assessment: guide to the procedures (ODPM 2003a), the government manual to developers, notes:

> Developers should also consider whether to consult the general public, and non-statutory bodies concerned with environmental issues, during the preparation of the environmental statement. Bodies of this kind may have particular knowledge and expertise to offer ... While developers are under no obligation to publicise their proposals before submitting a planning application, consultation with local amenity groups and with the general public can be useful in identifying key environmental issues, and may put the developer in a better position to modify the project in ways which would mitigate adverse effects and recognise local environmental concerns. It will also give the developer an early indication of the issues which are likely to be important at the formal application stage if, for instance, the proposal goes to public inquiry.

The good practice guide on preparing EISs (DoE 1995) adds:

> It is at the scoping stage that the developer should consider the most appropriate point at which to involve members of the public. Developers may be reluctant to make a public announcement about their proposals at an early stage, perhaps because of commercial concerns... There may also be occasions when public disclosure of development proposals in advance of a formal planning application may cause unnecessary blight. However, early announcement of plans for prospecting and site or route selection, and the provision of opportunities for environmental/amenity groups and local people to comment on environmental issues, may channel legitimate concerns into constructive criticism.

From this it is clear that in the UK the requirements for public participation have been implemented half-heartedly at best, and developers and the competent authorities have in turn generally limited themselves to the minimal legal requirements. An environmental consultant suggests:

> On the one hand assessment may be seen as a process in which all should participate; which involves the whole community in the design process and in which the statement merely becomes the statutory document required at the time the planning application is submitted. On the other hand assessment may be seen as a process in which the statement forms a critical milestone, the point at which the developer unveils his plans and gives his account of their likely environmental impacts. Discussion and debate ensue.
>
> Both models... are valid. The first may be seen as an ideal where a public spirited developer has the time, resources and ability to initiate a wide ranging programme of participation. It requires all participants to take a lively and rational interest in the proposal and preferably not take up an entrenched position at the outset. It would seem most suited to public sector projects, projects initiated by the not-for-profit sector and proposals which are unlikely to provoke much opposition in principle... relatively few projects requiring assessment will fulfil the last criterion. The second model is more suited to the private sector developer wrestling with the problems of commercial confidentiality and time constraints. Its acceptability appears to be endorsed by the latest [government guidelines on EIA] which, whilst emphasizing the value of scoping and the need for early consultation with the LPA and statutory consultees, acknowledges the possible need for confidentiality. The guidance on public participation is similarly cautious, balancing the desirability and potential benefits of early disclosure with commercial concerns. (McNab 1997)

6.3 Consultation with statutory consultees

Some of the most useful inputs to project decision-making are comments by statutory and other relevant consultees (Wende 2002, Wood & Jones 1997). The statutory consultees have accumulated a wide range of knowledge about environmental conditions in various parts of the country. They can give valuable feedback on the appropriateness of a project and its likely impacts. However, the consultees may have their own priorities, which may prejudice their response to the EIS.

Article 6(1) of Directive 85/337 (as amended) states:

> Member States shall take the measures necessary to ensure that the authorities likely to be concerned by the project by reasons of their specific environmental responsibilities are given an opportunity to express their opinion on the information supplied by the developer and other requests for development consent. To this end, Member States shall designate the authorities to be consulted...The information gathered...shall be forwarded to those authorities. Detailed arrangements for consultation shall be laid down by Member States.

In the UK, different statutory consultees have been designated for different types of development. For planning projects, for instance, the statutory consultees are any principal council to the area in which the land is situated (if not the LPA), the Countryside Agency or the Countryside Council for Wales, the Environment Agency, the Secretary of State for Wales (where the land is in Wales) and any body the LPA would be required to consult under Article 10 of the T&CP (General Development Procedures) Order 1995, for instance, the local highways authority if the project is likely to affect the road network.

In terms of best (but not mandatory) practice, the consultees should already have been consulted at the scoping stage. In addition, it is a legal requirement that the consultees should be consulted before a decision is made. Once the EIS is completed, copies can be sent to the consultees directly by the developer or by the competent authority. In practice, many competent authorities only send particular EIS chapters to the consultees, e.g. the chapter on archaeology to the archaeologist. However, this often limits the consultee's understanding of the project context and wider impacts; generally consultees should be sent a copy of the entire EIS.

6.4 EIA presentation

Although the EIA regulations specify the minimum contents required in an EIS, they do not give any standard for the presentation of this information. Past EISs have ranged from a three-page typed and stapled report to glossy brochures with computer graphics and multi-volume documents in purpose-designed binders. This section discusses the contents, organization, clarity of communication and presentation of an EIS.

6.4.1 *Contents and organization*

An EIS should be *comprehensive*. Its contents must at least fulfil the requirements of the relevant EIA legislation. As we shall discuss in Chapter 8, past EISs have not all fulfilled these requirements; however, the situation is improving rapidly, and LPAs are increasingly likely to require information on topics they feel have not been adequately discussed in an EIS. A good EIS will also go further than the minimum requirements if other significant impacts are identified. Most EISs are broadly organized into four sections: a non-technical summary, a discussion of relevant methods and issues, a description of the project and of the environmental baseline conditions, and a discussion of likely environmental impacts (which may include a discussion of baseline environmental conditions and predicted impacts, proposed mitigation measures and residual impacts). Ideally, an EIS should also include the main alternatives considered, and proposals for monitoring. It could include much or all of the information given in Appendix 5 of *Environmental impact assessment: guide to the procedures* (ODPM 2003a): see Table 3.5. Table 1.1 provides an example of a good EIS outline.

An EIS should *explain why some impacts are not dealt with*. It should include a "finding of no significant impact" section to explain why some impacts may be considered insignificant. If, for instance, the development is unlikely to affect the climate, a reason should be given explaining this conclusion. An EIS should *emphasize key points*. These should have been identified during the scoping exercise, but additional issues may arise during the course of the EIA. The EIS should set the context of the issues. The names of the developer, relevant consultants, relevant LPAs and consultees should be listed, along with a contact person for further information. The main relevant planning issues and legislation should be explained. The EIS should also indicate any references used, and give a bibliography at the end. Ideally, the cost of the EIS should be given.

The preparation of a *non-technical summary* is particularly important in an EIS, as this is often the only part of the document that the public and decision-makers will read. It should thus briefly cover all relevant impacts and emphasize the most important, and should ideally contain a list or a table that allows readers to identify them at a glance. Chapter 4 gave examples of a number of techniques for identifying and summarizing impacts.

An EIS should ideally be one *unified document*, with perhaps a second volume for appendices. A common problem with the organization of EISs stems from how environmental impacts are assessed. The developer (or the consultants co-ordinating the EIA) often subcontracts parts of the EIA to consultancies which specialize in those fields (e.g. ecological specialists, landscape consultants). These in turn prepare reports of varying lengths and styles, making a number of (possibly different) assumptions about the project and likely future environmental conditions, and proposing different and possibly conflicting mitigation measures. One way developers have attempted to circumvent this problem has been to summarize the impact predictions in a main text, and add the full reports as appendices to the main body of the EIS. Another has been to put a "company cover" on each report and present the EIS as a multi-volume document, each volume discussing a single type of impact. Both of these methods are problematic: the appendix method in essence discounts the great majority of findings, and the multi-volume method is cumbersome to read and carry. Neither method attempts to present findings in a cohesive manner, emphasizes crucial impacts or proposes a coherent package of mitigation and monitoring measures. A good EIS would incorporate the information from the subcontractors' reports into one coherent document which uses consistent assumptions and proposes consistent mitigation measures.

The EIS should be kept as *brief* as possible while still *presenting the necessary information*. The main text should include all the relevant discussion about impacts, and appendices should present only additional data and documentation. In the US, the length of an EIS is generally expected to be less than 150 pages. In the UK, the DoE (1995) recommends:

> For projects which involve a single site and relatively few areas of significant impact, it should be possible to produce a robust ES of around 50 pages. Where

more complex issues arise, the main body of the statement may extend to 100 pages or so. If it exceeds 150 pages it is likely to become cumbersome and difficult to assimilate and this should generally be regarded as a maximum . . . However, the quality of an ES will not be determined by its length. What is needed is a concise, objective analysis.

6.4.2 *Clarity of communication*

Weiss (1989) very well notes that an unreadable EIS is an environmental hazard:

> The issue is the quality of the document, its usefulness in support of the goals of environmental legislation, and, by implication, the quality of the environmental stewardship entrusted to the scientific community . . . An unreadable EIS not only hurts the environmental protection laws and, thus, the environment. It also turns the sincere environmental engineer into a kind of "polluter".

Weiss identifies three classes of error that mar the quality of EIS's communications:

1. strategic errors, "mistakes of planning, failure to understand why the EIS is written and for whom";
2. structural errors related to the EIS's organization;
3. tactical errors of poor editing.

An EIS has to communicate information to many audiences, from the decision-maker, to the environmental expert, to the lay person. Although it cannot fulfil all the expectations of all its readers, it can go a long way towards being a useful document for a wide audience. It should at least be *well written*, with good spelling and punctuation. It should have a clear structure, with easily visible titles and a logical flow of information. A table of contents, with page numbers marked, should be included before the main text, allowing easy access to information. Principal points should be clearly indicated, perhaps in a table at the front or back.

An EIS should *shun technical jargon*. Any jargon it does include should be explained in the text or in footnotes. All the following examples are from actual EISs:

> *Wrong*: It is believed that the aquiclude properties of the Brithdir seams have been reduced and there is a degree of groundwater communication between the Brithdir and the underlying Rhondda beds, although . . . numerous seepages do occur on the valley flanks with the retention regime dependent upon the nature of the superficial deposits.

> *Right*: The accepted method for evaluating the importance of a site for waterfowl (i.e. waders and wildfowl) is the "1% criterion". A site is considered to be of National Importance if it regularly holds at least 1% of the estimated British population of a species of waterfowl. "Regularly" in this context means counts (usually expressed as annual peak figures), averaged over the last 5 years.

The EIS should clearly *state any assumptions* on which impact predictions are based:

Wrong: As the proposed development will extend below any potential [archaeo-logical] remains, it should be possible to establish a method of working which could allow adequate archaeological examinations to take place.

Right: For each operation an assumption has been made of the type and number of plant involved. These are:
    Demolition: 2 pneumatic breakers, tracked loader
    Excavation: backacter excavator, tracked shovel...

The EIS should be *specific*. Although it is easier and more defensible to claim that an impact is significant or likely, the resulting EIS will be little more than a vague collection of possible future trends.

Wrong: The landscape will be protected by the flexibility of the proposed [monorail] to be positioned and designed to merge in both location and scale into and with the existing environment.

Right: From these [specified] sections of road, large numbers of proposed wind turbines would be visible on the skyline, where the towers would appear as either small or indistinct objects and the movement of rotors would attract the attention of road users. The change in the scenery caused by the proposals would constitute a major visual impact, mainly due to the density of visible wind turbine rotors.

Predicted impacts should be *quantified* if possible, perhaps with a range, and the use of non-quantified descriptions, such as severe or minimal, should be explained:

Wrong: The effect on residential properties will be minimal with the nearest properties... at least 200 m from the closest area of filling.

Right: Without the bypass, traffic in the town centre can be expected to increase by about 50–75% by the year 2008. With the bypass, however, the overall reduction to 65–75% of the 1986 level can be achieved.

Even better, predictions should give an *indication of the probability* that an impact will occur, and the degree of confidence with which the prediction can be made. In cases of uncertainty, the EIS should propose worst-case scenarios:

Right. In terms of traffic generation, the "worst case" scenario would be for 100% usage of the car park... For a more realistic analysis, a redistribution of 50% has been assumed.

Finally, an EIS should be *honest and unbiased*. A review of local authorities noted that "[a] number of respondents felt that the Environmental Statement concentrated too much on supporting the proposal rather than focusing on its impacts and was therefore not sufficiently objective" (Kenyan 1991). Ginger & Mohai (1993) suggest that lack of objectivity is a problem in the US as well, and that EISs are used to justify, not assess, decisions. Developers cannot be expected to conclude that their projects have such major environmental impacts that they should be stopped. However, it is unlikely that all major environmental issues will have been resolved by the time the statement is written.

> *Wrong*: The proposed site lies adjacent to lagoons, mud and sands which form four regional Special Sites of Scientific Interest [*sic*]. The loss of habitat for birds, is unlikely to be significant, owing to the availability of similar habitats in the vicinity.

Table 6.3 provides a simple example of a clear presentation of the environmental effects of a road development on adjacent areas. Table 6.4 provides an example of a summary table of environmental impacts.

6.4.3 Presentation

Although it would be good to report that EISs are read only for their contents and clarity, in reality, presentation can have a great influence on how they are received. EIAs are, indirectly, public relations exercises, and an EIS can be seen as a publicity document for the developer. Good presentation can convey a concern for the environment, a rigorous approach to the impact analysis and a positive attitude to the public. Bad presentation, in turn, suggests a lack of care, and perhaps a lack of financial backing. Similarly, good presentation can help to convey information clearly, whereas bad presentation can negatively affect even a well-organized EIS.

The presentation of an EIS will say much about the developer. The type of paper used – recycled or not, glossy or not, heavy- or lightweight – will affect the image projected, as will the choice of coloured or black-and-white diagrams and the use of dividers between chapters. The ultra-green company will opt for double-sided printing on recycled paper, while the luxury developer will use glossy, heavyweight paper with a distinctive binder. Generally, a strong binder that stands up well under heavy handling is most suitable for EISs. Unless the document is very thin, a spiral binder is likely to snap or bend open with continued handling; similarly, stapled documents are likely to tear. Multi-volume documents are difficult to keep together unless a box is provided.

The use of maps, graphs, photo-montages, diagrams and other forms of visual communication can greatly help the EIS presentation. As we noted in Chapter 4, a location map, a site layout of the project and a process diagram are virtually essential to a proper description of the development. Maps showing, for example, the extent of visual impacts, the location of designated areas or classes of agricultural land are a succinct and clear way of presenting such information. Graphs are often much more effective than tables or figures in conveying numerical information. Forms of visual communication break up the page, and add interest to an EIS. Increasingly some developers are also producing the EIS as a CD.

6.5 Review of EISs

The comprehensiveness and accuracy of EISs are matters of concern. As will be shown in Chapter 8, many EISs do not meet even the minimum regulatory requirements, much less provide comprehensive information on which to base

Table 6.3 Presentation of environmental effects

Feature	Effects	Affected	Timescale	Magnitude	Controversial	Probability	Mitigation	Significance
Pigeon House Road	Reduced risk of HGV traffic	Residents	Short term permanent	Local	No	High	None	Minor beneficial
	Perceived severance due to elevated structure	Residents	Short term reducing with time	Local	Potentially	Low	None	Minor adverse
Breman Grove	Reductions in traffic flow by about 80 per cent	Residents and children using the park	Short term permanent	Local	No	High	None	Minor beneficial
Beach Road	Reductions in traffic flow by about 80 per cent	School children	Short term permanent	Local	No	High	None	Minor beneficial

(*Source*: P. Tomlinson, Ove Arup.)

Table 6.4 Example of ES summary table showing relative weights given to significance of impacts (*Note*: only a selection of key issues given.)

Topic area	Description of impact	Geographical level of importance of issue *I N R D L*	Impact	Nature	Significance
Human beings	Disturbance to existing properties from traffic and noise	*N*	Adverse	St, R	Major
	Coalescence of existing settlements	*N*	Adverse	Lt, IR	Major
Flora and fanna	Loss of grassland of local nature conservation value	*D*	Adverse	Lt, IR	Minor
	Creation of new habitats	*D*	Beneficial	Lt, R	Minor
	Increased recreation pressure on SSSI	*R*	Adverse	Lt, R	Minor
Soil and geology	Loss of 300 acres agricultural soils (grade 3B)	*N*	Adverse	Lt, IR	Minor
Water	Increased rates of surface water run-off	*D*	Adverse	Lt, IR	Minor
	Reduction in groundwater recharge	*D*	Adverse	Lt, R	Minor

	Key:	I	International	St	Short-term
		N	National	Lt	Long-term
		R	Regional	R	Reversible
		D	District	IR	Irreversible
		L	Local		

(*Source*: DoE 1995.)

decisions. In some countries, for example the Netherlands, Canada, Malaysia and Indonesia, EIA Commissions have been established to review EISs and act as a quality assurance process. However, in the UK there are no mandatory requirements regarding the pre-decision review of EISs to ensure that they are comprehensive and accurate. A planning application cannot be judged invalid just because it is accompanied by an inadequate or incomplete EIS: a competent authority may only request further information, or refuse permission and risk an appeal.[2]

Many competent authorities do not have the full range of technical expertise needed to assess the adequacy and comprehensiveness of an EIS. Some authorities, especially those which receive few EISs, have consequently had difficulties in dealing with the technical complexities of EISs. In about 10–20 per cent of cases, consultants have been brought in to review the EISs. Other authorities have joined the Institute of Environmental Management and Assessment (IEMA), which reviews EISs. Others have been reluctant to buy outside expertise, especially at a time of restrictions on local spending (Fuller 1992, McDonic 1992). A technique advocated by the International Association for Impact Assessment (IAIA) (Partidario 1996), although not seen often in practice, is to involve parties other than just the competent authority in EIS review, especially the public.

In an attempt to fill the void previously left by the national government, several non-mandatory review criteria have been established which aim to:

- ensure that all relevant information has been analysed and presented;
- assess the validity and accuracy of information contained in the EIS;
- quickly become familiar with the proposed project and consider whether additional information is needed;
- assess the significance of the project's environmental effects;
- evaluate the need for mitigation and monitoring of environmental impacts; and
- advise on whether a project should be allowed to proceed (Tomlinson 1989).

Lee & Colley (1990) developed a hierarchical review framework. At the top of the hierarchy is a comprehensive mark (A = well-performed and complete, through to F = very unsatisfactory) for the entire report. This mark is based on marks given to four broad sub-headings: description of the development, local environment and baseline conditions; identification and evaluation of key impacts; alternatives and mitigation of impacts; and communication of results. Each of these, in turn, is based on two further layers of increasingly specific topics or questions. Lee and Colley's criteria have been used either directly or in a modified form (e.g. by the IEMA) to review a range of EISs in the UK. It is the most commonly used review method in the UK. Appendix 3 gives the Lee & Colley framework.

The European Commission has also published review criteria (CEC 2001a). These are similar to Lee & Colley's, but use seven sub-headings instead of four, include a longer list of specific questions, and judge the information based on relevance to the project context and importance for decision-making as well as presence/absence in the EIS.

The review criteria given in Appendix 4 are an amalgamation and extension of Lee and Colley's and the EC's criteria, developed by the Impacts Assessment Unit at Oxford Brookes University. See also Rodriguez-Bachiller with Glasson (2003) for an expert system approach to EIS review. It is unlikely that any EIS will fulfil all the criteria. Similarly, some criteria may not apply to all projects. However, they should act as a checklist of good practice for both those preparing and those reviewing EISs. Table 6.5 shows a number of possible ways of using these criteria. Example (a), which relates to minimum requirements, amplifies the presence or otherwise of key information. Example (b) includes a simple grading, which could be on the A–F scale used by Lee and Colley, for each criterion (only one of which is shown here). Example (c) takes the format of the EC criteria, which appraise the relevance of the information and then judge whether it is complete, adequate (not complete but need not prevent decision-making from proceeding) or inadequate for decision-making.

Table 6.5 Examples of possible uses for EIS review criteria

(a)

Criterion	Presence/absence (page number)	Information	Key information absent
Describes the proposed development, including its design, and size or scale	✓ (p. 5)	Location (in plans), existing operations, access	Working method, vehicle movements, restoration plans
Indicates the physical presence of the development	X		Site buildings (location, size), restoration

(b)

Criterion	Presence/absence (page number)	Comments	Grade
Explains the purposes and objectives of the development	✓ (p. 11)	Briefly in introduction, more details in Sec. 2	A
Gives the estimated duration of construction etc. phases	✓ (p. 12)	Not decommissioning	B

(c)

Criterion	Relevant? (Y/N)	Judgement (C/A/I)*	Comment
Considers the "no action" alternative, alternative processes, etc.	Y	A	Alternative sites discussed, but not alternative processes
If unexpectedly severe adverse impacts are identified, alternatives are reappraised	N		Impacts of sand/ gravel working well understood

*C = complete; A = adequate; I = inadequate.

6.6 Decisions on projects

6.6.1 EIA and project authorization

Decisions to authorize or reject projects are made at several levels:

> At the top of the tree are the relevant Secretaries of State...; below them are a host of Inspectors, sometimes called Reporters (Scotland); further down the list come Councillors, the elected members of district, county, unitary or metropolitan borough councils; and at the very bottom are chief or senior planning officers who deal with "delegated decisions".... [as] a rough guide, the larger the project the higher up the pyramid of decision makers the decision is made. (Weston 1997)

Where required by the competent authority, an EIS must be submitted with the application for authorization.[3] The decision on an application with an EIS must be made within a specified period (e.g. 16 weeks for a planning application), unless the developer agrees to a longer period. It is at this stage that the EIS review is undertaken. When making a decision, the competent authority is required to have regard to all the environmental information, i.e. the information contained in the EIS and other documents, any comment made by the statutory consultees and representations from members of the public, as well as to other material considerations. By any standards, making decisions on development projects is a complex undertaking. Decisions for projects requiring EIAs tend to be even more complex, because by definition they deal with larger, more complex projects, and probably a greater range of interest groups:

> The competition of interests is not simply between the developer and the consultees. It can also be a conflict between consultees, with the developer stuck in the middle hardly able to satisfy all parties and the "competent authority" left to establish a planning balance where no such balance can be struck. (Weston 1997)

Whereas in the early years the decision-making process for projects with EIA was accepted as being basically a black box, more recently attempts have been made to make the process more rigorous and transparent. Research by the University of Manchester (Wood & Jones 1997) and Oxford Brookes University (Weston et al. 1997) has focused on how environmental information is used in UK decision-making; this is discussed further in Chapter 8. Similar work carried out by Land Use Consultants resulted in a good practice guide (DoE 1994) on the evaluation of environmental information for planning projects; the advice, however, could relate equally well to other types of project. The good practice guide begins with a definition of evaluation:

... in the context of environmental assessment, there are a number of different stages or levels of evaluation. These are concerned with:

- checking the adequacy of the information supplied as part of the ES, or contributed from other sources;
- examining the magnitude, importance and significance of individual environmental impacts and their effects on specific areas of concern . . .;
- preparing an overall "weighing" of environmental and other material considerations in order to arrive at a basis for the planning decision.

The guide suggests that, after vetting the application and EIS, advertising the proposals and EIS, and relevant consultation, the LPA should carry out two stages of decision-making: an evaluation of the individual environmental impacts and their effects, and weighing the information to reach a decision. The evaluation of impacts and effects first involves verifying any factual statements in the EIS, perhaps by highlighting any statements of concern and discussing these with the developer. The nature and character of particular impacts can then be examined; either the EIS will already have provided such an analysis (e.g. in the form of Table 6.3 or 6.4) or the case-work officer could prepare such a table. Finally, the significance and importance of the impacts can be weighed up, taking into consideration such issues as the extent of the area affected, the scale and probability of the effects, the scope for mitigation and the importance of the issue.

Weighing up the information to reach a decision involves not only considering the views of different interest groups and the importance of the environmental issues, but also determining whether the proposed project is in accordance with the development plan. Environmental impacts can be divided into three groups: those which by themselves provide grounds for refusal or approval, those which in conjunction with others influence the decision, and those which are unlikely to influence the outcome of the decision. Then,

> decision-makers will usually be faced with a choice. The planning merits will depend upon a comparison of the advantages and disadvantages arising from the construction and operation of the development, with the consequences of maintaining the status quo – or "do-nothing" option. (DoE 1994)

In the case of a planning application, the planning officer's recommendations will then go to the planning committee, which makes the final decision.

The range of decision options are as for any application for project authorization: the competent authority can grant permission for the project (with or without conditions) or refuse permission. It can also suggest further mitigation measures following consultations, and will seek to negotiate these with the developer. If the development is refused, the developer can appeal against the decision. If the development is permitted, people or organizations can challenge the permission. The relevant Secretary of State can also "call in" an application, for a variety of reasons. A public inquiry may result.

But decision-making is not a clinical exercise. In a Canadian context, Ross (2000) explains how he and fellow panel member Mike Fanchuk wrote the report that explained their decision about whether to permit a pulp mill:

> Mike Fanchuk [is] a farmer from just north of the pulp mill site...During my work with Mike, we discussed when we would be satisfied with the report, and thus when we would be willing to sign it...I believe Mike's approach is the best I have ever encountered. He would only be willing to sign the report when he felt that, in future years, he would be pleased to tell his eight-year old granddaughter that he had served on the panel and authored the report. In academic terms, this intergenerational equity illustrates very well the principles of sustainable development...More importantly, however, it illustrates the basic human need to be proud of work one has done.

6.6.2 EIA and public inquiries

Weston (1997) compellingly discusses why all parties involved in EIA try to avoid public inquiries:

> By the time a project becomes the subject of a public inquiry the sides are drawn and the hearing becomes a focus for adversarial debate between opposing, expensive, experts directed and spurred on by advocates schooled in the art of cajoling witnesses into submission and contradictions. Such debates are seldom rational or in any other way related to the systematic, iterative and cooperative characteristics of good practice EIA. By the time the inquiry comes around, and all the investment has been made in expert witnesses and smooth talking barristers, it is far too late for all that. (Weston 1997)

Nevertheless, hundreds of projects involving EIA have gone to inquiry.

The environmental impact of proposals, especially traffic, landscape and amenity issues, will certainly be examined in detail during any inquiry. The EIA regulations allow inquiry inspectors and the Secretary of State to require (a) the submission of an EIS before a public inquiry, if they regard this as appropriate, and (b) further information from the developer if they consider the EIS is inadequate as it stands. In practice, before public inquiries involving EIAs the inspector generally receives a case file (including the EIS) which is examined to determine whether any further information is required. Pre-inquiry meetings may be held where the inspector may seek further information; these meetings may also assist the developer and competent authority to arrive at a list of agreed matters before the start of the inquiry; this can avoid unnecessary delays during it. At the inquiry, inspectors often ask for further information, and they may adjourn the inquiry if the information cannot be produced within the available time. The information contained in the EIS will be among the material considerations taken into account. However, an inadequate EIS is not a valid reason for preventing authorization, or even for delaying an inquiry.[4]

An analysis of 10 public inquiries involving projects for which EISs had been prepared (Jones & Wood 1995) suggested that in their recommendations most inspectors give "moderate" or "considerable" weight to the EIS and consultations on the EIS, and that environmental information is of "reasonable" importance to the decision whether to grant consent. However, a subsequent study of 54 decision letters from inspectors (Weston 1997) suggested that EIA has had little influence on the inquiry process: in about two-thirds of the cases, national or local land-use policies were the determining issues identified by the inspectors and the Secretary of State, and in the remaining cases other traditional planning matters predominated:

> The headings which dominate the decision letters of the Inspectors and Secretaries of State are the traditional planning material considerations such as amenity, various forms of risk, traffic and need, although some factors such as flora and fauna, noise and landscape do tend to be discussed separately. (Weston 1997)

6.6.3 Challenging a decision: judicial review

The UK planning system has no official provisions for an appeal against development consent. However, if permission is granted, a third party may challenge that decision on the grounds, for example, that no EIA was prepared when it should have been, or that the competent authority did not adequately consider the relevant environmental information. The only way to do this is through judicial review proceedings in the courts, or through the European Union (EU).

Judicial review proceedings in the UK courts first require that the third party shows it has "standing" to bring in the application, namely sufficient interest in the project by virtue of attributes specific to it or circumstances which differentiate it from all other parties (e.g. a financial or health interest). Establishing standing is one of the main difficulties in applying for judicial review.[5] If standing is established, the third party must then convince the court that the competent authority did not act according to the relevant EIA procedures. The court does not make its own decision about the merits of the case, but only reviews the way in which the competent authority arrived at its decision:

> The court will only quash a decision of the [competent authority] where it acted without jurisdiction or exceeded its jurisdiction or failed to comply with the rules of natural justice in a case where those rules apply or where there is an error of law on the face of the record or the decision is so unreasonable that no [competent authority] could have made it. (Atkinson & Ainsworth 1992)

Various possible scenarios emerge. A competent authority may fail to require an EIA for a Schedule 1 project, or may grant permission for such a project without considering the environmental information. In such a case, its decision would be void.

A competent authority may decide that a project does not require EIA because it is not in Schedule 1 or Schedule 2 with significant environmental effects. This was the issue in the case of *R v. Swale Borough Council and Medways Port Authority ex parte RSPB* (1991) concerning the construction of a storage area for cargo, which would require the infill of Lappel Bank, a mudflat important for its wading birds. The appellants argued that the project fell within either Schedule 1 or Schedule 2 with significant environmental effects; the LPA felt that an EIA was unnecessary. The courts held that a project falling within a schedule is a decision for the local authority to make, and open for review only if no reasonable local authority could have made it. Several court cases have also revolved around the level of detail needed in EISs of outline planning applications: the subtle rules emerging from these decisions are discussed by ODPM (2003b).

A competent authority's decision may have been made by an officer who does not have the authority to do so. A decision not to require an EIA was overturned in *R v. St. Edmundsbury Council, ex parte Walton* (1999) because the decision had been taken by an officer who did not have formal delegation.

A competent authority may make a decision in the absence of a formal EIA, but with environmental information available in other forms. This was the case in *R v. Poole Borough Council, ex parte Beebee and others* (1991) concerning a decision to develop part of Canford Heath. In this case the courts ruled that, despite the lack of an EIS and the attendant rigour and publicity, enough environmental information was available for the council to make an informed decision. In a similar Scottish appeal case against an LPA decision to refuse planning permission for an open-cast coal mine, the Reporter felt that an EIA would not have raised issues that would not have been raised by other means (Weston 1997). A different judgement may have been made if the competent authority had been shown to have made its decision before it had received all the relevant environmental information.[6]

A competent authority may not obtain all of the information needed to reach an appropriate decision. For instance, in *R v. Cornwall County Council ex parte Jill Harding* (2001) the planning authority gave permission for a development despite the fact that the EIS did not include information on bats, which site conditions favoured. Instead the authority imposed a condition that required the developer to carry out a bat survey prior to development. The inspector quashed this decision, ruling that the information on bats should have been part of the EIS.

In summary, judicial reviews of competent authority decisions have to date been limited by the issue of standing, and by the courts' relatively narrow inter-pretation of the duties of competent authorities under the EIA regulations. More recent court cases have strengthened and widened this interpretation, but it is very unlikely that the UK courts will play as active a role as those in the US did in relation to the NEPA.

6.6.4 *Challenging a decision: the European Commission*

Another avenue by which third parties can challenge a competent authority's decision to permit development, or not to require EIA, is the European Commission. Such cases need to show that the UK failed to fulfil its obligations as a Member State under the Treaty of Rome by not properly implementing EC legislation, in this case Directive 85/337. In such a case, Article 169 of the Treaty allows a declaration of non-compliance to be sought from the European Court of Justice. The issue of standing is not a problem here, since the European Commission can begin proceedings either on its own initiative or based on the written complaint of any person. To use this mechanism, the Commission must first state its case to the Member State and seek its observations. The Commission may then issue a "reasoned opinion". If the Member State fails to comply within the specified time, the case proceeds to the European Court of Justice.

Under Article 171 of the Treaty of Rome, if the European Court of Justice finds that a Member State has failed to fulfil an obligation under the Treaty, it may require the Member State to take the necessary measures to comply with the Court's judgement. Under Article 186, the EC may take interim measures to require a Member State to desist from certain actions until a decision is taken on the main action. However, to do so the Commission must show the need for urgent relief, and that irreparable damage to Community interests would result if these measures were not taken. Readers are referred to Atkinson & Ainsworth (1992), Buxton (1992) and Salter (1992a,b,c) for further information on procedures.

The latest Five-Year Review of the implementation of the Directive (CEC 2003) shows an increasing number of new complaints about impact assessment opened, rising from 81 in 1997 to 237 in 2001, with the highest numbers for Spain, Ireland, Germany and Italy. However, for almost, 1000 cases over the period 1997–2001, the Commission sent only 36 reasoned opinions to Member States for non-conformity or wrong application of the EIA Directive. This can be explained largely by (i) unfounded complaints and (ii) compliance by a Member State to a breach, resolving the issue before the Commission is obliged to send a reasoned opinion.

6.7 Summary

Active public participation, thorough consultation with relevant consultees, and good presentation are important aspects of a successful EIA process. All have been undervalued to date. The presentation of environmental information has improved, and statutory consultees are becoming increasingly familiar with the EIA process, but public participation is likely to remain a weak aspect of EIA in the UK until developers and competent authorities see the benefits exceeding the costs. However, as noted in Chapter 2 (Section 2.7), the transposition of the Aarhus Convention to provide for enhanced public participation in environmental decision-making may provide an opportunity for improvements in public participation in EIA (CEC 2001b).

A formal review of EIA is also rarely carried out, despite the availability of several non-mandatory review guidelines and government advice on the use of environmental information for decision-making. Such review procedures can contribute to the processing of the EIS as part of the decision-making stage. The links between the quality of an EIS and that of the planning decision is discussed in Chapter 8.

Several appeals against development consents or against competent authorities' failure to require EIA have been brought to the UK courts or the EC. The UK courts have been unwilling to overturn the decisions of competent authorities, and have generally given a relatively narrow interpretation of the duties of competent authorities under the EIA regulations. The EC, by contrast, has challenged the UK Government on its implementation of Directive 85/337 and on a number of specific decisions resulting from this implementation.

More positively, the next step in a good EIA procedure is the monitoring of the development's actual impacts and the comparison of actual and predicted impacts. This is discussed in the next chapter.

Notes

1. Although this section refers to public consultation and participation together as "public participation", the two are in fact separate. Consultation is in essence an exercise concerning a passive audience: views are solicited, but respondents have little active influence over any resulting decisions. In contrast, public participation involves an active role for the public, with some influence over any modifications to the project and over the ultimate decision.
2. Weston (1997) notes that LPAs need to be aware that they have the power to ask for further information, and that failure to use it could later be seen as tacit acceptance of the information provided. For instance, when deciding on an appeal for a Scottish quarry extension, the Reporter noted that it was significant that the LPA had not requested further information when they were processing the application, and had not objected to the EIS until the development came to appeal.
3. Where the project has already been built without authorization, the competent authority considers the environmental information when determining whether the project will be demolished or not.
4. For instance, in the case of a Scottish appeal regarding a proposed quarry extension (Scottish Office, P/PPA/SQ/336, 6 January 1992), the Reporter noted that: "The ES has been strongly criticised ... [it] does not demonstrate that a proper analysis of environmental impacts has been made ... Despite its shortcomings, the ES appears to me to comply broadly with the statutory requirements of the EA regulations."
5. An EC court case, for instance, ruled that Greenpeace had insufficient individual concerns to contest a decision to use regional funds to help build power stations in the Canary Islands (*Greenpeace v Commission of the European Communities* (1996) 8 Journal of Environmental Law 139). Similar judgements have been made in the UK context.
6. The UK is not alone in this. A 1994 German Federal Administrative Court ruling held that it was necessary for a plaintiff to demonstrate that a decision would have to be different had an EIA been carried out, before that decision could be quashed (Weston 1997).

References

Arnstein, S. R. 1971. A ladder of public participation in the USA. *Journal of the Royal Town Planning Institute*, April, 216–24.
Atkinson, N. & R. Ainsworth 1992. Environmental assessment and the local authority: facing the European imperative. *Environmental Policy and Practice* 2(2), 111–28.

Audit Commission 2000. *Listen Up! Effective Community Consultation* (http://www.audit-commission.gov.uk/publications/pdf/mpeffect.pdf).

Buxton, R. 1992. Scope for legal challenge. In *Environmental assessment and audit: a user's guide*, Ambit (ed.), 43–44. Gloucester: Ambit.

Canter, L. W. 1996. *Environmental impact assessment*, 2nd edn. New York: McGraw-Hill.

CEC (European Commission of Communities) 2001a. *Guidance on* EIA: EIS *review*. DG XI. Brussels: CEC.

CEC 2001b. *Proposal for a Directive of the European Parliament and of the Council providing for public participation in respect of drawing up of certain plans and programmes relating to the environment and amending Council Directives, 85/337/EEC and 96/61/EC.* COM(2000), 839 final, 18 January 2001. Brussels: DE Environment, EC.

CEC 2003. (Impacts Assessment Unit, Oxford Brookes University). Five years Report to the European Parliament and the Council on the Application and Effectiveness of the EIA Directive. Website of DG Environment, EC: http://europa.eu.int/comm/environment/eia/home.htm.

Clark, B. 1994. Improving public participation in environmental impact assessment. *Built Environment* 20(4), 294–308.

Dallas, W. G. 1984. Experiences of environmental impact assessment procedures in Ireland. In *Planning and Ecology*, R. D. Roberts & T. M. Roberts (eds), 389–95. London: Chapman & Hall.

DETR (Department of Environment, Transport and the Regions) 1999. *Quality of Life Counts*. London: HMSO.

DoE (Department of the Environment) 1994. *Evaluation of environmental information for planning projects: a good practice guide*. London: HMSO.

DoE 1995. *Preparation of environmental statements for planning projects that require environmental assessment*. London: HMSO.

Fuller, K. 1992. Working with assessment. In *Environmental assessment and audit: a user's guide*, Ambit (ed.), 14–15. Gloucester: Ambit.

Ginger, C. & P. Mohai 1993. The role of data in the EIS process: Evidence from the BLM wilderness review. *Environmental Impact Assessment Review* **13**, 109–39.

Hancock, T. 1992. Statement as an aid to consent. In *Environmental assessment and audit: a user's guide*, Ambit (ed.), 34–35. Gloucester: Ambit.

House of Lords 1981. *Environmental assessment of projects*, Select Committee on the European Communities, 11th Report, Session 1980–81, HMSO.

Hydro-Québec 1993. *Overview 21: communication and environmental assessment*. Quebec: Hydro-Québec.

IEMA (Institute of Environmental Management and Assessment) 2002. *Perspectives: Guidelines on participation in environmental decision-making*. Lincoln: IEMA.

Jones, C. E. & C. Wood 1995. The impact of environmental assessment in public inquiry decisions. *Journal of Planning and Environment Law*, October, 890–904.

Kenyan, R. C. 1991. Environmental assessment: an overview on behalf of the R.I.C.S. *Journal of Planning and Environment Law*, 419–22.

Lee, N. & R. Colley 1990. *Reviewing the quality of environmental statements*, Occasional Paper No. 24. Manchester: University of Manchester, EIA Centre.

McCormick, J. 1991. *British politics and the environment*. London: Earthscan.

McDonic, G. 1992. The first four years. In *Environmental assessment and audit: a user's guide*, Ambit (ed.), 2–3. Gloucester: Ambit.

McNab, A. 1997. Scoping and public participation. In *Planning and EIA in practice*, J. Weston (ed.), 60–77. Harlow: Longman.

Mollison, K. 1992. A discussion of public consultation in the EIA process with reference to Holland and Ireland (written for MSc Diploma course in Environmental Assessment and Management). Oxford: Oxford Polytechnic.

Morrison-Saunders, A., D. Annandale, J. Cappelluti 2001. Practioner perspectives on what influences EIA quality. *Impact Assessment and Project Appraisal* **19**(4), 321–25.

ODPM (Office of the Deputy Prime Minister) 2003a. *Environmental impact assessment: guide to the procedures.* London: HMSO.

ODPM 2003b. *Note on environmental impact assessment for local planning authorities.* London: HMSO.

Partidario, M. R. 1996. *16th Annual Meeting, International Association for Impact Assessment: Synthesis of Workshop Conclusions.* Estoril: IAIA.

Petts, J. 1999. Public participation and EIA. In *Handbook of environmental impact assessment,* J. Petts (ed.), vol. 1. Oxford: Blackwell Science.

Petts, J. 2003. Barriers to Deliberative Participation in EIA: Learning from Waste Policies, Plans and Projects. *Journal of Environmental Assessment Policy and Management* **5**(3), 269–94.

Read, R. 1997. Planning authority review. In *Planning and EIA in practice,* J. Weston (ed.). Harlow: Longman.

Rodriguez-Bachiller, A. with J. Glasson 2003. *Expert Systems and Geographical Information Systems for Impact Assessment.* London: Taylor and Francis.

Ross, W. A. 2000. Reflections of an environmental assessment panel member. *Impact Assessment and Project Appraisal* **18**(2), 91–98.

Salter, J. R. 1992a. Environmental assessment: the challenge from Brussels. *Journal of Planning and Environment Law,* January, 14–20.

Salter, J. R. 1992b. Environmental assessment – the need for transparency. *Journal of Planning and Environment Law,* March, 214–21.

Salter, J. R. 1992c. Environmental assessment – the question of implementation. *Journal of Planning and Environment Law,* April, 313–18.

Tomlinson, P. 1989. Environmental statements: guidance for review and audit. *The Planner* **75**(28), 12–15.

Turner, T. 1988. The legal eagles. *Amicus Journal,* Winter, 25–37.

Weiss, E. H. 1989. An unreadable EIS is an environmental hazard. *Environmental Professional* **11**, 236–40.

Wende, W. 2002. Evaluation of the effectiveness and quality of environmental impact assessment in the Federal Republic of Germany. *Impact Assessment and Project Appraisal* **20**(2), 93–99.

Westman, W. E. 1985. *Ecology, impact assessment and environmental planning.* New York: Wiley.

Weston, J. (ed.) 1997. *Planning and EIA in practice.* Harlow: Longman.

Weston, J., J. Glasson, R. Therivel, E. Weston, R. Frost 1997. *Environmental information and planning projects,* Working Paper No. 170. Oxford: Oxford Brookes University, School of Planning.

Williams, G. & A. Hill 1996. Are we failing at environment justice? How minority and low income populations are kept out of the public involvement process. *Proceedings of the 16th Annual Meeting of the International Association for Impact Assessment.* Estoril: IAIA.

Wood, C. M. & C. Jones 1997. The effect of environmental assessment on UK local authority planning decisions. *Urban Studies* **34**(8), 1237–57.

7 Monitoring and auditing: after the decision

7.1 Introduction

Major projects, such as roads, airports, power stations, petrochemical plants, mineral developments and holiday villages, have a life cycle with a number of stages (see Figure 1.5). The life cycle may cover a very long period (e.g. 50–60 years for the planning, construction, operation and decommissioning of a fossil-fuelled power station). EIA, as it is currently practised in the UK and in many other countries, relates primarily to the period *before* the decision. At its worst, it is a partial linear exercise related to one site, produced in-house by a developer, without any public participation. There is a danger of a short-sighted "build it and forget it" approach (Culhane 1993). However, EIA should not stop at the decision. It should be more than an auxiliary to the procedures to obtain a planning permission; rather it should be a means to obtain good environmental management *over the life* of the project. This means including monitoring and auditing in the EIA process.

The first section clarifies the definitions of and differences between monitoring and auditing, and outlines their potentially important roles in EIA. An approach to the better integration of monitoring into the process, drawing in particular on international practice, is then outlined. We then discuss approaches to environmental impact auditing, including a review of attempts to audit a range of EISs in a number of countries. The final section draws briefly on detailed monitoring and auditing studies of the local socio-economic impacts of the construction of the Sizewell B pressurized water reactor (PWR) nuclear power station in the UK.

7.2 The importance of monitoring and auditing in the EIA process

Monitoring involves the measuring and recording of physical, social and economic variables associated with development impacts (e.g. traffic flows, air quality, noise, employment levels). The activity seeks to provide information on the characteristics and functioning of variables in time and space, and in particular on the occurrence and magnitude of impacts. Monitoring can improve project management. It can be used, for example, as an early warning system, to identify harmful trends in a locality before it is too late to take remedial action. It can help to identify and correct unanticipated impacts. Monitoring can also provide an accepted database, which can be useful in mediation between interested parties. Thus, monitoring of the origins, pathways and destinations of, for example, dust in an industrial area may clarify where the responsibilities lie. Monitoring is also essential for successful environmental impact auditing, and can be one of the most effective guarantees of commitment to undertakings and to mitigation measures.

As noted by Buckley (1991), the term *environmental auditing* is currently used in two main ways. *Environmental impact auditing*, which is covered in this chapter, involves comparing the impacts predicted in an EIS with those that actually occur after implementation, in order to assess whether the impact prediction performs satisfactorily. The audit can be of both impact predictions (how good were the predictions?) and of mitigation measures and conditions attached to the development (is the mitigation effective? are the conditions being honoured?). This approach to auditing contrasts with *environmental management auditing*, which focuses on public and private corporate structures and programmes for environmental management and the associated risks and liabilities. We discuss this latter approach further in Chapter 11.

In total, monitoring and auditing can make important contributions to the better planning and EIA of future projects (Figure 7.1). Sadler (1988) writes of the need to

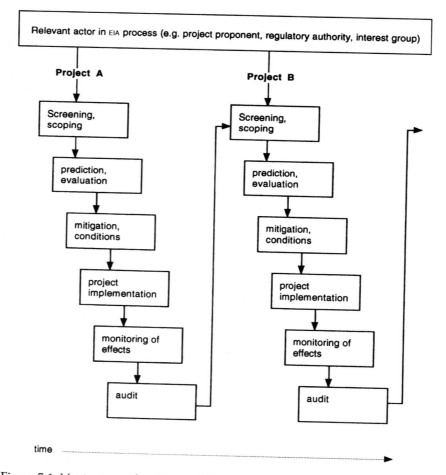

Figure 7.1 Monitoring and auditing and learning from experience in the EIA process. (Adapted from Bisset & Tomlinson 1988, Sadler 1988.)

introduce feedback in order to learn from experience; we must avoid the constant "reinventing of the wheel" in EIA. Monitoring and auditing of outcomes can contribute to an improvement in all aspects of the EIA process, from understanding baseline conditions to the framing of effective mitigating measures. In addition, Greene et al. (1985) note that monitoring and auditing should reduce time and resource commitments to EIA by allowing all participants to learn from past experience; they should also contribute to a general enhancing of the credibility of proponents, regulatory agencies and EIA processes. We are learning, and there is a considerable growth of interest in examining the effectiveness of the EIA process in practice. Unfortunately, there are a number of significant issues that have greatly limited the use of monitoring and auditing to date. These issues and possible ways forward for monitoring and auditing in practice are now discussed.

7.3 Monitoring in practice

7.3.1 Key elements

Monitoring implies the systematic collection of a potentially large quantity of information over a long period of time. Such information should include not only the traditional *indicators* (e.g. ambient air quality, noise levels, the size of a workforce) but also *causal underlying factors* (e.g. the *decisions* and *policies* of the local authority and developer). The causal factors determine the impacts and may have to be changed if there is a wish to modify impacts. *Opinions* about impacts are also important. Individual and group "social constructions of reality" (IAIA 1994) are often sidelined as "mere perceptions, or emotions", not to be weighted as heavily as facts. But such opinions can be very influential in determining the response to a project. To ignore or undervalue them may not be methodologically defensible and is likely to raise hostility. Monitoring should also analyse impact equity. The distribution of impacts will vary between groups and locations; some groups may be more vulnerable than others, as a result of factors such as age, race, gender and income. So a systematic attempt to identify opinions can be an important input into a monitoring study.

The information collected needs to be stored, analysed and communicated to relevant participants in the EIA process. A primary requirement, therefore, is to focus monitoring activity only on "those environmental parameters expected to experience a significant impact, together with those parameters for which the assessment methodology or basic data were not so well established as desired" (Lee & Wood 1980).

Monitoring is an integral part of EIA; baseline data, project descriptions, impact predictions and mitigation measures should be developed with monitoring implications in mind. An EIS should include a *monitoring programme* which has clear objectives, temporal and spatial controls, an adequate duration (e.g. covering the main stages of the project's implementation), practical methodologies, sufficient funding, clear responsibilities and open and regular reporting. Ideally, the monitoring activity should include a partnership between the parties involved; for example, the collection of information could involve the developer, local authority and local community. Monitoring programmes should also be adapted to the dynamic nature of the environment (Holling 1978).

7.3.2 *Mandatory or discretionary*

Unfortunately, monitoring is not a mandatory step in many EIA procedures, including those current in the UK. European Commission regulations do not specifically require monitoring. This omission was recognized in the review of Directive 85/337 (CEC 1993). The Commission is a strong advocate for the inclusion of a formal monitoring programme in an EIS, but EU Member States are normally more defensive and reactive. In consequence, the amended Directive does not include a mandatory monitoring requirement.[1] However, this has not deterred some Member States. For example, in the Netherlands the competent authority is required to monitor project implementation, based on information provided by the developer, and to make the monitoring information publicly available. If actual impacts exceed those predicted, the competent authority must take measures to reduce or mitigate these impacts. However, despite such legal provisions, practice has been limited and little post-EIA monitoring and evaluation has been carried out. See Arts (1998) for a comprehensive coverage of EIA follow-up in the Netherlands.

In other Member States, in the absence of mandatory procedures, it is usually difficult to persuade developers that it is in their interest to have a continuing approach to EIA. This is particularly the case where the proponent has a one-off project, and has less interest in learning from experience for application to future projects. Fortunately, we can turn to some examples of good practice in a few other countries. A brief summary of monitoring procedures in Canada is included in Chapter 10. In Western Australia (also see Chapter 10), the environmental consequences of developments are commonly monitored and reported. If it is shown that conditions are not being met, the government may take appropriate action. Interestingly, the EIS is called an "environmental review and management programme" (Morrison-Saunders 1996).

7.3.3 *The case of California*

The monitoring procedures used in California, for projects subject to the CEQA, are of particular interest (California Resources Agency 1988). Since January 1989, state and local agencies in California have been required to adopt a monitoring and/or reporting programme for mitigation measures and project changes which have been imposed as conditions to address significant environmental impacts. The aim is to provide a mechanism which will help to ensure that mitigation measures will be implemented in a timely manner in accordance with the terms of the project's approval. Monitoring refers to the observation and oversight of mitigation activities at a project site, whereas reporting refers to the communication of the monitoring results to the agency and public. If the implementation of a project is to be phased, the mitigation and subsequent reporting and monitoring may also have to be phased. If monitoring reveals that mitigation measures are ignored or are not completed, sanctions could be imposed; these can include, for example, "stop work" orders, fines and restitution. The components of a monitoring programme would normally include the following:

- a summary of the significant impacts identified in the Environmental Impact Report (EIR);
- the mitigation measures recommended for each significant impact;

- the monitoring requirements for each mitigation measure;
- the person or agency responsible for the monitoring of the mitigation measure;
- the timing and/or frequency of the monitoring;
- the agency responsible for ensuring compliance with the monitoring programme;
- the reporting requirements.

Figure 7.2 provides an extract from a monitoring programme for a woodwaste conversion facility at West Berkeley in California.

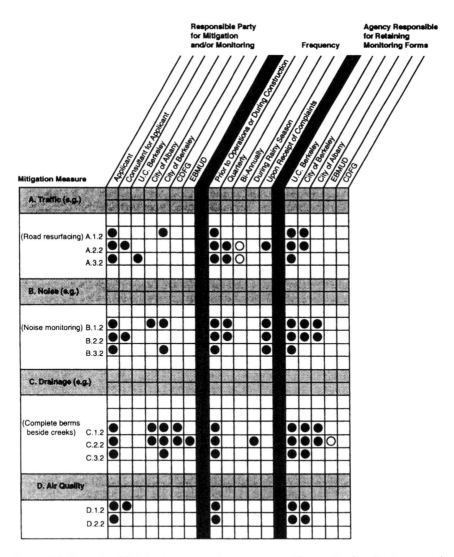

Figure 7.2 Example of Californian monitoring programme. (*Source*: Baseline Environmental Consulting 1989.)

7.3.4 The case of Hong Kong

In Hong Kong, a systematic, comprehensive environmental monitoring and auditing system was introduced in 1990 for major projects. A major impetus for action was the construction of the new $20-billion airport at Chap Lap Kok, which included the construction of not only the airport island, but also a railway, highways and crossings and a major Kowloon reclamation project. The environmental monitoring and audit manual includes three stages of an event-action plan: (1) trigger level, to provide an early warning; (2) action level, at which action is to be taken before an upper limit of impacts is reached and (3) target level, beyond which a predetermined plan response is initiated to avoid or rectify any problems. The approach does build monitoring much more into project decision-making, requiring proponents to agree monitoring and audit protocols and event-action plans in advance; however, enforcement has been problematic (Au & Sanvícens 1996).

Since April 1998 there have been EIA regulations in force which stipulate in detail when and how environmental monitoring and auditing should be done (EPD 1997, 1998). The regulations normally result in permit conditions relating to project approval. This has provided a statutory basis for follow-up work, and offences carry stiff penalties (up to $250,000, and six months imprisonment). A recent and fascinating innovation in the Hong Kong system is the use of the Internet for monitoring the effects of large projects and of compliance with the permit conditions. Under procedures introduced since 2000 major projects must set up a monitoring website (see: http://www.info.gov.hk/epd/eia). Some sites include webcams focused on parts of the project. There is public access to the websites, and concerned members of the public can report their views on project performance back to both the government and the developer (Hui & Ho 2002). Is this the shape of things to come?

7.3.5 UK experience

Although monitoring is not a mandatory requirement under UK EIA regulations, there is monitoring activity. A research study at Oxford Brookes University (see Frost 1997, Glasson 1994) has sought to provide an initial estimate of the extent of such activity using a "contents analysis" and a "practice analysis". The contents analysis of references to monitoring intentions uses a representative sample of almost 700 EISs and summaries of EISs (taken from the Institute of Environmental Assessment's *Digest of Environmental Statements*) (IEA 1993). For some EISs there was a clearly indicated monitoring section; for others, monitoring was covered in sections related to mitigation. In several cases there were generic monitoring proposals with, for example, a proposal to check that contractors are in compliance with contract specifications. Overall, approximately 30 per cent of the cases included at least one reference to impact monitoring. The maximum number of monitoring types was six, suggesting that impact monitoring is unlikely to be approaching comprehensiveness in even a select few cases. Table 7.1 shows the types of monitoring in EISs. Water quality monitoring was more frequently cited than air quality monitoring. Point of origin monitoring of air and aqueous emissions was also frequently cited. There was only very limited reference to the monitoring of non-biophysical (i.e. socio-economic) impacts. The type of monitoring varied between project types. For Combined Cycle Gas Turbine (CCGT) power stations, proposals were often made for monitoring air emissions, air quality and construction

Table 7.1 Types of impact monitoring in UK EISs

Type	% of total monitoring proposals
Water quality	16
Air emissions	15
Aqueous emissions	13
Noise	12
General	9
Others	7
Ecological	7
Archaeological	6
Air quality	5
Structural survey	4
Liaison group	3
Water levels	3
	100

(*Source*: Glasson 1994.)

noise; for landfill projects, the proposals were skewed towards the monitoring of leachate, landfill gas and water quality.

The practice analysis used a small representative sample of 17 projects, with EIS monitoring proposals, which had started. The LPAs were contacted to clarify monitoring arrangements including, for example, whether monitoring arrangements had been made operational under the terms of various consents (e.g. planning conditions, S106 agreements, Integrated Pollution Control (IPC) conditions, site licence conditions) or whether monitoring was being carried out voluntarily. The findings revealed that overall EISs tended to understate, on average by about 30 per cent, the amount of monitoring actually undertaken. This may be a response to planning conditions and agreements resulting from the decision-making process; it may also relate to other relevant licensing procedures, such as IPC. Whatever the case, the findings do suggest that monitoring proposals in EISs are carried out and are often more extensive than the, admittedly often limited, coverage in EIAs. The findings do not, of course, provide any information on the quality of the monitoring or about the accuracy of the predictions.

7.4 Auditing in practice

Auditing has developed a considerable variety of types. Tomlinson & Atkinson (1987a,b) have attempted to standardize *definitions* with a set of terms for seven different points of audit in the "standard" EIA process, as follows:

- decision point audit (draft EIS) – by regulatory authority in the planning approval process;
- decision point audit (final EIS) – also by regulatory authority in the planning approval process;
- implementation audit – to cover start-up; it could include scrutiny by the government and the public and focus on the proponent's compliance with mitigation and other imposed conditions;

- performance audit – to cover full operation; it could also include government and public scrutiny;
- predictive techniques audit – to compare actual with predicted impacts as a means of comparing the value of different predictive techniques;
- project impact audits – also to compare actual with predicted impacts and to provide feedback for improving project management and for future projects;
- procedures audit – external review (e.g. by the public) of the procedures used by the government and industry during the EIA processes.

These terms can and do overlap. The focus here is on project, performance and implementation audits. Whatever the focus, auditing faces a number of major *problems* as outlined in Table 7.2.

Such problems may partly explain the dismal record of the early set of Canadian EISs examined, from an ecological perspective, by Beanlands & Duinker (1983), for which accurate predictions appeared to be the exception rather than the rule. There are several examples, also from Canada, of situations where an EIA has failed to predict significant impacts. Berkes (1988) indicated how an EIA on the James Bay mega-HEP (1971–85) failed to pick up a sequence of interlinked impacts, which resulted in a significant increase in the mercury contamination of fish and in the mercury poisoning of native people. Dickman (1991) identified the failings of an EIA to pick up the impacts of increased lead and zinc mine tailings on the fish population

Table 7.2 Problems associated with post-auditing studies

Nature of impact predictions
Many EISs contain few testable predictions; instead, they simply identify issues of potential concern.
Many EIS predictions are vague, imprecise and qualitative.
Testable predictions often relate to relatively minor impacts, with major impacts being referred to only in qualitative terms.

Project modifications
Post-EIS project modifications invalidate many predictions.

Monitoring data
Monitoring data and techniques often prove inadequate for auditing purposes.
Pre-development baseline monitoring is often insufficient, if undertaken at all.
Most monitoring data are collected and provided by the project proponent, which may give rise to fears of possible bias in the provision of information.

Comprehensiveness
Many auditing studies are concerned only with certain types of impacts (e.g. biophysical but not socio-economic; operational but not construction-stage impacts) and are therefore not full-project EIA audits.

Clarity
Few published auditing studies are explicit about the criteria used to establish prediction accuracy; this lack of clarity hampers comparisons between different studies.

Interpretation
Most auditing studies pay little attention to examining the underlying causes of predictive errors: this needs to be addressed if monitoring and auditing work is to provide an effective feedback in the EIA process.

(*Source:* Chadwick & Glasson 1999.)

in Garrow Lake, Canada's most northerly hypersaline lake. Such outcomes are not unique to Canada. Canada is a leader in monitoring, and the incidence of such research may result in improved and better predictions than in most countries.

Findings from the early limited auditing activity in the UK were not too encouraging. A study of four major developments – the Sullom Voe (Shetlands) and Flotta (Orkneys) oil terminals, the Cow Green reservoir and the Redcar steelworks – suggested that 88 per cent of the predictions were not auditable. Of those that were auditable, fewer than half were accurate (Bisset 1984). Mills's (1992) monitoring study of the visual impacts of five recent UK major project developments (a trunk road, two wind farms, a power station and an opencast coal mine) revealed that there were often significant differences between what was stated in an EIS and what actually happened. Project descriptions changed fundamentally in some cases, landscape descriptions were restricted to land immediately surrounding the site and aesthetic considerations were often omitted. However, mitigation measures were generally carried out well.

More recent examples of auditing include the Toyota plant study (Ecotech Research and Consulting Ltd 1994), and various wind farm studies (Blandford, C. Associates 1994, ETSU 1994). The Toyota study took a wide perspective on environmental impacts; auditing revealed some underestimation of the impacts of employment and emissions, some overestimation of housing impacts and a reasonable identification of the impacts of construction traffic. The study by Blandford, C. Associates of the construction stage of three wind farms in Wales confirmed the predictions of low ecological impacts, but suggested that the visual impacts were greater than predicted, with visibility distance greater than the predicted 15 km. However, the latter finding related to a winter audit; visibility may be less in the haze of summer.

One of the most comprehensive nationwide auditing studies of the precision and accuracy of environmental impact predictions was carried out by Buckley (1991) in Australia. At the time of his study, he found that adequate monitoring data to test predictions were available for only 3 per cent of the up to 1,000 EISs produced between 1974 and 1982. In general, he found that testable predictions and monitoring data were available only for large, complex projects, which had often been the subject of public controversy, and whose monitoring was aimed primarily at testing compliance with standards rather than with impact predictions. Some examples of over 300 major and subsidiary predictions tested are illustrated in Table 7.3.

Table 7.3 Examples of auditing of environmental impact predictions

Component/ parameter	Type of development	Predicted impact	Actual impact	Accuracy/ precision
Surface water quality: salts, pH	Bauxite mine	No detectable increase in stream salinity	None detected	Correct
Noise	Bauxite mine	Blast noise <115 dBA	Only 90 per cent <115 dBA	Incorrect: 90 per cent accurate, worse
Workforce	Aluminium smelter	1,500 during construction	Up to 2,500	Incorrect: 60 per cent accurate, worse

(*Source*: Buckley 1991.)

Overall, Buckley found the average accuracy of quantified, critical, testable predictions was 44 ± 5 per cent standard error. The more severe the impact, the lower the accuracy. Inaccuracy was highest for predictions of groundwater seepage. Accuracy assessments are of course influenced by the degree of precision applied to a prediction in the first place. In this respect, the use of ranges, reflecting the probabilistic nature of many impact predictions, may be a sensible way forward and would certainly make compliance monitoring more straightforward and less subject to dispute. Buckley's national survey, showing less than 50 per cent accuracy, does not provide grounds for complacency. Indeed, as it was based on monitoring data provided by the operating corporations concerned, it may present a better result than would be generated from a wider trawl of EISs. On the other hand, we are learning from experience, and more recent EISs may contain better and more accurate predictions. In a more recent UK study, Marshall (2001) reviewed a set of 1,118 mitigation proposals from 41 EISs. He found that in 38 per cent of the cases (418 in total), the proposals were expressed in such a way that the proponent could not be held to be committed to their implementation. In such cases mitigation is of little value, and there may be major compliance issues.

There has not, until recently, been much emphasis in auditing studies on the important area of predictive techniques audit, and on the value of particular predictive techniques. Where there have been studies, they have tended to focus on identifying errors associated with predictive methods rather than on explaining the errors. There is a need to develop appropriate audit methodologies, and as more projects are implemented there should be more scope for such studies. The pioneering study by Wood on visibility, noise and air quality impacts, using GIS to audit and model EIA errors, provides an example of a way forward for such work (Wood 1999a,b, 2000).

7.5 A UK case study: monitoring and auditing the local socio-economic impacts of the Sizewell B PWR construction project

7.5.1 Background to the case study

Although monitoring and auditing impacts are not mandatory in EIA procedures in the UK, the physical and socio-economic effects of developments are not completely ignored. For example, a number of public agencies monitor particular pollutants. LPAs monitor some of the conditions attached to development permissions. However, there is no systematic approach to the monitoring and auditing of impact predictions and mitigation measures. This case study reports on one attempt to introduce a more systematic, although still very partial, approach to the subject.

In the 1970s and early 1980s, Britain had an active programme of nuclear power station construction. This included a commitment, since revised, to build a family of PWR stations. The first such station to be approved was Sizewell B in East Anglia. The approval was controversial, and followed the longest public inquiry in UK history. Construction started in 1987, and the project was completed in 1995. The IAU in the School of Planning at Oxford Brookes University had studied the impacts of a number of power stations and made contributions to EISs, with a focus on the socio-economic impacts. A proposal was made to the relevant public utility, the CEGB, that the construction of Sizewell B provided an invaluable opportunity to monitor in detail the project construction stage, and to check on the predictions made at the public inquiry and on the mitigating conditions attached to the project's approval. Although the predictions were not formally packaged in an EIS, but rather as a series of reports based on the inquiry, the research was extensive and comprehensive (DoEn 1986). The CEGB supported a monitoring study, which began in 1988. To the credit of the utility, which is now Nuclear Electric/British Energy following privatization, there was a continuing commitment to the monitoring study – despite the uncertainty about further PWR developments in Britain. Monitoring reports for the whole construction period and on the project's operation have now been completed (Glasson et al. 1989–97).

7.5.2 Operational characteristics of the monitoring study

It is important to clarify the *objectives of the monitoring study*, otherwise irrelevant information may be collected and resources wasted. Figure 7.3 outlines the scope of the study. The development under consideration is the construction stage of the Sizewell B PWR 1,200 MW nuclear power station. The focus is on the socio-economic impacts of the development, although with some limited consideration of physical impacts. The socio-economic element of EIA involves "the systematic advanced appraisal of the impacts on the day to day quality of life of people and communities

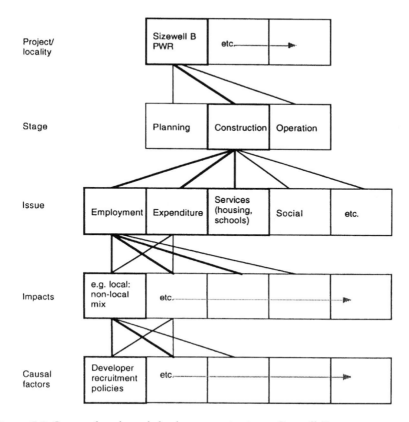

Figure 7.3 Scope of study and database organization – Sizewell B monitoring study. (*Source*: Glasson et al. 1989–97.)

when the environment is affected by development or policy change" (Bowles 1981). This involves a consideration of the impacts on employment, social structure, expenditure, services, etc. Although socio-economic studies have often been the poor relation in impact assessment studies to date, meriting no more than a chapter or two in EISs, they are important, not least because they consider the impacts of developments on people, who can answer back and object to developments.

The highest priority in the study has been to identify the impacts of the development on local employment; this emphasis reflects the pivotal role of employment impacts in the generation of other local impacts, particularly accommodation and local services. In addition to providing an updated and improved database to inform future assessments, assisting project management of the Sizewell B project in the local community and auditing impact predictions, the study also monitored and audited some of the conditions and undertakings associated with permission to proceed with the construction of the power station. These included undertakings on the use of rail and the routeing of road construction traffic, as well as conditions on the use of local labour and local firms, local liaison arrangements and (traffic) noise (DoEn 1986).

The monitoring study included the collection of a range of information, including statistical data (e.g. the mixture of local and non-local construction-stage workers, the housing tenure status and expenditure patterns of workers), decisions, opinions and perceptions of impacts. The spatial scope of the study extended to the commuting zone for construction workers (Figure 7.4). The study included information from the developer and the main contractors on site, from the relevant local authorities and other public agencies, from the local community and from the construction workers. The local upper-school Geography A-level students helped to collect data on the local perceptions of impacts via biennial questionnaire surveys in the town of Leiston, which is adjacent to the construction site. A major survey of the socio-economic characteristics and activities of a 20 per cent sample of the project workforce was also carried out every 2 years. The IAU team operated as the catalyst to bring the data together. There was a high level of support for the study, and the results are openly available in published annual monitoring reports and in summary broadsheets, which are available free to the local community (Glasson et al. 1989–97).

The study has highlighted a number of *methodological difficulties with monitoring and auditing*. The first relates to the disaggregation of project-related impacts from baseline trends. Data are available that indicate local trends in a number of variables, such as unemployment levels, traffic volumes and crime levels. But problems are encountered when we attempt to explain these local trends. To what extent are they due to (a) the construction project itself, (b) national and regional factors or (c) other local changes independent of the construction project? It is straightforward to isolate the role of national and regional factors, but the relative roles of the construction project and other local changes are very difficult to determine. "Controls" are used where possible to isolate the project-related impacts.

A second problem relates to the identification of the indirect, knock-on effects of a construction project. Indirect impacts – particularly on employment – may well be significant, but they are not easily observed or measured. For example, indirect employment effects may result from the replacement of employees leaving local employment to take up work on site. Are these local recruits replaced by their previous employers? If so, do these replacements come from other local employees, the local unemployed or in-migrant workers? It was feasible to obtain this sort of information.

Figure 7.4 Sizewell B commuting zone – monitoring study area. (*Source*: Glasson et al. 1989–97.)

Further indirect employment impacts may stem from local businesses gaining work as suppliers or contractors at Sizewell B. They may need to take on additional labour to meet their extra workload. The extent to which this has occurred is again very difficult to estimate, although surveys of local companies have provided some useful information on these issues (Glasson & Heaney 1993).

7.5.3 *Some findings from the studies*

A very brief summary of a number of the findings are outlined below and in Figures 7.5a and b.

Employment

An important prediction and condition was that at least 50 per cent of construction employment should go to local people (within daily commuting distance of the site). This has been the case, although, predictably in a rural area, local people have the largely semi-skilled or unskilled jobs. As the employment on site has increased, with a shift from civil engineering to mechanical and electrical engineering trades, the pressure on maintaining the 50 per cent proportion has increased. In 1989, a training centre was opened in the nearest local town, Leiston, to supply between 80 and 120 trainees from the local unemployed.

Local economy

A major project has an economic multiplier effect on a local economy. By the end of 1991, Sizewell B workers were spending about £500,000 per week in Suffolk and Norfolk, Nuclear Electric had placed orders worth over £40 million with local companies and a "good neighbour" policy was funding a range of community projects (including £1.9 million for a swimming pool in Leiston).

Housing

A major project, with a large in-migrant workforce, can also distort the local housing market. One mitigating measure at Sizewell B was the requirement of the developer to provide a large site hostel. A 600-bed hostel (subsequently increased to 900) was provided. It was very well used, accommodating in 1991 over 40 per cent of the in-migrants to the development, at an average occupancy rate of over 85 per cent, and it helped to reduce demand for accommodation in the locality.

Traffic and noise

The traffic generated by a large construction project can badly affect local towns and villages. To mitigate such impacts, there was a designated construction route to Sizewell B. The monitoring of traffic flows on designated and non-designated (control) routes indicated that this mitigation measure was working. Between 1988 and 1991, the amount of traffic rose substantially at the four monitoring points on the designated route, but much less so at most of the seven points not on that route. Construction noise on site has been a local issue. Monitoring has led to modifications in some construction methods, notably improvements to the railway sidings and changes in the piling methods used.

Crime

An increase in local crime is normally associated with the construction stage of major projects. The Leiston police division did see a significant increase in the

Employment impacts

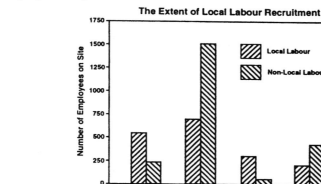

Actual and Predicted Growth of the Construction Workforce

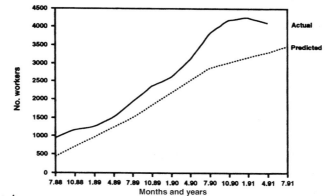

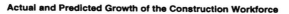

Social impacts

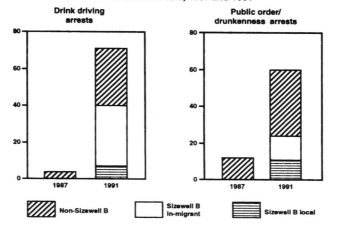

Figure 7.5(a) Brief summary of some findings from the Sizewell B PWR construction project monitoring and auditing study. (*Source*: Glasson et al. 1989–97.)

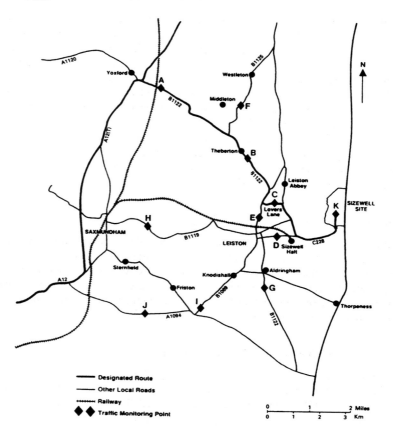

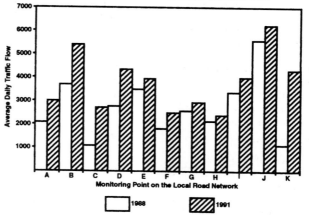

Figure 7.5(b) More findings from the Sizewell B PWR construction project monitoring and auditing study. (*Source*: Glasson et al. 1989–97.)

number of arrests in certain offence categories after the start of the project. However, local people not employed on the project were involved in most of the arrests, and in the increase in arrests, with the exception of drink-driving, for which Sizewell B employees (mainly in-migrants) accounted for most arrests and for most of the increase. However, the early diagnosis of the problems facilitated remedial action, including the introduction of a shuttle minibus service for workers, the provision of a large bar in the site hostel, the stressing at site-workers' induction courses of the problems of drink-driving, and the exclusion from the site (effectively the exclusion from Sizewell B jobs) of workers found guilty of serious misconduct or crime. Since the early stages of the project, worker-related crime has fallen substantially, and the police have considered the project workforce to be relatively trouble-free, with fewer serious offences than anticipated.

Residents' perceptions

Surveys of local residents in 1989 and 1991 revealed more negative than positive perceived impacts, increased traffic and disturbance by workers being seen as the main negative impacts. The main positive impacts of the project were seen to be the employment, additional trade and ameliorative measures associated with the project. The monitoring of complaints about the development revealed substantially fewer complaints over time, despite the rapid build-up of the project.

7.5.4 Learning from monitoring: Sizewell B and Sizewell C

Table 7.4 shows the nature and auditability of the Sizewell B socio-economic predictions. In contrast to the findings from previous post-auditing studies (see Dipper et al. 1998), a vast majority of the Sizewell B predictions were expressed in quantitative terms. The monitoring of impacts and the auditing of the predictions

Table 7.4 Nature and auditability of the Sizewell B predictions

	No. of predictions	% of total
Nature of prediction		
Quantitative		
Expressed in absolute terms	35	51
Expressed in % terms	21	30
Qualitative	11	16
Incorporates quantitative and qualitative elements	2	3
Total: all predictions	69	100
Auditability of predictions		
Auditable: monitoring data subject to no or little potential error	30	43
Auditable: but monitoring data subject to greater potential error	28	41
Not auditable	11	16
Total: all predictions	69	100

(*Source*: Chadwick & Glasson 1999.)

Table 7.5 Accuracy of auditable Sizewell B predictions

% error in prediction	No. of predictions	% of total
None: prediction correct or within predicted range	15	26
Less than 10%	9	16
10–20%	11	19
20–30%	5	9
30–40%	5	9
40–50%	2	3
Over 50%	8	14
Prediction incorrect, but % error cannot be calculated	3	5
Prediction cannot be audited	11	–
Total: all predictions	69	100

Notes: For quantified predictions, the predicted value was used as the denominator in the calculation of the % errors in the table. For non-quantified predictions, the % error could not be calculated and predictions were classified as either "correct" or "incorrect", based on assessment by the research team. (*Source*: Chadwick & Glasson 1999.)

and mitigation measures revealed (Table 7.5) that many of the predictions used in the Sizewell B public inquiry were reasonably accurate – although there was an underestimate of the build-up of construction employment and an overestimate of the secondary effects on the local economy. Predictions of traffic impacts, and on the local proportion of the construction workforce, were very close to the actual outcomes. Mitigation measures also appeared to have some effect. Overall, approximately 60 per cent of the predictions had errors of less than 20 per cent. Explanations of variations from the predictions included the inevitable project modification (particularly associated with new-technology projects, with few or no comparators at the time of prediction), and the very lengthy project authorization process (with a gap of almost 10 years between the predictions and peak construction). Other local issues have been revealed by monitoring, allowing some modifications to manage the project better in the community. Unfortunately, such systematic monitoring is still discretionary in the UK and very much dependent on the goodwill of developers.

Information gained from monitoring can also provide vital intelligence for the planning and assessment of future projects. This is particularly so when the subsequent project is of the same type, and in the same location, as that which has been monitored. Nuclear Electric applied for consent to build and operate a replica of Sizewell B, to be known as Sizewell C. A full EIS was produced for the project (Nuclear Electric 1993). Its prediction of the socio-economic impacts drew directly on the findings from the Sizewell B monitoring study. Figure 7.6 provides an overview of the cumulative employment impacts of the operational Sizewell B plus the construction of Sizewell C (with two reactors, C1 and C2). The regular peaks in the figure are the refuelling intervals. However, this proposed follow-on project fell victim to the abandonment of the UK nuclear power station programme.

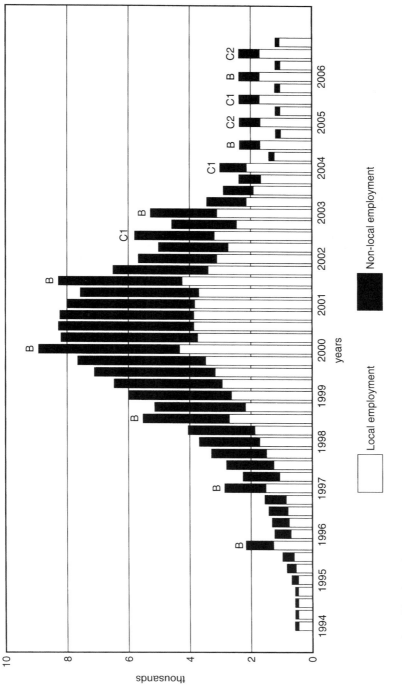

Figure 7.6 Predicted local employment impacts of Sizewell B operational station and Sizewell C construction project (with reactors C1 and C2). (*Source*: Nuclear Electric 1993, 1994.)

7.6 Summary

A mediation of the relationship between a project and its environment is needed throughout the life of a project. Environmental impact assessment is meant to establish the terms and conditions for project implementation; yet there is often little follow-through to this stage and even less follow-up after it. Arts (1998) concludes, after a thorough examination of "ex-post evaluation of EIA", that in practice it is lagging behind the practice of EIA itself. Few countries have made arrangements for some form of follow-up. In those that have, experience has not been too encouraging – reflecting deficiencies in often over-descriptive EISs, inadequate techniques for follow-up, organizational and resource limitations, and limited support from authorities and project proponents alike. Yet many projects have very long lives, and their impacts need to be monitored on a regular basis. Morrison-Saunders et al. (2001) show how this can bring positive outcomes for different stakeholders. Figure 7.7 shows the benefits not only to the proponent and the community (as exemplified by the Sizewell B case study), but also to the regulator – in the form of a better decision, and improvement of the EIA process. Such monitoring can improve project management and contribute to the auditing of both impact predictions and mitigating measures. Monitoring and auditing can provide essential feedback to improve the EIA process, yet

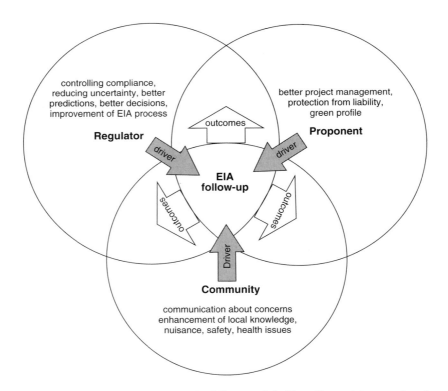

Figure 7.7 Outcomes of EIA follow-up for different stakeholders. (*Source*: Morrison-Saunders et al. 2001.)

this is still probably the weakest step of the process in many countries. Discretionary measures are not enough; monitoring and auditing need to be more fully integrated into EIA procedures on a mandatory basis.

Note

1. Early drafts of the EC Directive did include a requirement for an ex-post evaluation of EIA projects. Section 11 of the 1980 draft (CEC 1980) stated that the competent authority should check at set intervals whether the provisions which are attached to the planning permission are observed or adequate, or other provisions for environmental protection are observed, and whether additional measures are required to protect the environment against the project's impacts.

References

Arts, J. 1998. EIA *follow-up: on the role of ex-post evaluation in environmental impact assessment.* Groningen: Geo Press.

Au, E. & G. Sanvícens 1996. EIA follow up monitoring and management. EIA *Process Strengthening.* Canberra: Environmental Protection Agency.

Baseline Environmental Consulting 1989. *Mitigation monitoring and reporting plan for a woodwaste conversion facility, West Berkeley, California.* Emeryville, California: Baseline Environmental Consulting.

Beanlands, G. E. & P. Duinker 1983. *An ecological framework for environmental impact assessment.* Halifax, Nova Scotia: Dalhousie University, Institute for Resource and Environmental Studies.

Berkes, F. 1988. The intrinsic difficulty of predicting impacts: lessons from the James Bay hydro project. *Environmental Impact Assessment Review* 8, 201–20.

Bisset, R. 1984. Post development audits to investigate the accuracy of environmental impact predictions. *Zeitschift für Umweltpolitik* 7, 463–84.

Bisset, R. & P. Tomlinson 1988. Monitoring and auditing of impacts. In *Environmental impact assessment*, P. Wathern (ed.). London: Unwin Hyman.

Blandford, C. Associates 1994. *Wind turbine power station construction monitoring study.* Gwynedd: Countryside for Wales.

Bowles, R. T. 1981. *Social impact assessment in small communities.* London: Butterworth.

Buckley, R. 1991. Auditing the precision and accuracy of environmental impact predictions in Australia. *Environmental Monitoring Assessment* 18, 1–23.

California Resources Agency 1988. *California EIA monitor.* State of California.

CEC (1980). Proposal for a Council Directive Concerning the Assessment of the Environmental Effects of Certain Public and Private Projects on the Environment, 9 July 1980. *Official Journal of the EC*, L175, 40–49. Brussels: EC.

CEC (Commission of the European Communities) 1993. *Report from the Commission of the Implementation of Directive 85/337/EEC on the assessment of the effects of certain public and private projects on the environment*, vol. 13, *Annexes for all Member States*, COM(93), 28 final. Brussels: EC, Directorate-General XI.

Chadwick, A. & J. Glasson 1999. Auditing the Socio-Economic Impacts of a Major Construction Project: The Case of Sizewell B Nuclear Power Station. *Journal of Environmental Planning and Management* 42(6), 811–36.

Culhane, P. J. 1993. Post-EIS environmental auditing: a first Step to making rational environmental assessment a reality. *Environmental Professional* 5.

Dickman, M. 1991. Failure of environmental impact assessment to predict the impact of mine tailings on Canada's most northerly hypersaline lake. *Environmental Impact Assessment Review* 11, 171–80.

Dipper, B., C. Jones, C. Wood 1998. Monitoring and post-auditing in environmental impact assessment: a review. *Journal of Environmental Planning and Management* **41**(6), 731–47.

DoEn (Department of Energy) 1986. *Sizewell B Public Inquiry: Report by Sir Frank Layfield*. London: HMSO.

Ecotech Research and Consulting Ltd 1994. *Toyota impact study summary*, unpublished.

EPD (Environmental Protection Department) 1997. *Environmental Impact Assessment Ordinance and Technical Memorandum on Environmental Impact Assessment*. Hong Kong: EPD Hong Kong Government.

EPD Hong Kong Special Administrative Region Government 1998. *Guidelines for Development Projects in Hong Kong – Environmental Monitoring and Audit*. Hong Kong: EPDHK.

ETSU (Energy Technology Support Unit) 1994. *Cemmaes Wind Farm: sociological impact study*. Market Research Associates and Dulas Engineering, ETSU.

Frost, R. 1997. EIA monitoring and audit. In *Planning and EIA in practice*, J. Weston (ed.), 141–64. Harlow: Longman.

Glasson, J. 1994. Life after the decision: the importance of monitoring in EIA. *Built Environment* **20**(4), 309–20.

Glasson et al. 1989–97. *Local socio-economic impacts of the Sizewell B PWR Construction Project*. Oxford: Oxford Polytechnic/Oxford Brookes University, Impacts Assessment Unit.

Glasson, J. & D. Heaney 1993. Socio-economic impacts: the poor relations in British environmental impact statements. *Journal of Environmental Planning and Management* **36**(3), 335–43.

Greene, G., J. W. MacLaren, B. Sadler 1985. Workshop summary. In *Audit and evaluation in environmental assessment and management: Canadian and international experience*, 301–21. Banff: The Banff Centre.

Holling, C. S. (ed.) 1978. *Adaptive environmental assessment and management*. Chichester: Wiley.

Hui, S. Y. M. & M. W. Ho 2002. EIA Follow-up: Internet-based Reporting. In *Conference Proceedings of 22nd Annual Meeting IAIA June 15–21, The Hague*. IAIA.

IAIA (International Association for Impact Assessment) 1994. Report of special committee on guidelines on principles for social impact assessment. *Impact Assessment* **12**(2).

IEA 1993. *Digest of Environmental Statements*. London: Sweet & Maxwell.

Lee, N. & C. Wood 1980. *Methods of environmental impact assessment for use in project appraisal and physical planning, Occasional Paper no. 7*. University of Manchester, Department of Town and Country Planning.

Marshall, R. 2001. Mitigation linkage: EIA follow-up through the application of EMPs in transmission construction projects. In *Conference Proceedings of 21st Annual Meeting IAIA, 26 May–1 June, Cartagena, Colombia*, IAIA.

Mills, J. 1992. *Monitoring the visual impacts of major projects* (MSc dissertation in Environmental Assessment and Management, School of Planning, Oxford Brookes University).

Morrison-Saunders, A. 1996. Auditing the effectiveness of EA with respect to ongoing environmental management performance. In *Conference Proceedings 16th Annual Meeting IAIA June 17–23 1996*, M. Rosario Partidario (ed.), vol 1. IAIA, Lisbon 317–22.

Morrison-Saunders, A., J. Arts, J. Baker, P. Caldwell 2001. Roles and stakes in EIA follow-up. *Impact Assessment and Project Appraisal* **19**(4), 289–96.

Nuclear Electric 1993. *Proposed Sizewell C PWR Power Station – environmental statement*. Gloucester: Nuclear Electric.

Nuclear Electric 1994. *Suffolk coastal economy with and without Sizewell C*. Gloucester: Nuclear Electric.

Sadler, B. 1988. The evaluation of assessment: post-EIS research and process. In *Environmental Impact Assessment*, E. Wathern (ed.). London: Unwin Hyman.

Tomlinson, E. & S. F. Atkinson 1987a. Environmental audits: proposed terminology. *Environmental Monitoring and Assessment* **8**, 187–98.

Tomlinson, E. & S. F. Atkinson 1987b. Environmental audits: a literature review. *Environmental – Monitoring and Assessment* **8**, 239.

Wood, G. 1999a. Assessing techniques of assessment: post-development auditing of noise predictive schemes in environmental impact assessment. *Impact Assessment and Project Appraisal* **17**(3), 217–26.

Wood, G. 1999b. Post-development Auditing of EIA Predictive Techniques: A Spatial Analysis Approach. *Journal of Environmental Planning and Management* **42**(5), 671–89.

Wood, G. 2000. Is What You See What You Get? Post-development Auditing of Methods Used for Predicting the Zone of Visual Influence in EIA. *Environmental Impact Assessment Review* **20**(5), 537–56.

Part 3

Practice

LET'S MAKE SURE I'VE GOT THIS RIGHT. WE GET TO KEEP SOME LIZARDS AND BLUE BUTTERFLIES ON OUR HEATHLAND, AND IN RETURN YOU GET TO BUILD 3,200 NEW HOUSES ON OUR GREEN BELT . . . .

8 An overview of UK practice to date

8.1 Introduction

Part 3 considers EIA practice: what is done rather than what should be done. Chapter 8 provides an overview of the first 15 years or so of UK practice since EC Directive 85/337 became operational. We develop this further with reference to particular case studies in Chapter 9. The case studies seek to develop particular themes and aspects of the EIA process raised in this and in earlier chapters, for example, on the treatment of alternatives, of public participation and on the effect of divided consent procedures. The case studies are largely UK-based, and project-focused, although a case of SEA is also included. Chapter 10 discusses international practice in terms of "best practice" systems, emerging EIA systems and the role of international funding agencies in EIA, such as the World Bank.

These chapters can be set in the context of the important international study on EIA effectiveness, a major three-year study, whose results have been written up by Sadler (1996). Sadler suggests that EIA effectiveness can be tested at different stages in a cycle of EIA systems: (1) whether a given EIA policy is effectively translated into practice through the application of relevant processes and procedures, (2) whether the practice results in effective EIA performance through contributions to decision-making and (3) whether this performance then effectively feeds back into changes in the EIA policy by examining whether EIA realizes its purpose.

Sadler also notes that these questions and the attendant techniques for investigating them must be seen in the context of the decision-making framework in which the relevant EIA system operates. As was discussed in previous chapters, EIA in the UK can broadly be described as having been

- imposed on a reluctant government by the EC;
- implemented since then relatively punctually and thoroughly;
- based on a strong pre-existing planning system, but with inelegant "patching" where Directive 85/337 has required EIA for projects covered by other authorization systems, and where regulations have since been amended;
- often implemented through negotiations rather than through direct confrontations between the relevant interest groups, with the attendant weakening of many decisions but also relatively good implementation; and
- focused on qualitative rather than quantitative techniques, eschewing high-tech methods and leading to short, quite readable EISs.

Chapter 8 broadly addresses Sadler's first two points in sequence. Section 8.2 considers the number, type and location of projects for which EIAs have been carried out in the UK since mid-1988, as well as where the resulting EISs can be found. Section 8.3 discusses the stages of EIA before the submission of the EIS and application for authorization. Section 8.4 addresses what has, to date, been the most heavily studied aspect of EIA practice, the quality of EISs. Section 8.5 considers the post-submission stages of EIA, and how environmental information is used in decision-making by LPAs and inspectors. Finally, Section 8.6 discusses the costs and benefits of EIA as seen from various perspectives. Sadler's third point is partially addressed by government-published good-practice guides on EIA preparation and review (DETR 1997a and DoE 1994, 1995, 1996), which reflect a first cycle of limited policy changes by the UK government in response to early research findings regarding EIS and EIA effectiveness.

The information in this chapter was correct at the time of writing in 2004; it will obviously change as more EISs are carried out.

8.2   Number, type and location of EISs and projects

In the absence of a central government lead in maintaining a comprehensive database of EISs, several organizations have begun to establish such databases (e.g. IEA 1993, Wood & Bellanger 1998). This section considers how many EISs have been produced, for which projects and developers, and where. It concludes with a brief review of where collections of EISs are kept.

This analysis is complicated by several problems. First, some projects fall under more than one schedule classification, for example mineral extraction schemes (Schedule 2.2) that are later filled in with waste (Schedule 2.11), or industrial/residential developments (Schedule 2.10) that also have a leisure component (Schedule 2.12). Second, the mere description of a project is often not enough to identify the regulations under which its EIA was carried out. For instance, power stations may fall under Schedule 1.2 or 2.3a depending on size. Roads may come under highways or planning regulations depending on whether they are trunk roads or local highways. Third, many EISs do not mention when, by whom or for whom they were prepared. Fourth, locational analysis after 1995 is complicated by local government reorganization and many changes in the nature and boundaries of authorities in England, Scotland and Wales. All these factors affect the analysis. This chapter is based primarily on information from Wood & Bellanger (1998), but their findings are very similar to others (e.g. Wood 1996, 2003).

8.2.1   Number of EISs

Between the mid-1970s and the mid-1980s, approximately 20 EISs were prepared annually in the UK (Petts & Hills 1982). After the implementation of Directive 85/337, this number rose dramatically and, despite the recession, about 350 EISs per year were produced in the early 1990s; but, as can be seen from Figure 8.1, this number began to drop in the mid-1990s partly as a result of a fall in major development activity under the planning regulations. However, the numbers quickly recovered in the late 1990s and, as noted in Chapter 3, there has been over 600p.a. since the implementation of the amended Directive. This probably reflects many factors – more projects

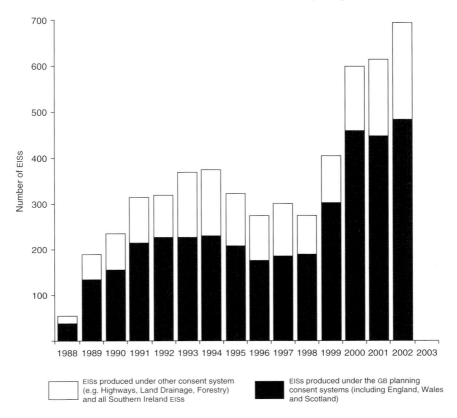

Figure 8.1 EISs prepared in the UK, 1988–2002. (*Source*: ODPM, DETR, Oxford Brookes University IAU statistics, Manchester University EIA Centre statistics.)

included in the amended Directive, a stronger UK economy and concern by developers and LPAs about certain court judgments involving the EIA Directive. By the end of 2003, approximately 6,000 EISs had been prepared, with approximately 70 per cent produced under the Planning Regulations for England, Wales and Scotland. The remainder are for projects in Northern Ireland and, more significantly, for projects under the other consent procedures (e.g. highways, forestry) as discussed in Chapter 3.

In parallel with the increase in the number of EISs, the participants in EIA have become increasingly familiar with the process. Surveys of UK local authorities carried out by Oxford Brookes University in the mid-1990s showed that over 80 per cent of LPAs even then had received at least one EIS. On average, strategic-level authorities (county and regional councils and national park authorities) had received 12 EISs and local-level authorities (district, borough, metropolitan boroughs and development corporations) had received four. Surveys of environmental consultants (e.g. Radcliff & Edward-Jones 1995, Weston 1995) found that about one-third of the consultancies surveyed had prepared 10 or more EISs. As noted, the total number of

EISs is now approximately 6,000, compared with approximately 2,500 by the end of 1995, and LPA and consultancy activity and experience with the process has continued to grow accordingly.

8.2.2 Types of projects

Figure 8.2 shows the types of projects for which EISs have been prepared. The largest numbers are for project types in waste,[1] urban/retail developments, roads, extraction schemes and energy projects (Wood & Bellanger 1998). Less than 10 per cent of these projects were Schedule 1 projects, primarily toxic waste disposal installations, power stations and motorways (Wood 2003).

Although these ratios have remained broadly steady over the years, some project types show clear trends and the incidence of EIS activity provides an interesting barometer of development activity and associated policies (Figures 8.3a–d). For instance, in response to privatization and the "dash for gas" the number of EISs for combined-cycle gas turbine power stations peaked at about ten per year (1989–93), falling back to about three per year (1994–98). EISs for roads increased steadily from about 30 in 1989 to about 80 in 1993 as a result of the roads programme, but fell back in the mid–late 1990s, given restrictions on government funding of road construction and policy trends towards traffic management. EISs for incinerators and wastewater treatment plants grew rapidly until 1992, as the government worked to meet EC quality standards, and then levelled off or fell back. The number of business park EISs dropped sharply in the early 1990s, as the recession affected speculative development (Frost & Wenham 1996, Wood & Bellanger 1998).

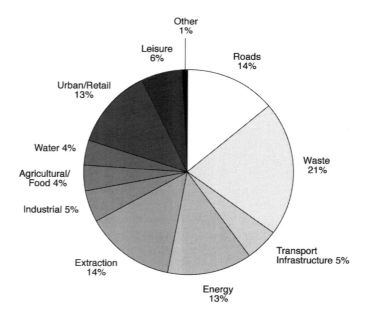

Figure 8.2 Trends in EISs for particular project types (1988–98). (*Source:* Wood & Bellanger 1998.)

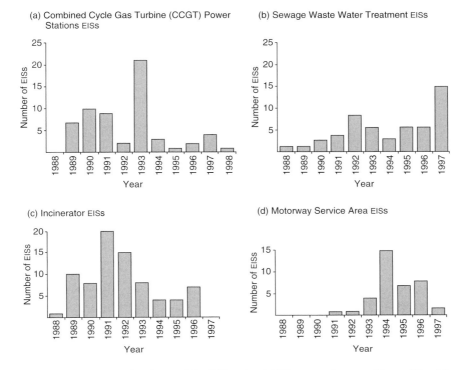

Figure 8.3 EISs prepared in the UK, July 1988 to April 1998: types of project. (*Source*: Wood & Bellanger 1998.)

In the first few years following the implementation of Directive 85/337, 40 per cent of EISs were produced for the public sector and 60 per cent for the private sector (Wood 1991). The percentage of private sector projects has increased somewhat owing to privatization, but much of this has been balanced by the heavy government investment in – and consequently EISs for – new roads. A particularly interesting subset is that of the 10 per cent of EIAs for which one agency acts as both the project proponent and the competent authority (e.g. the Highway Agency for roads, the Forestry Agency for afforestation).

8.2.3 Location of projects

Figure 8.4 shows the distribution of known EISs (1988–98) in England, Scotland and Wales by county or region. Generally, more is known about English and Scottish than about Welsh and Northern Irish EISs, so the figure may understate the situation in Wales and Ireland. By the end of 1998, the most EISs – about 170 – had been prepared for projects in Kent: the Channel Tunnel, the M20 "missing link" and the Dartford Crossing have spawned proposals for major secondary projects, including

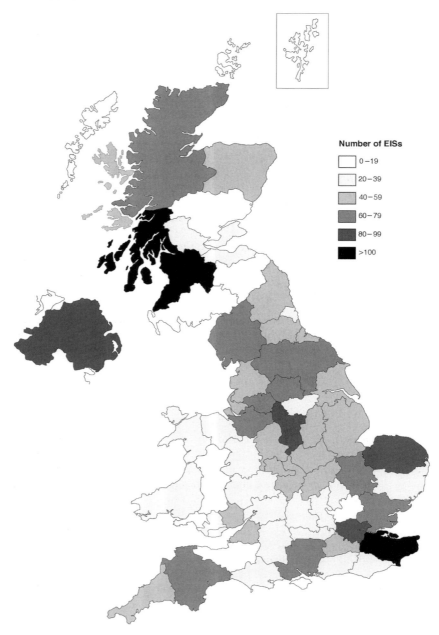

Figure 8.4 EISs prepared in the UK, July 1988 to April 1998: location of projects. (*Source*: Wood & Bellanger 1998.)

mixed-use, road and waste-disposal projects. The authorities in the major conurbations have also received a relatively large number of EISs. In Greater London (about 95) many of them were for road and rail schemes. In Greater Manchester (70) and the West Midlands (50), light rail and waste disposal also feature prominently. Many EISs (105) have been prepared in Strathclyde, largely for extraction, incinerator and leisure proposals. Recently several wind-farm EISs have also been submitted. The areas with the fewest EISs are in southern Scotland, mid-Wales, the northern Home Counties and Somerset.

The types of development vary considerably between regions, reflecting differences in the local economic bases. Thus, certain areas show concentrations of particular project types, for instance opencast coal schemes in Derbyshire and the northern English counties, power stations in Greater London and Humberside, wind farms in Wales and Cornwall, afforestation schemes in the Scottish Highlands and agricultural projects in Lincolnshire and Shropshire.

8.2.4 Sources of EISs

When an LPA receives an EIS, it is required to send a copy to the government office in its region, which then forwards it to the Office of Deputy Prime Minister (ODPM) library in London once the application has been dealt with. However, this process can be a long one. The ODPM library is open to the public by appointment; photo-copies can be made on the premises. In Wales, planning EISs are forwarded to the Welsh Assembly. In Scotland, all EISs are sent to the Scottish Assembly, while in Northern Ireland they are sent to the Northern Ireland DoE. Other government agencies, such as the Highways Agency and the Forestry Agency, also hold collections and lists of the EISs that fall under their jurisdiction. These collections are, however, generally not publicly available, although limited access for research purposes may be allowed.

The Institute of Environmental Management and Assessment (IEMA), based in Lincoln, has a collection of over 600 EISs, which are available by pre-arrangement with institute staff and can also be mailed on a loan basis to members. It has also published a *Digest of Environmental Statements* (IEA 1993), which provides compre-hensive summaries of 1,800 EISs. The EIA Centre at the University of Manchester keeps a large database of EISs and EIA-related literature: its collection of over 500 EISs is, like its database, open to the public, by appointment. Oxford Brookes University's collection of approximately 900 EISs is open to the public, by appointment, and photo-copies can be made on the premises. The University's IAU publishes an occasional directory of EISs, the latest being by Wood & Bellanger (1998). Other organizations, such as the Royal Society for the Protection of Birds, the Institute of Terrestrial Ecology, EN and the Campaign to Protect Rural England, as well as many environ-mental consultancies, also have limited collections of EISs, but these are generally kept by individuals within the organization for in-house use only, and are not available to the public.

The difficulty of finding out which EISs exist and their often prohibitive cost make the acquisition and analysis of EISs arduous. Various organizations, e.g. IEMA, the University of Manchester and Oxford Brookes University, have called for one central repository for all EISs in the UK.

8.3 The pre-submission EIA process

This is the first of three sections which discuss how EIAs are carried out in practice in the UK. It focuses on some of the pre-EIS submission stages of EIA, namely screening, scoping and pre-submission consultation.

8.3.1 Screening

Underpinning any analysis of the implementation of EIA in the UK are the requirements of the EC and UK government legislation. Under the original legislation, competent authorities in the UK were given wide discretion to determine which Schedule 2 projects require EIA within a framework of varying criteria and thresholds established by the 40-plus regulations and additional guidance. Generally this screening process worked quite well (CEC 1993). However, some specific problems did arise regarding screening in the UK. First, because of the largely discretionary system for screening, LPAs often – about half of the time – require an EIS to be submitted only after they receive a planning application (DoE 1996). For the same reason, screening requirements varied considerably between competent authorities. For instance, a 1991 search of 24 LPA returns, registers and files revealed 30 projects in 12 authorities for which EIAs could have been required but were not (DoE 1991). Most of these types of development (e.g. 6 mineral extraction schemes, 2 landfills and 17 mixed-use developments) had been subject to EIA in other local authorities. The decision not to require an EIA had mostly been taken by junior members of staff who had never considered the need for an EIA, or who thought (incorrectly) that no EIA was required if the land was designated for the type of use specified in the development plan, or if the site was being extended or redeveloped rather than newly developed. Similarly, different DoE regional offices have given different decisions on appeals for what are essentially very similar developments (Gosling 1990).

The new screening criteria established by the amendments to the Directive seek to reduce these problems post-1999 (see Sections 3.4 and 4.3). An ODPM-sponsored study, undertaken by the Oxford Brookes University's IAU (IAU 2003), has provided some research evidence on the nature and characteristics of LPA screening decision-making under the T&CP (EIA) Regulations 1999. The research, based on survey responses from over 100 LPAs in England and Wales in 2002, sought information on frequency of screening activity, on main considerations in LPA decision whether or not an EIA should be undertaken and on the importance of different screening criteria (with regard to the most recent project screened). For screening activity in general, and for LPAs in total, the indicative thresholds were identified as the main consideration in screening decision-making (44 per cent of LPAs). Schedule 3 and Circular 2.99 considerations (project is a major development of more than local importance, is in a sensitive location or will have complex/hazardous environmental effects), which require a greater degree of professional judgement, were noted second most frequently (35 per cent of LPAs). Tables 8.1 and 8.2 relate to views on the LPAs' most recent single screening decision. Using regulations and thresholds is seen as the most effective approach overall, but professional judgement is an equally important approach amongst the more experienced LPAs (Table 8.2). The most important factors in the screening decision are the size and scale of the project with nearly 87 per cent of LPAs indicating that these are "important" or "very important". Proximity to sensitive environmental receptors (76 per cent) and the nature of the project (74 per cent) are the next most important factors. At the other extreme, only 15 per cent indicated that risk of accidents was important or very important. The main constraints on screening decision-making were identified as lack of resources (45 per cent), time-frame constraints (44 per cent), lack of clarity of the regulations (33 per cent) and uncertainty (over baseline data, project characteristics, etc.) (32 per cent). Overall, the findings show that whilst thresholds are clearly important

Table 8.1 Most effective approach in most recent screening decision

Screening approach considered most effective	LPA (%) N = 56	≤5 EIAs (n = 26)	>5 EIAs (n = 30)
Consultation with own organization	3.6	7.7	0.0
Community consultation	12.5	11.5	13.3
Asked for screening direction from SoS	0.0	0.0	0.0
Followed screening guidance in local development plan	1.8	3.8	0.0
Followed guidance in other plans/policies	7.1	11.5	3.3
Used regulations and thresholds as guide	35.7	38.5	33.3
Consulted examples of similar projects	1.8	0.0	3.3
Used professional judgement/experience	26.8	19.2	33.3
Used checklist to identify possible impacts	1.8	0.0	3.3
Used other formal technique	1.8	3.8	0.0
Own standard approach	0.0	0.0	0.0
Likely controversy of project	0.0	0.0	0.0
Other	7.1	3.8	10.0

(*Source:* IAU 2003.)

Table 8.2 Importance of issues in most recent screening decision

Issue (n = 97)	Very imp. (%)	Imp. (%)
Size/scale of project	47	40
Proximity to receptor	44	43
Nature of project	42	32
Traffic/access impacts	33	32
Ecological impacts	32	31
Emissions	31	31
Landscape impacts	26	35
Cumulative impacts	20	26
Economic impacts	6	24
Social impacts	5	22
Controversy/concern	9	16
Risk of accidents	5	10
Other	3	1

(*Source:* IAU 2003.)

in the screening decision, they are often conditioned by professional judgement. In other words, in themselves, they do not provide sufficient justification for a screening decision (Weston 2000); as stated in Circular 02/99, "The fundamental test to be applied in each case is whether *that* particular type of development and its specific impacts are likely, in *that* particular location, to result in significant effects on the environment" (DETR 1999b).

Finally, EIA in the UK is an all-or-nothing process: either an EIA is needed or it is not. This is in contrast with some other countries (e.g. Peru and China, Chapter 10), where a brief environmental study is carried out to determine whether a full-scale EIA is needed. In the UK, where the provision of environmental information with planning applications was the norm before the implementation of Directive 85/337, many developers still voluntarily submit environmental documents without specifying whether these are EISs or not. Some competent authorities treat these documents as EISs, with the attendant requirements for consultation and publicity, but in other cases they simply treat them as additional information (Hughes & Wood 1996).

8.3.2 Scoping and pre-submission consultation

Competent authorities also have much discretion to determine the scope of EIAs. As we discussed in Chapter 3, the original Directive 85/337's Annex III was interpreted in UK legislation as being in part mandatory and in part discretionary. Table 8.3 shows the type of information included in early EISs, based on a survey of 100 EISs prepared before 1990 (Jones et al. 1991). It shows that although the mandatory

Table 8.3 EISs prepared in the UK July 1988 to December 1989: comprehensiveness

Type of information	EISs including information (%)
Specified information	
Description of proposed development	93
Data to identify and assess the main environmental effects	76
Description of likely significant effects	88
With reference to	
human beings	75
flora/fauna	85
soil/geology	51
water	65
air	54
climate	24
landscape	91
interaction between them	14
material assets/cultural heritage	48
Description of mitigating measures	75
Non-technical summary	67
Additional information	
Physical characteristics of proposed development	
construction	51
operation	74
Residues and emissions from the development	
With reference to	
water	63
air	54
soil	29
noise	68
vibration	17
light	9
heat	2
radiation	1
Outline of main alternatives studied	34
Forecasting methods used	45
Difficulties in compiling information	4

(*Source:* Jones et al. 1991.)

requirements of the legislation were generally carried out, the discretionary elements (e.g. the consideration of alternatives, forecasting methods, secondary and indirect impacts, scoping) were, understandably, carried out less often.

Although early scoping discussions between the developer, the consultants carrying out the EIA work, the competent authority and relevant consultees are advised in government guidance and are increasingly considered vital for effective EIA (Jones 1995, Sadler 1996), in practice, pre-submission consultation has been carried out sporadically. For instance, a survey of environmental consultants (Weston 1995) showed that only 3 per cent had been asked to prepare their EISs before site identification, and 28 per cent before detailed design. LPAs were consulted by the developer before EIS submission in between 30 and 70 per cent of cases, although this has increased (DoE 1996, Lee et al. 1994, Leu et al. 1993, Radcliff & Edward-Jones 1995, Weston 1995). A survey by Weston (1995) showed that EN was consulted before EIS submission in about 50 per cent of the cases, the (then) National Rivers Authority in about 40 per cent, the (then) Countryside Commission in about 25 per cent, and Her Majesty's Inspectorate of Pollution (HMIP) only rarely. Other studies (DoE 1996, Pritchard et al. 1995) also showed that very limited consultation with statutory or non-statutory consultees or the public occurred at this stage, although where extensive consultation had been carried out, project design was often modified significantly before the submission of the planning application (Pritchard et al. 1995).

However, even early consultation does not necessarily mean that the consultees will be satisfied with the outcome. For instance, in some cases groups have lodged objections to planning applications despite having been consulted. In particular, consultees from whom the developer has requested information before EIS submission may expect the EIS to cover more than just the data that they have provided (DoE 1996). Similarly, consultation may be widespread but may avoid organizations that could be hostile to the project (Pritchard et al. 1995).

As noted in Chapters 2 and 3, the amended Directive and subsequent UK regulations have raised the profile of scoping in the EIA process. The ODPM study noted earlier (IAU 2003, Wood 2003) also carried out research on the nature and characteristics of scoping activities by LPAs, consultants and statutory consultees. Nearly 75 per cent of the LPAs had been involved in producing scoping opinions. All three sets of stakeholders ranked very high the preliminary assessment of characteristics of the site, consideration of mitigation and consideration of impact magnitude in formulating the scoping opinion/report. Similarly, all ranked professional judgement and consultation within own organization as key approaches to impact identification; use of legal regulations and thresholds were also very important for LPAs and consultancies, but much less so for statutory consultees. Table 8.4 shows the issues of most concern in the most recent scoping project for each group of participants in the process. There is considerable similarity in emphasis between the LPAs and consultancies, with traffic/transport, landscape/visual and flora/fauna issues ranking particularly high. For statutory consultees, there are some similarities, but other environmental issues (e.g. other emissions, waste disposal) come more into play. Climatic factors and risk of accidents (health and safety) rank surprisingly low, reflecting perhaps uncertainty about what are seen as more long-term and less predictable issues.

Considerable experience has been gained with screening and scoping. After initial hiccups, the screening process now seems to be relatively well accepted and has been refined after the 1999 amendments. Scoping is generally considered to be a very valuable

Table 8.4 Ranking of issues of major concern in most recent project scoping opinion/report (% ranking of major concern)

	LPAs (n = 78)	Consultancies (n = 98)	Statutory consultees (n = 28)
Social issues	10	20	11
Culture/heritage	19	29	18
Economic	21	22	4
Flora/fauna	46	50	68
Soil	18	20	21
Air quality	35	21	14
Noise and vibration	42	45	11
Other omissions	26	31	39
Climatic factors	6	5	7
Waste disposal	13	19	32
Water resources	37	31	43
Geo-technical issues	18	21	29
Landscape/visual	58	57	32
Traffic/transportation	47	55	11
Risk of accidents	5	9	7
Inter-relationships of above	18	17	43
Others	3	9	4

(*Source:* IAU 2003, Wood 2003.)

and cost-effective part of EIA by all those concerned, and again, following the 1999 amendments, has increased in significance in the UK EIA process.

8.4 EIS quality

As we mentioned in Section 8.1, the preparation of high-quality EISs is one component of an effective translation of EIA policy into practice. Two schools of thought exist about the standards that should be required of an initial EIS. Some argue that developers should be encouraged to submit EISs of the highest standard from the outset. This reduces the need for costly interaction between developer and competent authority (Ferrary 1994), provides a better basis for public participation (Sheate 1994), places the onus appropriately on the developer and increases the chance of effective EIA overall. Others argue that it is the entirety of environmental information that is important, and that the advice of statutory consultees, the comments of the general public and the expertise of the competent authority can substantially overcome the limitations of a poor EIS (Braun 1993). This view is also supported by planning inspectors at appeal and judicial review cases.

Environmental impact statement quality in the UK is affected by the limited legal basis for EIA and by the facts that planning applications cannot be rejected if the EIS is inadequate, that (to date) some crucial steps of the EIA process (e.g. public participation, the consideration of alternatives (until 1999) and monitoring) are weak and often not mandatory and that developers undertake EIAs for their own projects. This section first considers the quality of EISs produced in the UK, based on several academic studies. It continues with a brief discussion of other perceptions of EIS quality, since competent authorities, statutory consultees and developers require different things from EIA and may thus have different views of EIA quality. It concludes with a discussion of factors that may influence EIS quality.

8.4.1 Academic studies of EIS quality

Academic studies of EIS quality can broadly be classified as aggregated or disaggregated. Aggregated studies consider the quality of a number of EISs overall, where the EISs either represent the total population of EISs or a specific subgroup (e.g. type of project). Disaggregated approaches focus on the quality of the treatment of individual EIS topic areas (e.g. landscape or noise), or performance with respect to certain EIS components (e.g. baseline data, the consideration of alternatives) or their presentation (Lee et al. 1994).

Researchers from the University of Manchester have studied aggregated approaches to EIS quality over the years, using the Lee & Colley criteria (1990) (see Appendix 3). Based on these criteria, EISs have been divided into "satisfactory" (i.e. marks of A, B or C) and "unsatisfactory" (D or below). Table 8.5 summarizes some of the findings. It shows EIS quality to be increasing after dismal beginnings, with about half of recent EISs being satisfactory.

A study carried out by Oxford Brookes University's IAU for the DoE compared 25 EISs prepared before 1991 with matched[2] EISs prepared after 1991, on the basis of different sets of criteria, including simple "regulatory requirements" and comprehensive criteria devised by the IAU. For the simple regulatory requirements, 44 per cent of the post-1991 EISs fulfilled all the nine criteria used, compared with 36 per cent of the pre-1991 EISs. A more detailed analysis indicated that 92 per cent of the post-1991 EISs fulfilled six or more of the criteria, compared with 64 per cent of the pre-1991 criteria. However, this review framework (Table 8.6), with simple yes/no grading and a very limited list of criteria, could be regarded as providing a crude and perhaps over-harsh review of quality. Using the more comprehensive range of criteria established by the IAU (Table 8.7 and Section 6.5), the quality of EISs rose from just unsatisfactory (D) before 1991 to just satisfactory (C) after 1991. The percentage of satisfactory EISs[3] increased from 36 to 60.

Other studies have focused on specific project types: for instance Kobus & Lee (1993) and Pritchard et al. (1995) reviewed EISs for extractive industry projects,

Table 8.5 Aggregated EIS quality, % satisfactory*

	*Authors and year of study***				
	Lee & Colley (1990)	Wood & Jones (1991)	Lee & Brown (1992)	Lee et al. (1994)	Jones (1995)
Year(s) EISs were prepared					
1988–89	25	37	34	17	
1989–90			48		
1990–91			60	47	
1988–93					"just over half"

* Satisfactory means marks of A, B or C based on the Lee & Colley criteria (1990 or 1992).
** No. of EISs analysed: Lee & Colley, 12; Wood & Jones, 24; Lee & Brown, 83; Lee et al., 47; Jones, 40.

Table 8.6 Disaggregated EIS quality based on simple "regulatory requirements", % covered

Criterion	25 pre-1991 EISs	25 post-1991 EISs
1. Describes the proposed development, including its design and size or scale.	76	84
2. Defines the land areas taken up by the development site and any associated works, and shows their location on a map.	76	92
3. Describes the uses to which this land will be put and demarcates the land use areas.	68	92
4. Considers direct and indirect effects of the project and any consequential development.	60	80
5. Investigates these impacts in so far as they affect human beings, flora, fauna, soil, water, air, climate, landscape, interactions between the above, material assets and cultural heritage.	56	76
6. Considers the mitigation of all significant negative impacts.	68	92
7. Mitigation measures include the modification of the project, the replacement of facilities and the creation of new resources.	60	92
8. There is a non-technical summary, which contains at least a brief description of the project and environment, the main mitigation measures and a description of any remaining impacts.	64	80
9. The summary presents the main findings of the assessment and covers all the main issues raised.	52	72
All criteria	36	44

(*Source*: DoE 1996.)

Table 8.7 Disaggregated EIS quality based on IAU criteria (average marks)

Criterion	25 pre-1991 EISs	25 post-1991 EISs
1. Description of the development	C/D	C
2. Description of the environment	C/D	C
3. Scoping, consultation and impact identification	D	C/D
4. Prediction and evaluation of impacts	D	C
5. Alternatives	E	D
6. Mitigation and monitoring	C/D	C/D
7. Non-technical summary	D	C/D
8. Organization and presentation of information	C/D	C
Overall mark	D	C
% satisfactory (A–C)	36	60
% marginal (C–D)	12	4
% unsatisfactory (D–F)	52	36

(*Source*: DoE 1996.)

Prenton-Jones for pig and poultry developments (Weston 1996), Radcliff & Edward-Jones (1995) for clinical waste incinerators and Davison (1992) and Zambellas (1995) for roads. These studies also broadly suggest that EIS quality is not very good, but improving.

In terms of disaggregated approaches, Lee & Dancey (1993) analysed 83 EISs and found 60 per cent to be satisfactory in terms of their description of the development, local environment and baseline conditions, 36 per cent in terms of identification of key impacts, 47 per cent in terms of alternatives and mitigation, and 49 per cent in terms of communication and presentation of results. Nelson (1994), Pritchard et al. (1995) and Jones (1995) made broadly similar findings. Tables 8.6 and 8.7 for the Oxford Brookes University DoE study noted earlier show the results of an analysis of the 25 matched pairs of EISs (1996) based on, respectively, the simple "regulatory requirements" and the more comprehensive IAU criteria. Coverage of each of the regulatory requirement criteria improved over time, from an average of about two-thirds before 1991 to more than 80 per cent since 1991. Based on the IAU criteria, quality in general also rose significantly between 1988 and 1990 and between 1992 and 1994, with improvements in each of the eight main categories of assessment. The EISs' description of monitoring and mitigation improved only marginally, but the other categories generally improved by about half of a mark (e.g. from D to C/D). A particular improvement was seen in the approach to alternatives. Of the 25 EIS pairs, 15 showed an improvement in quality, while nine became worse (DoE 1996).

Other studies have analysed the quality of specific EIS environmental components, for instance, landscape/visual impacts (e.g. Mills 1994) and socio-economic impacts (e.g. Hall 1994). These show findings similar to those discussed above. In sum, although both aggregated and disaggregated studies by academics show a continued and pleasing improvement in EIS quality, there must still be concern that many EISs, from between one-third to one-half depending on the criteria used, are still not satisfactory, and in several cases poor.

8.4.2 Quality for whom?

These findings must, however, be considered in the wider context of "quality for whom?" Academics may find that an EIS is of a certain quality, but the relevant planners or consultees may perceive it quite differently. For instance, the DoE (1996) study, Radcliff & Edward-Jones (1995), and Jones (1995) found little agreement about EIS quality between planners, consultees and the researchers; the only consistent trend was that consultees were more critical of EIS quality than planners were.

In interviews conducted by the IAU (DoE 1996), planning officers thought EISs were intended to gain planning permission and minimize the implication of impacts. Just over 40 per cent felt that EIS quality had improved, although this improvement was usually only marginal. Most of the others felt that this was difficult to assess when individual officers see so few EISs and when those they do see tend to be for different types of project, which raises different issues. A lack of adequate scoping and discussion of alternatives was felt to be the major problem. EISs were seen to be getting "better but also bigger". Some officers linked EIS quality with the reputation of the consultants producing them, and believed that the use of experienced and reputable consultants is the best way to achieve good quality EISs.

Statutory consultees differed about whether EIS quality was improving. They generally felt that an EIS's objectivity and clear presentation were important and

Table 8.8 Changing perceptions of participants of quality of EISs in relation to particular criteria (%)

	Good		Marginal		Poor	
	Post-1991	*Pre-1991*	*Post-1991*	*Pre-1991*	*Post-1991*	*Pre-1991*
Comprehensiveness	31	55	31	27	38	18
Objectivity	18	41	37	35	45	24
Clarity of information	25	55	56	38	19	7

(*Source:* DoE 1996.)

were improving, yet still wanting. Table 8.8 from the DoE study (1996) indicated that participants (LPAs, developers, consultants and consultees) generally thought the key EIS criteria of comprehensiveness, objectivity and clear information were improving. Yet it is interesting to note that only about 40 per cent of interviewees regarded the objectivity of the more recent EISs as good.

Developers and consultants link EIS quality with ability to achieve planning permission. Consultants felt that developers are increasingly recognizing the need for environmental protection and are starting to bring in consultants early in project planning, so that a project can be designed around that need. One reason for this improvement may be that pressure groups are becoming more experienced with EIA, and thus have higher expectations of the process (DoE 1996).

8.4.3 Determinants of EIS quality

Several factors affect EIS quality including the type and size of a project, and the nature and experience of various participants in the EIA process. Certain types of *project* have been associated with higher quality EISs. For instance, Schedule 1 projects, which generally have a high profile and attract substantial attention and resources, are likely to have better EISs. Better EISs have been linked with projects coming under the electricity and pipeline EIA regulations, the Scottish EIA regulations (Lee & Brown 1992) and the post-1993 highways regulations (Zambellas 1995) and, within the planning regulations, with wind farms, waste-disposal and treatment plants, sand and gravel extraction schemes and opencast coal projects (DoE 1996). Larger projects generally have more satisfactory EISs than smaller projects, as is shown in Table 8.9.

Regarding the *nature and experience of the participants* in the EIA process, EISs produced in-house by developers are generally of poorer quality than those produced by outside consultants: the DoE (1996) study, for instance, showed that EISs prepared in-house had an average mark of D/E, while those prepared by consultants averaged C/D, and those prepared by both B/C. Lee & Brown's (1992) analysis of 83 EISs concluded that 57 per cent of those prepared by environmental consultants were satisfactory, compared with only 17 per cent of those prepared in-house. Similarly, EISs prepared by independent applicants tend to be better (C/D) than those prepared by local authorities for their own projects (D/E) (DoE 1996).

The experience of the developer, consultant and competent authority also affects EIS quality. For instance, Lee & Brown (1992) showed that of EISs prepared by developers (without consultants) who had already submitted at least one EIS 27 per cent were satisfactory, compared with 8 per cent of those prepared by developers with no

Table 8.9 Project size vs. EIS quality, % satisfactory*

	*Lee & Brown 1992***	DoE 1996**
Project size***		
Small	20	50
Medium	35	54
Large, very large	50–65	64

* Satisfactory means marks of A, B or C based on the Lee & Colley (1990 or 1992) criteria.
** Lee & Brown reviewed 83 EISs, DoE 50.
*** Small is defined as <75% of the threshold size used to determine whether EIA is needed (DoE 1989), large as >125% of the threshold size, and medium as between the two.

prior experience; Kobus & Lee (1993) cited 43 and 14 per cent respectively. A study by Lee & Dancey (1993) showed that of EISs prepared by authors with prior experience of four or more, 68 per cent were satisfactory compared with 24 per cent of those with no prior experience. The DoE (1996) study showed that of the EISs prepared by consultants with experience of five or fewer, about 50 per cent were satisfactory, compared with about 85 per cent of those prepared by consultants with experience of eight or more. EISs prepared for local authorities with no prior EIS experience were just over one-third satisfactory, compared with two-thirds for local authorities with experience of eight or more (DoE 1996).

Other determinants of EISs' quality include the availability of EIA guidance and legislation, more guidance (e.g. DoE 1995, DoT 1993, local authority guides such as those of Kent and Essex (Essex Planning Officers' Association 2001)) leading to better EISs; the stage in project planning at which the development application and EIA are submitted, EISs for detailed planning applications generally being better than those for outline applications; and issues related to the interaction between the parties involved in the EIA process, including commitment to EIA, the resources allocated to the EIA and communication between the parties. There are also more EISs in the public domain to provide evidence of good practice.

Environmental impact statement length also shows some correlation with EIS quality. For instance, Lee & Brown (1992) showed that the percentage of satisfactory EISs rose from 10 for EISs less than 25 pages long to 78 for those more than 100 pages long. In the DoE (1996) study, quality was shown to rise from an average of E/F for EISs of less than 20 pages to C for those of over 50 pages. However, as EISs became much longer than 150 pages, quality became more variable: although the very large EISs may contain more information, their length seems to be a symptom of poor organization and coordination. Cashmore et al. (2002) had similar findings from a study of EISs in a case study in Greece, with a correlation between length and quality – the average length of a satisfactory EIS being 203 pages and of an unsatisfactory EIS being 63 pages.

8.5 The post-submission EIA process

After a competent authority receives an EIS and application for project authorization, it must review it, consult with statutory consultees and the public and come to a decision about the project. This section covers these points in turn.

8.5.1 Review

Interviews with local authority planners show that planning officers see little difference between projects subject to EIA and other projects of similar complexity and controversy: once an application is lodged, the development process takes over. Competent authorities usually review EISs using their own knowledge and experience to pinpoint limitations and errors: the review is carried out primarily by reading through the EIS, consulting with other officers in the competent authority, consulting externally and comparing the EIS with the relevant regulations.

Despite the ready availability of the Lee & Colley review criteria (1990), only about one-third of local authorities use any form of review methods at all, and then usually as indicative criteria, to identify areas for further investigation, rather than in a formal way. About 10–20 per cent of EISs are sent for review by external consultants or by the IEA; but even when outside consultants are hired to appraise an EIS, it is doubtful whether the appraisal will be wholly unbiased if the consultants might otherwise be in competition with each other. There are also problems involved in getting feedback from the reviewing consultants quickly enough, given the tight timetable for making a project determination. An innovative approach being used by some developers requires consultants who are bidding to carry out an EIA to include as part of their bid an "independent" peer reviewer who will guarantee the quality of the consultants' work.

Various studies (e.g. Jones 1995, Kobus & Lee 1993, Lee et al. 1994, Weston 1995) suggest that LPAs require additional EIA information in about two-thirds of cases. This is usually done informally, without invoking the regulations. The 1999 regulations do provide for some tightening-up of EIA procedures in relation to EIA information. Where an EIA is required, the developer must now produce an EIS that includes at least the information listed in Part 2 of Schedule IV of the regulations and any relevant information listed in Part 1 of Schedule IV as they can be "reasonably required to assess the environmental effects of the project" (DETR 1999a). The Circular 2/99 reinforces the point with an important change of emphasis: ". . . if a developer fails to provide enough information to complete the ES, the application can be determined only by a refusal" (Regulation 3, DETR 1999b). As Weston (2002) notes, . . . "This effectively removes the ability LPAs had under the 1988 regulations to accept ESs that did not include the 'minimum requirements'".

8.5.2 Post-submission consultation and public participation

Competent authorities generally rely heavily on statutory and non-statutory consultees to review the different elements of an EIS, and in most cases the comments received are "substantial or very substantial" (Kobus & Lee 1993). More recent evidence (Kreuser & Hammersley 1999) confirms that planning officers continue to "place great reliance on the consultees to review, verify and summarise at least parts of ESs". Where the EIS contains insufficient information about a specific environmental component, competent authorities often put the developer and consultee in direct contact with each other rather than formally require further information themselves (DoE 1996).

Although there were early problems when EISs were not sent to the consultees (DoE 1991, Lee & Brown 1992), these seem to have mostly been ironed out (Jones 1995). In particular, EN and the Countryside Agency seem to participate quite actively at this stage of EIA, as well as local interest groups. However, some LPAs are not consulting all statutory consultees despite regulatory requirements to do so, possibly because they feel that a proposed project does not affect their area of interest (Pritchard et al. 1995). The Environment Agency also has little reason to carry out extensive consultation as part of the EIS process:

> HMIP, the NRA and the waste regulatory authorities (which have since been merged to form the Environment Agency) require impact assessments to be supplied with pollution permit applications. Therefore in their role as statutory planning EIS consultees, HMIP and the NRA are unlikely to waste time complaining about the poorly detailed designs given in a planning EIS, if they will be receiving another type of EIA document which precisely covers their area of concern. The Didcot B case study showed that even though HMIP considered the EIS to be satisfactory, they later demanded major design changes. (In the case of the Hamilton Oil gas terminal project in Liverpool Bay) HMIP raised no objectives to the EIS, but then rejected the IPC authorization on the grounds that design neither met the requirements of BATNEEC nor represented the BPEO. (Bird 1996)

This problem of duplicate authorization procedures and the issues relating to discussion between EIA participants will be discussed further in case studies in Chapter 9. The CEC (1993) feels that the UK situation:

> is satisfactory concerning the publication of ESs and their availability for consultation once they have been submitted. Copies can, in most cases, be obtained from either the developer or the competent authority concerned. Where the information was available to the EIA Centre, just under half of 290 ESs were available free of charge, with 18% available for purchase at £20 or less, and the remaining 33% available at more than £20. In most cases copies of ESs are available, particularly in the specific locality where an application for consent is submitted. However, in a few cases copies of ESs are only available for consultation, but not for purchase by the public.

In general, however, public participation in the UK EIA system is often very partial, limited to a short period following EIS submission. The public concerned is not defined, and notice of an EIS is only given locally (CEC 2003). For other wider aspects of public consultation and participation, the reader is referred to Section 6.2.

8.5.3 Decision-making

As we noted in Chapter 1, one of the main purposes of EIA is to help to make better decisions, and it is therefore important to assess the performance of EIA to date in relation to this purpose. It is also important to remember that all decisions involve trade-offs. Wood (2003) identifies some of these, including trade-offs in the EIA process between simplification and complexity, urgency and the need for better information, facts and values, forecasts and evaluation, and certainty and uncertainty. There are also trade-offs of a more substantive nature, in particular between the socio-economic and biophysical impacts of projects – sometimes reduced to the "jobs vs. the environment" dilemma. Box 8.1 illustrates the trade-off issue in relation

Box 8.1 Socio-economic and biophysical impact trade-offs – the example of Newbury Bypass, UK

MIXED FEELINGS GREET GO-AHEAD TO NEWBURY

The Government's shock decision to approve the A34 Newbury Bypass has been greeted with relief by district and county planners, who feel that traffic congestion is paralysing the Berkshire town. Environmental campaigners, however, have reacted angrily to ex-transport secretary Brian Mawhinney's announcement last Wednesday – his final one before being replaced by Sir George Young. Protests on a similar scale to those at Twyford Down are now expected at the site, where construction work could begin before the end of the year.

Last December, Mawhinney said he would delay any decision on the controversial proposal for a year to consider alternatives, and his sudden announcement took both local authorities and environmentalists by surprise. "I had no doubt that the current situation on the A34 was intolerable and that there was strong economic justification for a bypass" he explained. "But I wanted to be sure of the environmental balance between the principal route alternatives and to confirm that the route proposed was the best solution to the problems of congestion in Newbury."

Peter Gilmour, community officer at Newbury Borough Council, called the decision "a triumph for common sense and local democracy". Council planning officers and members had backed the scheme "root and branch", he said – while 13,000 of the town's 27,000 population had signed a petition of support. According to council research 50,000 vehicles a day travel through Newbury, an estimated two thirds of which are "simply passing through the town", said Gilmour. Incidents of asthma are growing among local people, while traffic delays of half-an-hour are common. "People are reluctant to come into the town to shop, while some local industries have moved away because they are unable to transport goods effectively," Gilmour continued. "It's all very well protesters taking the moral high ground, but we have a moral duty to the welfare of people in the town."

Berkshire County Council also welcomed the announcement, which follows years of uncertainty and two public inquiries to decide on the western route. The council's environment committee has been lobbying Mawhinney to back the scheme – arguing that the delay was making transport planning very difficult. "The council's transport strategy generally has been to move away from major roads but we feel this is one of the exceptions", said county environment officer Keith Reed.

He stressed that the council is concerned about potential environmental damage from the scheme, which protesters say will destroy parts of four SSSIs, partly cross a civil war battlefield and pass near to the 14th century Grade 1 listed Donnington Castle and the Watermill Theatre at Bagnor. "We will be pressing the DoT to carry out more archaeological studies, and calling for the use of porous asphalt to cut down on traffic noise levels", said Reed. Environmental campaigners remain furious, nonetheless. Tony Juniper of Friends of the Earth said Mawhinney's decision "makes a mockery of the 'great transport debate'. Local people have been preparing to start local dialogue this month on alternatives to the plan, but will now have their efforts thrown back in their face". "This road must be stopped at all costs. It is one of the most destructive schemes in the national roads programme and will mobilise massive countrywide opposition", Juniper said.

Graham Wynne, RSPB conservation director, said the bypass will destroy part of Snelsmore Common – home to a special community of plants and animals including the rare nightjar. "This decision flies in the face of the Government's stated commitment to a sustainable transport policy. No-one denies that Newbury needs a solution to its traffic problems, but it should not be regardless of the environmental cost." The European Commission is also considering legal action to stop the scheme, which protesters say will break various environmental directives.

(*Source: Planning Week* July 1995.)

to the UK's Newbury Bypass, which generated direct action by aggrieved parties, who sought to influence the project decision.

Some impacts may be more tradable in decision-making than others. Sippe (1994) provides an illustration, for both socio-economic and biophysical categories, of negotiable and non-negotiable impacts (Table 8.10). Sadler (1996) identifies such trade-offs as the core of decision-making for sustainable development.

In the UK there is an important decision-making stage linked most normally to a planning approval process by the competent authority, and involving the consideration of the EIS and associated information. The EIS may have an impact on a planning officer's report, on a planning committee's decision and on modifications and conditions to the project before and after submission. But the impact of EIA on decision-making may be much wider than this, influencing, for example, the alternatives under consideration, project design and redesign, and the range of mitigation measures and monitoring procedures (Glasson 1999). Indeed, the very presence of an effective EIA system may lead to the withdrawal of unsound projects and the deterrence of the initiation of environmentally damaging projects.

In Chapter 3 the various participants in the EIA process were identified. These participants will have varying perspectives on EIA in decision-making. A local planning officer may be concerned with the *centrality* of EIA in decision-making (does it make a difference?), central government might be concerned about *consistency* in application to development proposals across the country; pressure groups may also be concerned with these criteria, but also with *fairness* (in providing opportunities for participation) and *integration* in the project cycle and approval process (to what extent is EIA easily bypassed?). A number of studies have attempted to determine whether EIA and associated consultations have influenced decisions about whether and how to authorize a project.

Table 8.10 Judging environmental acceptability – trade-offs

	Non-negotiable impacts	*Negotiable impacts**
Ecological (physical and biological systems components)	Degrades essential life support systems	No degradation beyond carrying capacity
	Degrades conservation estate	No degradation of productive systems
	Adversely affects ecological integrity	Wise use of natural resources
	Loss of biodiversity	
Social (humans as individuals or in social groupings)	Loss of human life	Community benefits and costs and where they are borne
	Reduces public health and safety unacceptably	Reasonable apportionment of costs and benefits
	Unreasonably degrades quality of life where people live	Reasonable apportionment of inter-generational equity
		Compatibility with defined environmental policy goals

* In terms of net environmental benefits.
(*Source*: Sippe 1994.)

Early surveys of local planning officers (Kobus & Lee 1993, Lee et al. 1994) suggested that EISs were important in the decision in about half of the cases. Interviews with a wider range of interest groups (DoE 1996) found that about 20 per cent of respondents felt that the EIS had "much" influence on the decision, more than 50 per cent felt that it had "some" influence, and the remaining 20–30 per cent felt that it had little or no influence. Jones (1995) found that about one-third of planning officers, developers and public interest groups felt that the EIS influenced the decision, compared with almost half of environmental consultants and only a very small proportion of consultees. For planning decisions, it is the members of the planning committees who make the final decision. Interviews suggest that they are not generally interested in reading the EIS, but instead rely on the officer's report to summarize the main issues (DoE 1996). According to Wood & Jones (1997), planning committees followed officers' recommendations in 97 per cent of the cases they studied.

The consultations related to the EIS are generally seen to be at least as important as the EIS itself (Jones 1995, Kobus & Lee 1993, Lee et al. 1994, Wood & Jones 1997). Figure 8.5 clearly illustrates this point. On the other hand, many interviewees from non-statutory bodies felt excluded from the decision-making process, and one national non-statutory wildlife body complained that if the then Nature Conservancy Council or the Countryside Commission did not object then their own objections went largely ignored (DoE 1996).

While studies of early EISs (e.g. Kobus & Lee 1993, Lee et al. 1994) suggested that material considerations were slightly more important than environmental considerations in the final decision on a project's authorization, a later study

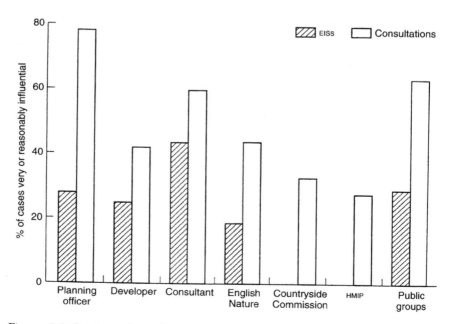

Figure 8.5 Opinions about the influence of EISs and consultations on decisions. *Note*: Consultees include equivalents in Scotland and Wales, in each case. (*Source*: Wood & Jones 1997.)

(Jones 1995) suggested that the environment was the principal factor influencing the decision, with planning policies given slightly less weight. Wood & Jones (1997) report that the environment was seen to be the overriding factor influencing the decisions in 37 per cent of the cases they studied. However, only in a very few cases would the final decision have been different in the absence of an EIS. Nor does the introduction of the requirements of the amended Directive appear to have markedly increased the influence of EIA. From a study of planning appeal cases, Weston (2002) concludes:

(Whilst) . . . Procedurally EIA is much stronger today in the UK than it was in the early years of its implementation – Yet the influence that EIA has on the actual decisions made by LPAs and planning inspectors remains weak, as those decisions are based on a complex web of factors that had evolved long before EIA was introduced. . . . Local authorities in England and Wales deal with around 450,000 applications for planning consent per year. The decisions on those applications are made on the basis of "material considerations" including the local development plan, national planning policy guidance and the results of a formal consultation process. EIA cases make up less than 0.1% of those applications and for the most part the actual decision-making process for those cases will be little different to the other 99.9%.

Overall, in the UK, project applications with EISs are not treated much differently from those without EISs. Although environmental issues are addressed more formally, in a discrete document, the final decision-making process is not changed much by EIA. The main procedural difference brought about by EIA is the need to consult people about the EIS, and the broader scope for public participation (not often used in practice) that it brings. However, the result of the entire EIA process is a modification of projects due to EIA, possibly additional or different conditions on the project, and perhaps a more comprehensive consideration of environmental issues by the competent authority.

8.6 Costs and benefits of EIA

Much of the early resistance to the imposition of EIA was based on the idea that it would cause additional expense and delay in the planning process. EIA proponents refuted this by claiming that the benefits of EIA would well outweigh its costs. This chapter concludes with a discussion of the costs and benefits of EIA to various parties in the UK.

8.6.1 Costs of EIA

Generally, EIA has probably slightly increased the cost to *developers* of obtaining planning permission. An EIS generally costs between 0.000025 and 5 per cent of project costs (Coles et al. 1992). Weston's (1995) survey of consultants showed that consultancies received on average £34,000 for preparing a whole EIS, £40,000 for several EIS sections, and £14,750 for one section: this itself highlights the variability of the costs involved. Another study of 20 EISs showed EIS preparation to vary from 22 person-days at a cost of £5,000 to 3–4 person-months with additional work contracted out (DoE 1996). Pritchard et al.'s (1995) study of eight EIAs found that developers felt that "the preparation of the ES had cost them too much time and money, and that the large amounts of work involved in EA often yielded few tangible benefits in terms of the actual planning decision reached". In its Consultation Paper on the amended Directive, the DETR suggested £35,000 as an appropriate median figure for the cost of undertaking an EIA under the new Regulations (DETR 1997b).

In terms of the delay caused to planning decisions, various studies (e.g. DoE 1991, Lee et al. 1994, Tarling 1991) have shown that the mean time to decide planning applications with EISs was about 40 weeks; but there were wide variations. This is considerably more than for applications without an EIS (DoE 1996), but then the projects with an EIS also tend to be larger, more complex and more politically sensitive. An early study (Coles et al. 1992) found that, on average, the entire EIA process, from the notification of intent for the project to the decision, took 62 weeks, the EIS preparation taking 25 weeks. Although some consultants feel that EIA slows down the decision-making process, imposes additional costs on developers and is a means through which LPAs can make unreasonable demands on developers to provide detailed information on issues "which are not strictly relevant to the planning decision" (Weston 1995), others feel that EIA does not necessarily slow things down: "The more organised approach makes it more efficient and in some cases it allows issues to be picked up earlier. The EIS can thus speed up the system" (DoE 1996). An EIA may well shorten the planning application stage but lengthen the period before the EIS is submitted.

There has been some concern that competition and cost-cutting by consultancies, an increase in "cowboy" consultancies and the tendency for developers to accept the lowest bid for preparing an EIS may affect the quality of the resulting EIAs by limiting the consultants' time, expertise or equipment. Consultants note that "on all but the largest developments there is always a limited budget – an EA expands to fill the available budget, and then some" (Radcliff & Edward-Jones 1995). However, Fuller (1992) argues that this may not be helpful to a developer in the long run:

> A poor-quality statement is often a major contributory factor to delays in the system, as additional information has to be sought on issues not addressed, or only poorly addressed, in the original...Therefore, reducing the cost of an environmental assessment below the level required for a thorough job is often a false economy.

The cost of EIA to *competent authorities* is much more difficult to measure and has until now been based on interviews rather than on a more systematic methodology. An early study (Lee & Brown 1992) found that about half the officers interviewed felt that the EIS had not influenced how long it took to reach a decision; the rest were about evenly split between those who felt that the EIA had speeded up or slowed down the process. In subsequent interviews (DoE 1996), many planning officers felt that dealing with the EIS and the planning application were one and the same and "just part of the job". Estimates for reviewing the EIS and associated consultation ranged from five hours to 6–8 months of staff time. Planning officers handling EIS cases tend to be development control team leaders and above, so staff costs would generally be higher than for standard planning applications. Where LPAs had engaged consultants to help them appraise an EIS, the cost of such review was between £1,000 and £10,000 for half the cases, the remaining being broadly split evenly between more than £10,000 and less than £1,000 (Leu et al. 1993).

In 20 case studies, the time spent by *consultees* on EIA ranged from four hours to one-and-a-half days for statutory consultees, and from one hour to two weeks for non-statutory consultees. Although some consultees, like planning officers, argued that "this is what we are here for", others suggested that they needed to prioritize what developments they got involved in because of time and resource constraints (DoE 1996).

8.6.2 *Benefits of* EIA

The benefits of EIA are mostly unquantifiable, so a direct comparison with the costs of EIA is not possible. Perhaps the clearest way to gauge whether EIA helps to reduce a project's environmental impacts is to determine whether a project was modified as a result of EIA. Early studies on EIA effectiveness (e.g. Kobus & Lee 1993, Tarling 1991) showed that modifications to the project as a result of the EIA process were required in almost half the cases, with most modifications regarded as significant. Jones's (1995) study of 40 EISs prepared before March 1993 showed that modifications before EIS submission were made in one-third of cases, modification after EIS submission in about one-sixth of cases, modifications before and after the EIS submission in one-fifth of cases.

Environmental impact assessment can have other benefits in addition to project modification. A survey of *environmental consultants* (Weston 1995) showed that about three-quarters of them felt that EIA had brought about at least some improvements in environmental protection, primarily through the incorporation of mitigation measures early in project design and the higher regard given to environmental issues. However, other consultants felt that the system is "often a sham with EISs full of platitudes". Jones et al. (1998) found that only one-fifth of developers and consultants felt that there had been no benefits associated with EIA.

Competent authorities generally feel that projects and the environment benefit greatly from EIA (Jones 1995, Lee et al. 1994). EIA is seen as a way to focus the mind, highlight important issues, reduce uncertainty, consider environmental impacts in a systematic manner, save time by removing the need for planning officers to collect the information themselves and identify problems early and direct them to the right people (DoE 1996, Jones 1995, Pritchard et al. 1995). One planning officer noted:

> when the system first appeared I was rather sceptical because I believed we had always taken all these matters into account. Now I am a big fan of the process. It enables me to focus on the detail of individual aspects at an early stage. (DoE 1996)

Consultees broadly agree that EIA creates a more structured approach to handling planning applications, and that an EIS gives them "something to work from rather than having to dig around for information ourselves". However, when issues are not covered in the EIS, consultees are left in the same position as with non-EIA applications: some of their objections are not because the impacts are bad but because they have not been given any information on the impacts or any explanation of why a particular impact has been left out of the assessment. Consultees feel that an EIA can give them data on sites that they would not otherwise be able to afford to collect themselves, and that it can help parties involved in an otherwise too often confrontational planning system to reach common ground (DoE 1996).

8.7 Summary

All the parties involved agree that EIA as practised in the UK helps to improve projects and protect the environment, although the system could be much stronger: EIA is thus at least partly achieving its main aims. There are time and money costs involved, but there are also tangible benefits in the form of project modifications and more informed decision-making. When asked whether EIA was a net benefit or cost, "the overwhelming response from both planning officers and developers/consultants was that it had been a benefit. Only a small percentage of both respondents felt that EIA had been a drawback" (Jones 1995). Some stages in EIA – particularly early scoping, good consultation of all the relevant parties and the preparation of a clear and unbiased EIS – are consistently cited as leading to particularly clear and cost-effective benefits (DoE 1996, Kobus & Lee 1993). More recent research by Weston (2002) reinforces this view. He suggests that there appears to be some agreement that – "LPAs and developers have grown to regard EIA as a positive procedural part of the planning process. Yet one consultant stated that they still 'come across clients who believe that an ES can be produced from scratch in a month and should contain the least possible information'". Chapter 9 provides a set of primarily UK case studies which seek to exemplify some of the issues of and responses to particular aspects of the EIA process. Suggestions for future directions in EIA in the UK and beyond are discussed in Chapter 11.

Notes

1. This was made up largely of landfill/raise projects (10 per cent), wastewater or sewage treatment schemes (4 per cent) and incinerators (3 per cent).
2. In terms of type and size of project, location, local planning authority and developer.
3. "Satisfactory" for the IAU criteria is very similar to "satisfactory" for the Lee & Colley criteria.

References

Bird, A. 1996. *Auditing environmental impact statements using information held in public registers of environmental information*, Working Paper 165. Oxford: Oxford Brookes University, School of Planning.

Braun, C. 1993. EIA in the 1990s: the view from Marsham Street. EIA in the UK: *Evaluation and prospect conference*, 6 April. Manchester: University of Manchester.

Cashmore, M., C. Epaminondas, D. Cobb 2002. An Evaluation of the Quality of Environmental Impact Statements in Thessaloniki, Greece, in *Journal of Environmental Assessment Policy and Management* 4(4), 371–96.

CEC (Commission of the European Communities) 1993. *Report from the Commission of the Implementation of Directive 85/337/EEC on the assessment of the effects of certain public and private projects on the environment*, vol. 13, *Annexes for all Member States*, COM(93), 28 final. Brussels: EC, Directorate-General XI.

CEC 2003 (Impacts Assessment Unit, Oxford Brookes University). *Five Years Report to the European Parliament and the Council on the Application and Effectiveness of the EIA Directive.* Website of DG Environment, EC: http://europa.eu.int/comm/environment/eia/ home.htm.

Coles, T. F., K. G. Fuller, M. Slater 1992. Practical experience of environmental assessment in the UK. *Proceedings of Advances in Environmental Assessment Conference.* London: IBC Technical Services.

Davison, J. B. R. 1992. *An evaluation of the quality of Department of Transport environmental statements* (MSc dissertation, Oxford Brookes University).

DETR 1997a. *Mitigation measures in environmental assessment.* London: HMSO.

DETR 1997b. *Consultation paper: implementation of the EC Directive (97/11/EC) – determining the need for environmental assessment.* London: DETR.

DETR (Department of Environment, Transport and the Regions) 1999a. *Town and Country Planning (Environmental Impact Assessment) Regulations.* London: HMSO.

DETR 1999b. *Town and Country Planning (Environmental Impact Assessment) Regulations.* Circular 2/99 London: HMSO.

DoE (Department of the Environment) 1989. *Environmental assessment: a good practice guide.* London: HMSO.

DoE 1991. *Monitoring environmental assessment and planning.* London: HMSO.

DoE 1994. *Evaluation of environmental information for planning projects: a good practice guide.* London: HMSO.

DoE 1995. *Preparation of environmental statements for planning projects that require environmental assessment: a good practice guide.* London: HMSO.

DoE 1996. *Changes in the quality of environmental impact statements.* London: HMSO.

DoT (Department of Transport) 1993. Design Manual for Roads and Bridges, vol. 11, *Environmental Assessment.* London: HMSO.

Essex Planning Officers' Association 2001. *The Essex Guide to Environmental Impact Assessment.* Chelmsford: Essex County Council.

Ferrary, C. 1994. Environmental assessment: our client's perspective. Environmental Assessment: RTPI Conference, 20 April. Andover.

Frost, R. & D. Wenham 1996. *Directory of environmental impact statements July 1988–September 1995.* Oxford: Oxford Brookes University, School of Planning.

Fuller, K. 1992. Working with assessment. In *Environmental assessment and audit: a user's guide*, 14–15. Gloucester: Ambit.

Glasson, J. 1999. Environment Impact Assessment – Impact on Decisions. In *Handbook of Environmental Impact Assessment*, J. Petts (ed.), vol. 1. Oxford: Blackwell Science.

Gosling, J. 1990. *The Town and Country (Assessment of Environmental Effects) Regulations 1988: the first year of application.* Proposal for a working paper, Department of Land Management and Development, University of Reading.

Hall, E. 1994. *The environment versus people? A study of the treatment of social effects in environmental impact assessment* (MSc dissertation, Oxford Brookes University).

Hughes, J. & C. Wood 1996. Formal and informal environmental assessment reports. *Land Use Policy* **13**(2), 101–13.

IAU (Impacts Assessment Unit) 2003. *Screening Decision Making under the Town and Country Planning (EIA) (England and Wales) Regulations 1999.* IAU: Oxford Brookes University.

IEA (Institute of Environmental Assessment) 1993. *Digest of environmental statements.* London: Sweet & Maxwell.

Jones, C. E. 1995. The effect of environmental assessment on planning decisions, *Report*, special edition (October), 5–7.

Jones, C. E., N. Lee, C. Wood 1991. UK *environmental statements 1988–1990: an analysis*, Occasional Paper no. 29. EIA Centre, University of Manchester.

Jones, C., C. Wood, B. Dipper 1998. Environmental assessment in the UK planning process. *Town Planning Review* **69**, 315–19.

Kobus, D. & N. Lee 1993. The role of environmental assessment in the planning and authorisation of extractive industry projects. *Project Appraisal* **8**(3), 147–56.

Kreuser, P. & R. Hammersley (1999). Assessing the assessments: British planning authorities and the review of environmental statements. *Journal of Environmental Assessment Policy and Management* **1**, 369–88.

Lee, N. & D. Brown 1992. Quality control in environmental assessment. *Project Appraisal* **7**(1), 41–45.

Lee, N. & R. Colley 1990 (updated 1992). *Reviewing the quality of environmental statements*, Occasional Paper no. 24. University of Manchester.

Lee, N. & R. Dancey 1993. The quality of environmental impact statements in Ireland and the United Kingdom: a comparative analysis. *Project Appraisal* **8**(1), 31–36.

Lee, N., F. Walsh, G. Reeder 1994. Assessing the performance of the EA process. *Project Appraisal* **9**(3), 161–72.

Leu, W.-S., W. P. Williams, A. W. Bark 1993. An evaluation of the implementation of environmental assessment by UK local authorities. *Project Appraisal* **10**(2), 91–102.

Mills, J. 1994. The adequacy of visual impact assessments in environmental impact statements. In *Issues in environmental impact assessment*, Working Paper no. 144. School of Planning, Oxford Brookes University, 4–16.

Nelson, P. 1994. Better guidance for better EIA. *Built Environment* **20**(4), 280–93.

Petts, J. & P. Hills 1982. *Environmental assessment in the UK*. Nottingham: University of Nottingham, Institute of Planning Studies.

Pritchard, G., C. Wood, C. E. Jones 1995. The effect of environmental assessment on extractive industry planning decisions. *Mineral Planning* **65** (December), 14–16.

Radcliff, A. & G. Edward-Jones 1995. The quality of the environmental assessment process: a case study on clinical waste incinerators in the UK. *Project Appraisal* **10**(1), 31–38.

Sadler, B. 1996. *Environmental assessment in a changing world: evaluating practice to improve performance*. Final report of the international study on the effectiveness of environmental assessment, Canadian Environmental Assessment Agency.

Sheate, W. 1994. *Making an impact: a guide to EIA law and policy*. London: Cameron May.

Sippe, R. 1994. *Policy and environmental assessment in Western Australia: objectives, options, operations and outcomes*. Paper for International Workshop, Directorate General for Environmental Protection, Ministry of Housing, Spatial Planning and the Environment, The Hague, Netherlands.

Tarling, J. P. 1991. *A comparison of environmental assessment procedures and experience in the UK and the Netherlands* (MSc dissertation, University of Stirling).

Weston, J. 1995. Consultants in the EIA process. *Environmental Policy and Practice* **5**(3), 131–34.

Weston, J. 1996. Quality of statement is down on the farm. *Planning* **1182**, 6–7.

Weston, J. 2000. EIA, decision-making theory and screening and scoping in UK practice. *Journal of Environmental Planning and Management* **43**(2), 185–203.

Weston, J. 2002. From Poole to Fulham: a changing culture in UK environmental impact decision making? *Journal of Environmental Planning and Management* **45**(3), 425–43.

Wood, C. 1991. *Environmental impact assessment in the United Kingdom*. Paper presented at the ACSP-AESOP Joint International Planning Congress, Oxford Polytechnic, Oxford, July.

Wood, C. 1996. Progress on ESA since 1985 – a UK overview. In the proceedings of the IBC Conference on Advances in Environmental Impact Assessment, 9 July. London: IBC UK Conferences.

Wood, C. 2003. *Environmental impact assessment: a comparative review*, 2nd edn. Harlow: Prentice Hall.

Wood, C. & C. Jones 1991. *Monitoring environmental assessment and planning*, DoE Planning and Research Programme. London: HMSO.

Wood, C. & C. Jones 1997. The effect of environmental assessment on local planning authorities. *Urban Studies* **34**(8), 1237–57.

Wood, G. & C. Bellanger (1998). *Directory of Environmental Impact Statements July 1988–April 1998*. Oxford: IAU, Oxford Brookes University.

Zambellas, L. 1995. *Changes in the quality of environmental statements for roads* (MSc dissertation, Oxford Brookes University).

9 Case studies of EIA in practice

9.1 Introduction

This chapter builds on the analysis in Chapter 8 by examining a number of case studies of EIA in practice. The selected case studies mainly involve EIA at the project level, although an example of the application of SEA is also included. Links between EIA and other types of assessment are also examined, including a case study of the so-called "appropriate assessment" process required under the European Union Habitats Directive. The selected case studies are largely UK-based and cover a range of project and development types, including energy (offshore wind energy, gas-fired power stations and overhead electricity transmission lines), transport (road projects), waste (municipal waste incinerator and wastewater treatment works) and infrastructure (port development and flood defence works). Some of the case studies involve developments in areas of particular environmental sensitivity, such as designated priority habitats under the EU Habitats Directive.

The case studies have been selected to illustrate particular themes or issues relevant to EIA practice, and some are linked to specific stages of the EIA process. These are:

- project definition in EIA and the effect of divided consent procedures on EIA (Wilton power station, Section 9.2);
- treatment of alternatives, and of indirect and consequential effects (M6–M56 motorway link road, Section 9.3);
- links between EIA and other EU Directives and the process of appropriate assessment (N21 link road, Section 9.4);
- approaches to public participation in EIA (Portsmouth incinerator, Section 9.5);
- mitigation in EIA (Cairngorm mountain railway, Section 9.6);
- assessment of cumulative impacts (Humber Estuary schemes, Section 9.7);
- strategic environmental assessment (SEA of UK government plans for future offshore wind energy development, Section 9.8); and
- EIA in developing countries (EIA of refugees in Guinea, Section 9.9).

It is not claimed that the selected case studies represent examples of best EIA practice – indeed two of the cases have been the subject of formal complaints to the European Commission regarding the inadequate assessment of environmental impacts. However, the examples do include some innovative or novel approaches towards particular issues – e.g. towards the assessment of cumulative effects (Humber Estuary) and the treatment of public participation and risk communication (Portsmouth incinerator). In other cases, practice appears to be somewhat in advance of existing legislative requirements – e.g. the SEA of UK government plans for offshore wind energy development was undertaken prior to the implementation of the EU SEA Directive. But the case studies also draw attention to some of the practical difficulties encountered in EIA, and the limitations of the process in practice. This reinforces some of the criticisms of UK EIA practice made in Chapter 8.

The selected case studies are largely based on original research either by the authors or by colleagues in the IAU at Oxford Brookes University (the exceptions are the power station case study in Section 9.2, which was researched by William Sheate, Reader in EIA at Imperial College, University of London, and published in 1995; the Cairngorm mountain railway study in Section 9.6; and the Guinea refugees study in Section 9.9).

9.2 Wilton power station case study – project definition in EIA

9.2.1 *Introduction*

This case study, originally documented by Sheate (1995), illustrates the problems of project definition in EIA, particularly in cases in which consent procedures for different elements of an overall scheme are divided. The case highlights the failure of the EIA process to fully assess the impacts of a proposed UK power station development. In particular, the EIA process failed to identify prior to the power station consent decision the environmental implications of the extensive electricity transmission lines required to service the new development.

The case study highlights a basic problem within EIA for UK energy sector projects caused by the splitting of consent procedures for electricity generation and transmission. This situation arose after the privatization of the UK electricity supply industry by the 1989 Electricity Act. The case illustrates how the division of consent procedures for individual components of the same overall project can result in conflicts with the EIA Directive's requirement to assess the direct, indirect and secondary effects of development projects. Although the case study relates to early EIA practice in the UK and EU (in the early 1990s), the issues raised remain largely unresolved and are still relevant to current practice.

9.2.2 *The Wilton power station project*

Early in 1991, newspaper reports began to identify the environmental conse‑ quences of proposed high voltage electricity transmission lines necessary to connect a new power station on Teesside, North East England to the National Grid system. To many, it was astonishing that these impacts had not been identified at the time the power station itself was proposed. Close inspection of the environmental statement (ES) produced for the power station revealed that such issues had barely been identified at the time and therefore did not feature in the consent process for the power station. Following considerable public uproar over the proposed power lines, in April 1991 the Council for the Protection of Rural England (CPRE) lodged a formal complaint with . . . the European Commission in Brussels against the UK Secretary of State for Energy. (Sheate 1995)

This state of affairs could hardly be regarded as an example of good EIA practice, but how did it come to occur?

The complaint concerned the EIA for a large new gas-fired power station at Wilton, near Middlesbrough on Teesside, proposed by a consortium known as Teesside Power Limited. CPRE argued that consent had been granted for the power station without the full environmental impacts of the proposal having been considered. Because of its size (a generating capacity of 1875 MW), the power station was an Annex I project and EIA was therefore mandatory. However, the overall "project" consisted of a number of linked components, in addition to the power station itself. These included:

- a new natural-gas pipeline;
- a gas reception and processing facility;
- a combined heat and power (CHP) fuel pipeline from the processing facility to the CHP facility; and
- new 400 kv overhead transmission lines and system upgrades (75–85 km in length, running from the power station site to Shipton, near York).

It was the implications of the transmission connections required to service the new power station that were of particular concern, although the other project components also had the potential for environmental impacts. Cleveland CC, in whose area the power station was located, expressed the view that a full assessment of the implications of all project components should be undertaken before the consent decision on the power station was taken. The County Planning Officer commented:

> My council wanted the power station [consent decision] deferred until all the implications could be fully considered. But the Secretary of State [for Energy – the consenting authority for schemes of this type at the time] wasn't prepared to do this. The result is that different features of the scheme, which includes pipelines and a gas cleaning plant as well as the main station and its transmission lines, come up at different stages with different approval procedures. An overall view hasn't been possible.

Despite these concerns, consent for the power station was granted by the Secretary of State for Energy in November 1990. The consent decision was based on the information contained in the environmental statement for the power station, and without the benefit of a public inquiry. However, crucially the power station ES did not include a description or assessment of the effects of the other elements of the overall "project", including the pipelines, gas processing facility and transmission lines. These elements of the project were seen to be the responsibility of other companies under separate consent and EIA procedures. The extensive nature of the transmission connections required to service the new power station was not therefore apparent from an examination of the power station ES. Although separate EIA procedures were in place for these other project components, so that their environmental impacts would subsequently be considered, CPRE in its complaint to the EC argued that, under the EIA Directive, the EIA for the power station should have included the main environmental effects of its associated developments. The failure to do so resulted in a piecemeal approach to EIA which, it was argued, contravenes the requirement in the EIA Directive that all direct, indirect and secondary effects of a project should be assessed prior to consent being granted.

Sheate (1995) summarizes the argument made by CPRE:

> Concern was expressed that the Secretary of State did not see fit to require further information on these aspects [the impacts of associated developments], as he is entitled to do under the UK's own implementing legislation. [The developer] had successfully received consent for the power station even though the major impacts on the environment of the electricity transmission lines, the gas pipeline, the gas processing facility and the CHP pipeline did not feature in the accompanying documentation provided to the Secretary of State for Energy. Since the relevant information was not available to the Secretary of State – nor did he request such information – it was argued that his decision might not have been the same had all the relevant information been available to him. Since the information was not contained in the ES, neither the public nor interest groups had been alerted to these consequential impacts, which might otherwise have caused a public inquiry to be held where the issue would inevitably have been aired.

In the UK, responsibility for new transmission lines rests with the National Grid Company (NGC), not the developer of the power station. NGC has an obligation to connect a new electricity generator into the national grid, and – if significant environmental impacts are likely – it must undertake its own EIA for new overhead lines and major upgrades of existing lines. EIA does therefore take place for new transmission lines (and for other types of associated development). However, this EIA process occurs after the power station has been given consent, and it is therefore unable to influence the decision over whether the power station should have been built in the first place, either in that location or somewhere closer to the existing transmission network, hence minimizing the adverse visual impacts of new overhead lines.

Essentially, the Wilton case revolved around the way in which "projects" are defined for the purposes of EIA. The ES for Wilton power station referred to the "overall project" as including both the power station and its associated developments, such as transmission lines. However, because of the fact that different elements of this overall project were subject to separate consent procedures, the project was divided into separate "sub-projects", with the environmental impacts of each being assessed separately and at different time periods depending on the timescale of the various consent procedures. CPRE argued that, under the Directive, it was not appropriate to assess the impacts of associated developments in isolation from (and after) the main development, including the implications of the latter's location.

In their response to CPRE's complaint, the European Commission agreed in principle that, in such cases, combined assessment was necessary and that splitting of a project in this way was contrary to the EIA Directive:

> I can confirm that it remains the Commission's view that, as a general principle, when it is proposed to construct a power plant together with any power lines either a) which will need to be constructed in order to enable the proposed plant to function, or b) which it is proposed to construct in connection with the proposals to construct the power plant, *combined assessment of the effects of the construction of both the plant and the power lines in question will be necessary* under Articles 3 and 5 of Directive 85/337/EEC when any such power lines are likely to

have a significant impact on the environment. (Letter from European Commission to CPRE, 11 November 1993; italics added).

The UK government had argued that the proposed transmission lines in the Wilton case were not required primarily to service the new power station, since the proposed upgrading would allow NGC to increase exports of electricity from Scotland to England. However, evidence presented by NGC to the subsequent public inquiry into the power line proposals appeared to contradict this view. Although the upgrading would have some wider benefits for NGC, it was clear that the primary justification for the proposals, and indeed for the specific routes proposed, was the needs of Wilton power station. The government also argued that the Directive allowed for separate EIA procedures for power stations and transmission lines, since the former tend to be Annex I projects whilst the latter fall within Annex II of the Directive (for which EIA is required only if significant effects are likely). However, the Directive requires an assessment of *direct and indirect effects*, which cannot be ensured for a power station scheme unless the transmission implications are included within the EIA. The Commission's response clearly supported this interpretation of the Directive.

Despite this clarification of the purpose and intention of the Directive, the EC decided against taking infringement action against the UK government in this case. Earlier, action had been taken by the EC in connection with EIA for the Channel Tunnel rail link and Kings Cross terminal "project". In that case, the EC had argued that these two projects were indivisible, because of the effect of each on the choice of site or route of the other:

> The effect of dividing the London–Channel Tunnel project into the rail link on the one hand, and the terminal on the other, leads to the circumvention of Directive 85/337/EEC, since the siting of the rail link in London is no longer capable of being assessed and – for instance by the choice of another site for the terminal – its effects minimised during the consideration of the rail link route. Terminal and link are, because of the impact of the choice of the terminal site on the link, or the link on the site, indissociable. The intention to assess the link once the assessment of the impact of the terminal is [completed] does not therefore make acceptable the assessment of the terminal . . . , which failed, contrary to Article 3 of the Directive, to take into account the effects of its siting on the choice of [route for] the rail link. (Letter from the Environment Commissioner to the UK Government, 17 October 1991).

The same argument seems to apply in the Wilton example. The power station and transmission lines are also indivisible, since the power lines would not be required were it not for the new power station, and the location of the power station is critical to any subsequent decisions on the route of the power lines.

9.2.3 *The Lackenby–Shipton power lines public inquiry*

As noted above, the installation of power lines to service the new power station was the responsibility of a separate developer (the NGC) and was subject to separate – later – consent and EIA procedures. In the event, five alternative routing proposals were considered concurrently at a public inquiry, which started in May 1992 – some

18 months after consent had been given for the power station; indeed, by this stage, construction of the power station had already begun.

The proposed power line routes started in Lackenby, adjacent to the Wilton power station site, and then proceeded south to Picton, via alternative southern and northern routes. From Picton, alternative western, eastern and central routes ran south to Shipton, north-west of York (Figure 9.1). The total length of new power lines and

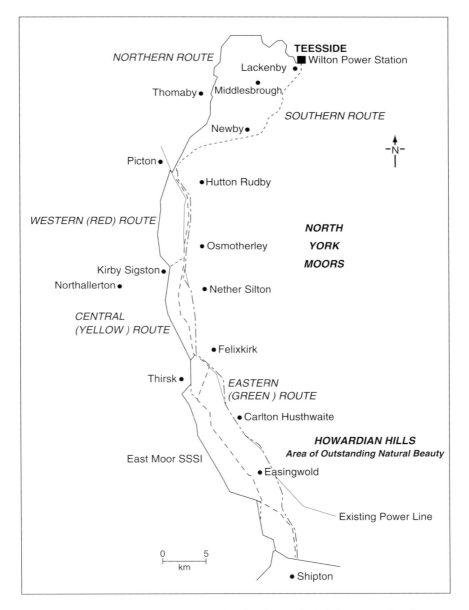

Figure 9.1 Alternative route options considered at the North Yorkshire power lines Inquiry.

system upgrades required was between 75 and 85 km, depending on the route options selected. The NGC itself expressed a preference for the shorter southern route from the power station to Picton, and for the western route option from Picton to Shipton. All of the proposed routes passed through or adjacent to (and visible from) important protected landscapes, including the North York Moors National Park and the Howardian Hills AONB. Key objectors to the proposals at the public inquiry included the local authorities through which the proposed routes ran (North Yorkshire County Council, Cleveland CC and others), the Country Landowners Association, the National Farmers Union and CPRE, as well as many individuals including farmers and local residents. The principal issues considered at the inquiry included the visual impact of the pylons and overhead lines, potential health risks from electromagnetic radiation, issues of need and alternatives and effects on farming operations.

CPRE argued at the inquiry that the visual impacts of the proposals were unacceptable and should have been foreseen at a much earlier stage. It urged that the inquiry inspectors "should not feel obliged to grant consent for the power lines simply because consent for the power station had already been granted and it was already being built" (Sheate 1995). It also invited the inspectors to comment on the inadequacy of the existing EIA procedures in such cases, in which consent for electricity generation is divided from consent for electricity transmission.

The inquiry ended in December 1992, and after a long delay the inspectors' report was finally published in May 1994. The report recommended approval of NGC's preferred route options – the southern route from Lackenby to Picton and the western route from Picton to Shipton, subject to various detailed modifications to minimize the environmental impacts (e.g. around East Moor SSSI). However, the inspectors agreed with CPRE's views on the EIA procedures in such cases:

> It seems to us that to site power stations without taking into account all relevant factors, including transmission to the areas of consumption, is likely to lead to the extension of high voltage power lines through areas currently not affected and the reinforcement of lines in areas already affected. It is not disputed that in the view of the scale and form of the towers these lines are inevitably highly intrusive and damaging to almost any landscape and as a result are unwelcome.

> It appears to us that there is a strong case for consideration to be given to the introduction of procedures to ensure that consents for future power stations take account of the resulting transmission requirements, and the environmental impacts of any necessary extension or reinforcement of the National Grid, between the proposed generating plant and areas of consumption. (Inspectors' conclusions, 23 September 1993).

The failure of the EIA for Wilton power station to address the implications of transmission connections resulted in a situation in which "Teesside Power Limited [the developer] neither had to demonstrate the full implications of the siting and development of the power station, nor to bear the full economic and environmental costs" (Sheate 1995). This is because there are limits on the costs that can be recouped by NGC for the provision of transmission connections to individual generating projects. This means that NGC is under pressure to develop the cheapest options, since any additional costs incurred to minimize the environmental impact of power lines – such as taking a longer route through less sensitive areas or placing all or part of the route underground – will be borne by NGC rather than the power station developer.

9.2.4 Lessons for EIA

Sheate (1995) argues that the Wilton power station case provides powerful evidence that, at the time, the procedures for consent approval in the electricity supply industry ran counter to the letter and spirit of the EIA Directive. According to Sheate, the situation could have been remedied by an amendment to the Electricity and Pipeline Works EIA Regulations. The suggested amendment read as follows:

> An environmental statement shall include information regarding the overall implications for, and impact of, power transmission lines and other infrastructure associated with the generating station where these are likely to have significant effects on the environment. (CPRE, letter to DTI, 22 February 1993)

The effect would be that, in cases where power lines or other associated infrastructure are likely to have a significant effect on the environment, these impacts should be material considerations in whether consent for the power station should be given and the Secretary of State for Energy (now Trade and Industry) should be aware of these before giving consent.

> The consequence of such an amendment would be to ensure that power station proponents were forced to consider the transmission implications of their proposals and that they would form part of the EIA and of any subsequent public inquiry. It would begin to reduce the difficulties which arise over the definition of projects and programmes. (Sheate 1995)

The case also highlights a wider problem within the EIA Directive concerning its ambiguous definition of the term "project". As we have seen, this is a particular issue for projects in the electricity supply industry, but it also applies to other infrastructure projects, especially road schemes. It has resulted in a number of complaints to the EC about whether a larger project can be split into a number of smaller schemes for the purposes of consenting and (therefore) EIA. The problem is that EIA in the UK – as in most EU Member States – has been implemented as part of existing consent procedures, and if these are divided for a project, then so is the requirement for EIA. This so-called "salami-slicing" of projects runs counter to the purposes of the Directive, which states "effects on the environment [should be taken] into account at the earliest possible stage in all the technical, planning and decision-making processes" (Preamble to Directive 85/337/EEC). As the case study illustrates, this purpose cannot be achieved if EIA is applied only to individual project components rather than to the project as a whole.

The issues raised by this case study remain relevant to current EIA practice. The issue of ambiguous project definition was not resolved in the revised EU EIA Directive that came into force in 1999, and consent procedures for electricity generation and transmission projects in the UK remain divided. The extent to which the implementation of the EU SEA Directive may help to resolve the problem of "salami-slicing" of projects is also currently unclear.

9.3 M6–M56 motorway link road – the treatment of alternatives and indirect effects

9.3.1 Introduction

This case study of the EIA for a proposed motorway scheme in North West England raises important issues in relation to the treatment of alternatives in EIA, and the assessment of indirect and consequential effects. EIA for road schemes in the UK has been subject to significant advances since the implementation of the original EIA Directive in the late 1980s, and this example relates to relatively early UK practice. The environmental statement for the scheme was submitted towards the end of 1992, just prior to the emergence of revised guidance on road scheme EIA (itself since superseded) (DoT 1993). The case also pre-dates the stronger line on the consideration of alternatives in the revised EIA Directive. Nevertheless, the case study raises a number of issues with continuing relevance to EIA practice, both for road schemes and for other infrastructure projects, particularly those of a linear nature.

9.3.2 The proposed development

The proposed scheme, known as the A556(M), involved the construction of a new motorway link between the existing M6 and M56 motorways, in the Cheshire countryside south of Manchester, in North West England (Figure 9.2). The existing link was provided by the A556 trunk road, which was mainly a four-lane single carriageway. This route, which served much of the motorway traffic between the cities of Manchester and Birmingham, was seen as problematic. It was characterized by a high volume of traffic, much of it consisting of heavy-goods vehicles, peak-hour congestion and a poor accident record. The do-minimum option and on-line improvements to the existing road were rejected in favour of a replacement link road. This was to be a three-lane dual carriageway for most of its 10-km length. The existing A556 route would lose its trunk road status when the new scheme was completed. The scheme was proposed by the DoT, which at the time was the authority responsible for trunk road and motorway construction in the UK.

The land surrounding the existing A556 was rural and predominantly agricultural. The relevant local authority, Cheshire CC, had identified three areas of special landscape value in the vicinity, at Tatton Park to the east, the Bollin Valley to the north and Tabley Park to the south-west (Figure 9.2). Ecologically designated sites included the important site of Rostherne Mere, near the existing junction with the M56 motorway. This was designated an SSSI, a national nature reserve and a wetland of international importance under the Ramsar Convention on account of the waterfowl which nest at the site (DoE/DoT 1995). There were also six designated sites of biological importance, as well as many patches of undesignated woodland, unimproved grassland, lane-side hedges and ponds of local importance; wildlife corridors connected these features (DoT 1992).

9.3.3 Background to the proposal and the key planning stages

The proposed scheme had originally entered the National Trunk Roads Programme in the late 1980s as a stand-alone scheme. However, it later became part of a much longer route, known as the Greater Manchester Western and Northern Relief Road

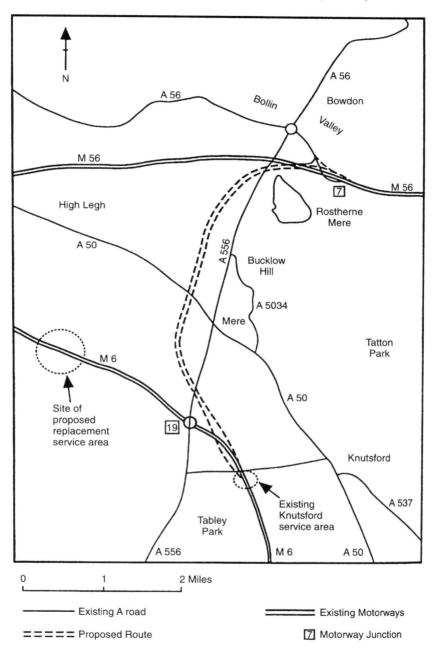

Figure 9.2 The proposed M6–M56 motorway link road. (*Source*: Based on DoT/Allott & Lomax (DoT 1992).)

(GMWNRR). This route, divided into three main parts, involved the construction of new relief roads around the northern and western fringes of the Greater Manchester conurbation. The A556(M) scheme represented Stage I of this overall route. Six alternative routes for the A556 scheme, including an on-line improvement of the existing road, were considered by the DoT before the public consultation stage, although only one route option was presented at the consultation stage at the end of 1989.

Following public consultation on the proposals, the DoT announced their preferred route for the scheme at the end of 1990 and detailed design and assessment of this route then began. The environmental statement was published in October 1992, together with the draft line orders for the scheme (for a road scheme, this is the equivalent of the planning application). Modifications to the draft orders and an addendum to the environmental statement were published in February 1993. A public inquiry into the draft orders was held during the latter part of 1993. Supporters of the scheme at the inquiry included the local authorities in the area, Cheshire CC and Macclesfield Borough Council, as well as the CPRE and the Mere Residents Association (representing residents along the existing A556 route). Objectors included adjacent local authorities (such as Trafford Metropolitan Borough Council), environmental bodies such as EN and FoE, and many local residents and members of the public. Approval for the scheme, confirming the draft line orders in modified form, was given in July 1995 (DoE/DoT 1995). A separate inquiry on the compulsory purchase orders necessitated by the scheme was held in the spring of 1997.

9.3.4 The scope and format of the environmental statement

The environmental statement for the scheme in fact comprised four separately bound volumes rather than a single document (DoT 1992). The main document described the proposed scheme and summarized its main impacts, while three further technical volumes dealt with a range of specific impacts in greater depth. The impacts addressed in the ES appear to have been based largely on those listed in the relevant DoT guidance at the time (DoT 1983). However, certain additional effects not identified in the guidance, including impacts on water and drainage, were also examined as part of the EIA for the scheme and were included in the ES.

More detailed assessments, prepared by various consultants, were incorporated into three additional, separately bound documents, which formed part of the ES. These included an agricultural assessment, an ecological survey and assessment, an archaeological assessment, an air quality report and a road traffic and construction noise report. Detailed technical reports in support of the conclusions in the ES were not provided for certain types of impact, including effects on water and drainage, landscape, severance and construction-stage impacts. However, supporting documents on water and drainage impacts were provided by the DoT at the subsequent public inquiry.

9.3.5 The consideration of alternatives

The environmental statement reveals that six alternative schemes had been considered by the DoT prior to the public consultation stage in 1989. These included a do-minimum option (involving an online improvement of the existing

A556 route) and five alternative off-line routes, including the preferred scheme. Compared with the preferred scheme, these alternative routes were characterized by a more westerly alignment and/or a more westerly location for the new interchange with the M56 motorway. Those alternatives not considered included the do-nothing option and alternative modes, such as public transport or park-and-ride. The ES stated that, as the major strategic route for motorway traffic between the Midlands and Manchester, "only a new road was judged to be appropriate or effective in coping with the forecast growth in demand for traffic movement" (DoT 1992).

Five of the six alternatives under consideration were rejected by the DoT before the public consultation, including the do-minimum option and all the more westerly route options. Consequently, only one option was presented at the public consultation stage in late 1989. Such single-option consultation, although not the norm with road schemes, is not unusual (NAO 1994). The route – in somewhat modified form – was confirmed as the SoS's preferred route at the end of 1990. Further modifications were made during the process of detailed route design prior to the submission of draft line orders for the scheme and the environmental statement in October 1992. The ES for the scheme was therefore submitted almost 2 years after the announcement of the preferred route and 3 years after the public consultation stage. This sequence of events encourages the belief that the crucial decisions about the general alignment of the route had been taken long before the appearance of the ES, and indeed before the public consultation stage. It is therefore not surprising that the treatment of alternatives in the environmental statement was far from satisfactory.

The relevant DoT guidance at the time, contained in Departmental Standard HD 18/88 (DoT 1989), indicated that a road scheme ES should include a brief description of the alternatives considered at the public consultation stage and the reasons for the choice of the preferred route. Therefore, since only one option for the A556(M) scheme was presented for public consultation, it would appear that the ES did not need to address the issue of alternatives at all. Notwithstanding this, a brief description of all six of the original options considered before the public consultation stage was included in the ES. The reasons for the choice of the preferred scheme were also outlined, although this did not amount to a systematic comparison of the various options. For example, the comparison of the environmental impacts of the different options occupied only one page of the ES, most of the discussion focusing on the relative economic, traffic and safety implications of the schemes.

The ES stated that "an assessment of the five do-something options revealed that there was little to choose between them in environmental terms", except that a more serious visual impact arose out of the routes involving a more westerly location for the interchange with the M56, which would be sited in a conspicuous location. This conclusion may or may not be true, but the ES did not contain the detailed information on the environmental effects of each option to support such a statement. Indeed, the more serious impact of the westerly M56 interchange locations was questioned by English Nature, one of the statutory consultees, in its comments on a draft version of the ES:

> The choice of the most easterly option [for the interchange] has been based on traffic, operational, safety and economic grounds. There appears to have been

> no consideration of the considerably greater impact of the chosen location on Rostherne Mere [a Ramsar site, national nature reserve and SSSI]. (English Nature, in DoT 1992)

English Nature argued that the impacts of the scheme on the important site of Rostherne Mere, including visual intrusion and increased air pollution and noise levels, had not been adequately addressed in the ES. Other objectors made similar comments at the subsequent public inquiry into the proposals (see DoE/DoT 1995). Subsequent to public consultation and the announcement of the preferred route, significant alterations were made to the proposed route, which, the DoT argued, reinforced its selection. However, rather unhelpfully, the ES did not clearly identify these alterations, nor did it justify them, either in environmental or any other terms. Further modifications to the proposals were made shortly after the submission of the ES; these were described in an addendum to the ES, published in February 1993.

Although the ES contained only a very limited treatment of alternatives, the issue of alternatives was one of the main pre-occupations of the subsequent public inquiry held during 1993. No fewer than 12 main alternatives proposed by objectors were considered at the inquiry (DoE/DoT 1995). Most of them (seven) involved route realignments, ranging from minor adjustments to the proposed scheme to entirely different route corridors. The other alternatives involved minor modifications to the side-road orders or other design changes, such as the placing of part of the route in a cutting rather than on an embankment. Four of these alternatives were subsequently accepted by the Secretaries of State in their decision on the scheme, following recommendations by the inquiry inspector. These included a slight westerly realignment of the route near its junction with the M6, to avoid the Mere Estate; putting the northern section of the route in a cutting rather than on a high embankment, to reduce the visual impact to and from Rostherne Mere and the Bollin Valley; bringing together the northern and southern carriageways along part of the route; and providing a replacement bridge, to retain access along an important side road severed by the scheme (DoE/DoT 1995).

9.3.6 A strategic level of assessment

As noted above, at the time of the publication of draft orders the proposed scheme was part of a much longer route, known as the GMWNRR. The GMWNRR was divided into three main stages, each of which was to be subjected to separate planning, EIA and consent procedures. The A556(M) scheme represented Stage I of the GMWNRR. Stage II involved the construction of a motorway link between the M56 and the M62 to the west of the Manchester conurbation, and Stage III continued the route along the M62 corridor around the northern perimeter of the conurbation. Public consultation on Stage II of the route took place in October 1992, at the time of the publication of draft orders for the A556 scheme. Despite this background, the ES for the scheme made no explicit reference to the existence of the proposed GMWNRR, or to the relationship of the proposed scheme to Stages II and III of the route. No strategic assessment of the environmental consequences of the whole route appears to have taken place in this case. Although Stage II of the GMWNRR was subsequently abandoned by the DoT, following overwhelming public opposition to the proposals,

the failure of the ES even to mention t
unfortunate.

9.3.7 Indirect and consequential effects

The proposed scheme, as described in the
have important implications for an existin;
the M6 at Knutsford. The scheme propos.
facing slip roads onto the M6, with the rest
be open to either northbound or southboun
left a gap of almost 40 miles between the nea
way, and might have been expected to result :
in the Knutsford area. The need for such a rep
environmental effects of its development and ι
for the A556(M) scheme.

However, at an earlier stage in the develo_ ..... or the scheme, the DoT had
identified a replacement site for such a service area and submitted a notice of
proposed development on the site. This site was east of Arley Hall, some three
miles north-west of the existing Knutsford service area (see Figure 9.2). The DoT
asked its consultants to include the site in their ecological survey and impact
assessment carried out during the design and assessment of the A556 scheme.
However, before the publication of draft orders for the scheme, the Department
abandoned its plans to develop the Arley Hall site. This was because of changes to
the planning regime for MSA provision introduced at this time. These transferred
responsibility for identifying new MSA sites, seeking planning permission and
acquiring the necessary land from the DoT to the private sector (see Sheate &
Sullivan 1993). As a result, the ES for the A556(M) scheme did not identify the
Arley Hall site or any other site for a replacement MSA. This means that a major
form of consequential development resulting directly from the scheme, and with
potentially significant environmental effects, was not addressed in the ES. Here we
find another example of the effect of divided consent procedures on EIA, similar to
that discussed at length in Section 9.2.

A somewhat ironic postscript is that shortly after the submission of the ES, the
scheme proposals were subject to further modifications which involved the
retention of access to the existing Knutsford MSA for M6 traffic. These changes
therefore removed the need for a replacement MSA site. Whether environmental
statements for motorway schemes should or could discuss the need for the provision or replacement MSAs – and the environmental impacts of such provision – is
open to debate. What is clear is that the present arrangements do not require any
consideration of such consequential development. The removal of responsibility
for MSA provision from the DoT reinforced the separation between the planning
and environmental assessments of motorway proposals and their associated service
areas.

9.3.8 Conclusions

The quality of environmental statements for road schemes has undoubtedly
improved substantially compared with the early examples produced in the years
immediately after the implementation of the original EIA Directive. However, there

...out the quality of the wider EIA process for major road ...the environmental statement at a time when many of the ...the route have already been made and the limited treatment of ...of indirect and consequential impacts are well illustrated by the

N21 link road, Republic of Ireland – assessing impacts on an EU priority wildlife habitat

9.4.1 Introduction

This case study, researched by Weston & Smith (1999), concerns a proposed road improvement scheme in County Kerry, Republic of Ireland. The proposed route of the road passed through part of a European protected habitat, a residual alluvial forest known as Ballyseedy Wood. Although the proposal was not subject to EIA (largely because the ecological status of the site was not known at the time), it was later subjected to a related procedure known as "appropriate assessment" which operates under the EU Habitats Directive. The Habitats Directive requires that projects likely to have a detrimental impact on a European priority habitat must be subject to an assessment of that impact. This assessment involves a series of sequential tests that must be passed for the project to be allowed to proceed. The case study examines the nature and interpretation of these tests, and demonstrates the high level of protection currently afforded to designated habitats in the EU (Weston & Smith 1999).

9.4.2 The proposals

Improvements to the N21 main road into Tralee, County Kerry, in the west of Ireland, had been an objective of the local authority, Kerry CC, since the late 1960s. However, it was not until the prospect of European funding for these improvements emerged during the mid-1990s that substantial progress was made in advancing the scheme. The route was included in the Irish Government's Operational Transport Programme for Ireland (OTP) in 1994. In the same year, the OTP was adopted for co-funding by the European Commission as part of the EU's Community Support Framework for Ireland. Under this Support Framework, the EU agreed to provide 85 per cent of the funding for the proposed improvements to the N21 link.

The proposed project comprised improvements to 12.5 km of the existing N21 highway between Castleisland and Tralee, including a short (2.4 km) new section of dual carriageway between Ballycarty and Tralee (Figure 9.3). The dual carriageway section of the scheme ran to the south of the existing highway and through Ballyseedy Wood, which was later discovered to be a priority habitat under the EU Habitats Directive. Following the announcement of European co-funding in July 1994, Kerry CC, as the local highways authority, began design work on the proposed scheme.

9.4.3 The planning and EIA process

The N21 road improvement scheme was an Annex II project, for which EIA is required only if there are likely to be significant environmental effects. Like most EU

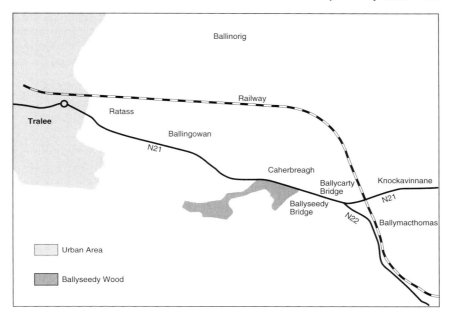

Figure 9.3 Map of the existing N21/N22 road network.

Member States, Ireland uses a series of size thresholds to help determine whether Annex II projects should be subject to EIA. In the case of road schemes of the type proposed (rural dual carriageways), EIA is required for schemes in excess of 8 km in length or if there are considered to be significant environmental effects. In such cases, EIA is carried out by the local highways authority and submitted to the DoE. After a period of consultation and a public inquiry (if one is held), the Minister for the Environment makes a decision on the application. For Annex II schemes falling below the size threshold (less than 8 km in length) and not considered likely to cause significant environmental impacts, EIA is not carried out and the proposal is dealt with under normal planning legislation. This allows for a period of public consultation, with the final decision as to whether to approve the scheme resting with the relevant local authority. For road schemes, it is the developer of the project, the CC as local highways authority, who determines whether or not EIA is required.

After commissioning a report from environmental consultants into the proposed scheme, Kerry CC decided that an EIA was not required in this case. This decision was based on the length of the dual carriageway section of the scheme (at 2.4 km, well below the 8-km size threshold for such schemes) and on the belief that there were unlikely to be any significant environmental effects. However, following the publication of the proposals, the Council received a number of objections, mainly regarding the impact of the new dual carriageway on Ballyseedy Wood. After considering these objections and the subsequent report to the Council on the scheme prepared by the authority's officers, the Council's elected members decided to proceed with the proposals. However, this was not the end of the authorization process, since the Compulsory Purchase Orders (CPOs) necessary for the scheme to proceed still had to be served and considered at a public inquiry (the CPO inquiry). Under the Irish system, members of the public and interested parties are allowed to give evidence at the CPO inquiry on environmental issues. However, the inquiry and the subsequent decision (including any alterations to the alignment of the route) must be based solely on land acquisition, rights of way and access issues. The CPO inquiry for the N21 scheme was held in March 1996.

Following the Council's decision to go ahead with the scheme, a local organization objecting to the proposals commissioned an ecological assessment of Ballyseedy Wood. This assessment concluded that the wood comprised an area of residual alluvial forest that, although currently lacking protected status, complied with the description of a priority habitat as set out in the EU Habitats Directive of 1992. These conclusions were accepted both by the relevant Irish national authorities and Kerry CC, and the site was subsequently proposed as a Special Area of Conservation under the terms of the Directive. The revelation of the important ecological status of Ballyseedy Wood resulted in a number of formal complaints being submitted to the EC concerning its co-funding of the proposed scheme. It was argued that the Commission should re-consider its decision to co-finance the project, given its potentially damaging impacts upon a habitat of recognized European-wide importance. As a result of these complaints, the EC commissioned an independent study to provide advice on whether there was a need to re-consider the co-funding of the scheme.

9.4.4 The EU Habitats Directive

The EU Habitats Directive (92/43/EEC) requires all Member States to designate sites hosting important habitat types and species as Special Areas of Conservation. Together with Special Protection Areas (SPAs) designated under the Birds Directive (79/409/EEC), it is intended that these sites will form a network of European protected habitats known as "Natura 2000". The Habitats Directive is designed to protect the integrity of this European-wide network of sites, and includes provisions for the safeguarding of Natura 2000 sites and priority habitats.

In cases in which a project is likely to have a significant impact on a protected site, the Directive states that there must be an "appropriate assessment of the implications for the site in view of [its] conservation objectives". Under the terms of the Directive, consent can only be granted for such a project if, as a result of this appropriate assessment, either (a) it is concluded that the integrity of the site will not be adversely affected, or (b) where an adverse effect is anticipated, there is shown to be an absence of alternative solutions and imperative reasons of over-riding public interest that the project should go ahead. The overall intention of the Directive is "to prevent the loss of existing priority habitat sites whenever possible by requiring alternative solutions to be adopted" (Weston & Smith 1999). Projects that have a negative impact on the integrity of a priority habitat, but which are able to satisfy both the absence of alternatives and overriding-reasons tests, can go ahead. However, in such cases, the developer must provide compensatory measures to replace the loss of priority habitat. Huggett (2003) discusses the development of such measures in relation to a range of port-development proposals in the UK.

The tests set out in the Directive are not absolute and require a degree of interpretation. For example, a literal interpretation of the "absence of alternative solutions" test could be taken to imply that any alternative that is less damaging to the protected habitat than the proposed scheme should be selected, regardless of cost or impacts on other interests. European case law provides some indication as to the appropriate interpretation of the Directive's tests, and this is reviewed by Weston & Smith (1999). They conclude that both the "absence of alternatives" and "reasons of public interest" tests should be interpreted stringently, in view of the intention of the Directive to provide a significant level of protection to priority habitats. The application of these tests to the N21 link road proposal is now explored.

9.4.5 *Appropriate assessment of the N21 link road proposals*

Under Article 6 of the Habitats Directive, the appropriate assessment of the N21 project involved a series of sequential tests. These concerned:

- the impact of the scheme on the integrity of the priority habitat;
- the presence or absence of alternative solutions; and
- the existence of imperative reasons of overriding public interest that the scheme should go ahead.

Each of these tests is examined in turn, drawing on the results of the independent study commissioned by the EC, as reported in Weston & Smith (1999).

Impact on the integrity of the priority habitat

Ballyseedy Wood covers a total of 41 h, although the priority habitat that was the subject of assessment represented only a very small part of this overall area. A small area in the northern corner of the wood accorded with the definition in the Habitats Directive of a residual alluvial forest. This northern part of the wood consists of alder and ash and is subject to regular flooding. Wet woodland of this type is the least common type of Irish forest, and the surviving examples tend to be small in area; the priority habitat area covered less than half a hectare. It was this northern edge of the wood that was to be lost to the proposed dual carriageway, including the areas of greatest ecological interest.

The direct land take of the proposed scheme involved the loss of only 3 per cent of the total area of the priority habitat. However, this does not necessarily mean that the effect on the integrity of the habitat would be insignificant. The independent study concluded that:

> On the basis of the assessments carried out by a number of environmental consultants and the evidence presented to the CPO inquiry, the loss of habitat could not be objectively assessed as being of no significance to the integrity of the habitat as a whole. There will be change caused to the habitat as a result of the removal of trees, the change in the hydrological regime and the re-routing of the river [another element of the scheme]. Evidence to the CPO inquiry suggested that areas of the wood, outside of the land take, would also be affected by this change. As part of EU policy the precautionary principle also needs to be applied to the assessment of the impact of a project on a priority habitat. In applying that principle it must be concluded that there is a risk that the integrity of the [habitat] will be significantly diminished by the proposed road. (Weston & Smith 1999)

It was therefore concluded that the impact on the priority habitat would be negative. Nevertheless, the project could still go ahead if it satisfied both of the remaining two tests. The first of these concerned the absence of alternative solutions.

The absence of alternative solutions

The County Council's identification of the preferred route alignment for the proposed scheme, and possible alternative routes, was based partly on a constraint mapping exercise in which areas with various environmental constraints (such as archaeological remains) were identified. Cost factors also featured in the choice of the preferred route, and a form of CBA was carried out. The Council prepared a Design Report, which set out the need for the scheme and the alternatives considered. This reveals that the Council had identified and investigated a number of alternatives. Weston & Smith (1999) identify a total of six main alternatives to the proposed scheme, including a do-minimum option. Most of the alternatives considered would completely avoid Ballyseedy Wood, generally by following more northerly route alignments. However, other adverse impacts would arise from some of these alternative schemes, such as the demolition of residential properties, farm severance and the relocation of Ballyseedy Monument, a local war memorial. Notwithstanding these impacts, a number of viable alternative route options were clearly available to meet the objectives of the proposed scheme.

The CC tested the various route options against their ability to provide the best solution "in terms of human safety, capacity and economic viability". However, there appears to have been no systematic attempt to test the alternatives against the need to avoid the loss of priority habitat, even after the Council became aware of the importance of Ballyseedy Wood. The conclusion of the independent study was that "[the] alternatives were not examined to the same rigour or degree as the preferred route and appear to have been rejected without clearly defined and quantified justification" (Weston & Smith 1999).

An issue that appears not to have been considered by the CC in its route selection is the need to serve those areas where future development growth is planned. In this case this was the northern edge of Tralee, which was the location of a new Regional Technical College and of allocated industrial areas. This suggests that a more northerly route alignment for the dual carriageway – which would have avoided the impacts on Ballyseedy Wood – may have been better placed than the proposed scheme to accommodate the growth in traffic generated by these planned developments. The proposed scheme would involve traffic serving these planned growth areas passing through the town centre of Tralee. This would add to existing traffic problems in the town, and may have resulted in the time benefits derived from the improved N21 being lost because of increased congestion in Tralee. The independent study comments:

> It is surprising therefore that an alternative alignment for both the N22 and N21, which links the infrastructure to the areas of Tralee where future development is planned, has not been more fully investigated. A northern route proposed by private individuals, which could be of dual carriageway standard, was not adequately assessed in terms of the strategic objectives of the Operational [Transport] Programme or in terms of its benefits such as avoiding Ballyseedy Wood and maintaining the existing distinctive quality of the area around the

Ballyseedy Monument. There are other possible alignments that appear not to have been fully considered, such as routes south or north of the railway line [which runs to the north of the existing N21]. Although the Council's Design Report rejects such routes because of the problems of crossing the railway line, farm severance and the impact on property, there appear to have been insufficient investigations and assessment on which to base such an outright rejection of such options.

Overall, there is little to suggest that the Council's alternatives have been tested to the same degree as the preferred option. The alternatives considered were not subjected to detailed costings, surveys, time-saving considerations, their ecological impacts or indeed their ability to meet the strategic objectives of Structural Funding. In the absence of the rigorous testing of all alternatives against clear objectively determined criteria it cannot, in this case, be concluded that the objectives of the [OTP] cannot be achieved with an alternative solution to that which would damage the priority habitat. (Weston & Smith 1999)

The second test, an absence of alternative solutions, was therefore failed. Under the terms of the Habitats Directive, the project could not therefore proceed, since a number of viable alternative solutions were shown to have been available in this case. It was not therefore necessary to carry out the third test, the existence of imperative reasons of overriding public interest in favour of the scheme. However, it is useful to do so, since this demonstrates how this test is applied in practice.

Public interest issues

The third test involves the balancing of the loss of priority habitat against other imperative public interest issues. Public interest issues would outweigh the loss of habitat if they resulted in "far greater adverse impacts than does the loss of habitat" (Weston & Smith 1999). So, for example, if only a minor impact on the habitat is anticipated and the alternative options would result in extreme economic or other public interest disbenefits, then the public interest issues could be said to outweigh the loss of habitat. Conversely, if the impact on the habitat was great or uncertain, and the impact on the public interest issues was small, then the interests of the habitat would take precedence.

European case law provides some guidance on the type of public interest issues that can be considered to be "imperative" reasons. Examples include the public interest of economic and social cohesion, human health, public safety and other environmental concerns. However,

> for such public interest reasons to out-weigh the loss of habitat they must be of a similar scale in importance [as the protection of the priority habitat] – that is of interest to the [European] Community as a whole – and be demonstrable. (Weston & Smith 1999)

In the case of the N21 scheme, it was the view of the CC that a number of public interest issues were relevant and that, when combined, the sum total of these issues outweighed the loss of the priority habitat at Ballyseedy Wood. The public interest issues arising in the case included:

- the strategic objectives of the wider OTP (of which the scheme was a component);
- the cost of alternative solutions;
- loss of family homes;
- road safety issues;
- heritage impacts on the Ballyseedy Monument;
- farm severance; and
- impacts on archaeology.

The independent study into the scheme concluded that none of these issues could be regarded as both imperative (that is, of equal importance as the loss of habitat) and overriding (that is, sufficiently damaging to override the protection of the habitat), and therefore this third test was also failed. The reasons for this conclusion included:

- *The existence of alternative solutions.* The fact that a range of alternative route options were available made it difficult to argue that the public interest issues arising in the case were unavoidable. For example, there was considerable local concern about the need to relocate the Ballyseedy Monument, a local war memorial, should one of the alternative routes be adopted. However, the need to relocate the Monument arose only with one of the six main alternatives considered and could therefore be avoided by the adoption of one of the other alternative solutions.
- *The alternatives would not necessarily result in greater adverse impacts than the proposed scheme.* For example, there was no evidence that, apart from one of the route options, any alternative solution would result in the loss of more family homes than the proposed scheme.
- *Some of the public interest issues were not demonstrable, due to a lack of data.* For example, no quantified data was produced on the road safety implications of alternative routes, compared with the proposed scheme. Evidence from the CC at the CPO inquiry suggested that all alternatives examined by the Council were equally safe. Also, it was not possible to argue that the proposed scheme was necessarily the most cost-effective, since the costs of all the alternative route options had not been worked out in detail. Indeed, it was suggested "that an alternative route may be cheaper to construct because of the decreased disruption to existing road users, the impact of construction on properties in the existing corridor and the reduction of some mitigation costs" (Weston & Smith 1999). Another issue raised was the impact of alternative routes on farm severance. Again, however, "there is little hard evidence to show that this is an area which has either been examined in any great detail, been quantified in any way or has been comparatively assessed against the [proposed] scheme" (Weston & Smith 1999).
- *The loss of habitat was a superior interest compared to most of the public interest issues raised.* Most of the public interest issues arising in the case were not equivalent in importance to the loss of priority habitat, and could not therefore be regarded as "overriding" interests. Examples include the loss of family homes and farm severance. Although important issues at a local scale, these cannot be seen as equal in importance to the need to protect the priority habitat, given the status of the latter in the EU Habitats Directive. Similarly, in relation to archaeological impacts, in order to be of "overriding" public interest, the archaeological feature affected would need to rank higher than the priority habitat on a European

scale. There was no evidence that such impacts would arise with any of the alternative routes.

One public interest issue that appeared to be of greater importance was the need to relocate the Ballyseedy Monument, which arose with one of the alternative route options.

> The Council and the local community generally consider the relocation of the Monument to be unacceptable as it is considered one of the most important modern monuments in Ireland. The Monument, however, has no national or local statutory protection, whereas [Ballyseedy Wood] has statutory protection [at European level] through the [Habitats] Directive . . . On that basis, the relocation of the Monument, while clearly a very important public interest issue, cannot be seen as an "overriding" public interest in terms of the presumption established by the Directive to protect the priority habitat. (Weston & Smith 1999)

Having failed all three of the tests required under the "appropriate assessment" process, EU funding for the proposed N21 link road scheme was withdrawn.

9.4.6 *Conclusions*

The process of appropriate assessment under the Habitats Directive, once a negative impact on the priority habitat has been established, is an exacting one. In particular, few projects are likely to have a genuine absence of viable alternatives, especially if the search for possible alternatives is widely defined. Also, to outweigh the loss of priority habitat, public interest reasons must be of equal or greater weight than the protection of priority habitats at European level. This means that issues of only local or even national importance would not be sufficient. Finally, as illustrated above, the absence of alternatives and imperative reasons tests are inextricably linked. "While there remains the possibility of alternative solutions there are unlikely to be 'imperative reasons of overriding public interest' to justify the preferred solution" (Weston & Smith 1999).

9.5 Portsmouth incinerator – new approaches to public participation in EIA

9.5.1 *Introduction*

This case study involves an innovative approach to public participation within the EIA process for a proposed municipal waste incinerator in Portsmouth, Hampshire, UK. The approach adopted by the developer in this case provided an opportunity for members of the public to take part in structured discussions about the project proposals and their environmental impacts before the submission of the planning application and environmental statement. This approach to extended public participation, beyond that required in the EU EIA Directive, has been used in a number of cases in the UK waste sector in recent years, not only at project level (as in this case) but also at more strategic levels in the development of local waste management strategies and plans (Petts 2003, 1995).

The increasing use of these methods reflects the perceived inadequacy of more traditional forms of public participation in the highly contentious arena of waste facility planning. However, questions remain about the effectiveness of such methods in providing genuine opportunities for the public and other interested stakeholders to participate in the EIA and wider development processes. The case described here is based largely on research carried out by Chris Snary as part of his PhD studies with the IAU at Oxford Brookes University (previously documented as Snary 2002), with additional material from Petts (2003, 1995).

9.5.2 Public participation and EIA

The wider context to the case study is the almost universal opposition towards proposed waste management facilities amongst those who live near proposed sites. Such public opposition is often dismissed simply as a NIMBY reaction or as being based on unjustified and irrational fears about potential impacts, particularly in relation to emissions and associated health risks. This is contrasted with the scientifically based technical assessments of impact and risk carried out by EIA practitioners. However, Snary (2002) points out that recent studies indicate that public opposition to such facilities is often based on a much wider range of considerations, including "concern about the appropriateness of the waste management option, the trustworthiness of the waste industry and the perceived fairness of the decision-making process".

Reflecting this improved understanding of the nature of public opposition, a number of commentators have called for better communication with the public at all stages of the waste management facility planning process (ETSU 1996, IWM 1995, Petts 1999). Such communication can take a variety of forms, ranging from a one-way flow of information from developer to public, through different levels of consultation and participation (in which there is a two-way exchange of views between the public and the developer and/or consenting authority, and the public's views are a legitimate input into the decision-making process). All of these types of communication are seen to be important components in the planning and EIA process for incinerators and other waste facilities, as Snary (2002) explains:

> Concerns about health risks require comprehensive *information* on the [predicted] emissions and a *consultation* process through which the public's views can affect the decision-making process. Concerns about the ability of the waste industry and regulators to manage risk competently require *participation* in a process through which their concerns can be openly addressed and conditions of competency discussed. Debate concerning fundamental policy issues and the legitimacy of the waste planning process [also] requires a public *participation* process through which a consensus may be built [at the plan-making stage of the waste incinerator planning process].

The search for improved methods of public participation is also linked to the growing social distrust of science and experts noted by a number of commentators (see, for example, House of Lords 2000, Petts 2003, Weston 2003).

9.5.3 Background to the proposed scheme

Hampshire is a county on the south coast of England, with a population of around 1.6 million. By the end of the 1980s, the county was faced with the problem of increasing volumes of household waste, set against a background of an ageing stock of incinerator plants (which failed to meet the latest emission standards) and growing difficulties in finding new and environmentally acceptable landfill sites. In response, the CC's Waste Management Plan prepared at the time (1989) advocated an integrated approach to waste management, supporting recycling and waste minimization initiatives and emphasizing the need for a reduced reliance on landfill. Government financial regimes in operation at the time (the Non-Fossil Fuel Obligation) also provided cost incentives for the development of energy-from-waste schemes rather than landfill. It was also recognized that significant economies of scale could be obtained by developing a single large plant in the county rather than several smaller ones. As a result, following a tendering process, an application was submitted at the end of 1991 for a large energy-from-waste incinerator in Portsmouth, in the south of the county on a site selected by the CC (Petts 1995). The capacity of the plant was 400,000 tonnes per annum, which represented two-thirds of the household waste arising in the county (Snary 2002). The proposed location was on the site of one of the county's redundant incinerators, which had been closed in 1991 after failing to meet the latest emission standards.

The proposal met with much local opposition, from both local residents and ultimately the relevant local authority, Portsmouth City Council. Objections focused on a number of environmental issues, including the health risks posed by emissions from the plant; visual, noise and traffic impacts; and the close proximity of the site to residential areas. Policy concerns were also raised, in particular that, by concentrating on incineration as the preferred waste option, the promotion of recycling and waste minimization in the county would be adversely affected. In the event, the CC (the consenting authority for this type of project at the time) decided that it could not support the application, on the grounds that the proposal was too large and did not form part of a more integrated waste management strategy for the area (Snary 2002).

The failure to gain approval for the proposed scheme resulted in a change of approach from the County Council, as Petts (1995) explains:

> By the summer of 1992 the County Council had failed to gain approval for the plant and was facing an urgent task to find a solution to the waste disposal problem. The traditional approach had failed. While the [county's waste management] plan which had supported the need for [energy-from-waste] had been subject to public consultation with relatively little adverse comment, this was now regarded as too passive a process and it seemed that the real concerns and priorities of the community had not been recognised by the County [Council]. There had not been strong support of the need for an integrated approach to waste management and there had been little recognition of the need to "sell" [energy-from-waste] to the public. The proponents had been overly optimistic about their ability to push the project through with the standard, information-based approach to public consultation.

Faced with these problems, the CC embarked on the development of a more integrated and publicly acceptable household waste management strategy (Snary 2002). The Council's new approach involved an extensive proactive public involvement programme, launched in 1993, to examine the various options for dealing with household waste in the county. The aim was to attempt to establish "a broad base of public support for a strategy which could be translated into new facilities" (Petts

1995). As part of this process, Community Advisory Forums were established in the three constituent parts of the county, based on the model of citizens' panels. Membership included a mix of people with different interests and backgrounds, including those with little prior knowledge of waste issues. At the end of the process (which lasted for six months), the forums presented their conclusions to the CC. The broad consensus reached was that:

- greater efforts should be made in waste reduction and recycling;
- energy-from-waste schemes would be needed as part of an integrated waste management strategy, but there was considerable concern about their environmental effects and the monitoring of plant; and
- landfill was the least preferred option (Petts 1995).

The public participation exercise in Hampshire resulted in the inclusion in the county's revised waste strategy (published in 1994) of plans to build three smaller energy-from-waste incinerators (each with a capacity of 100,000–165,000 tonnes), rather than the single large incinerator originally proposed. The new plants were to be located in Portsmouth (on the same site as the earlier application), Chineham, near Basingstoke, and Marchwood, near Southampton. EIA work for these proposed developments began in 1998 (Petts 2003). It is the first of these plants that is the focus of this case study.

9.5.4 The contact group process

The EIA process for each of the three proposed incinerators in Hampshire involved a method of public participation known as the "contact group" process. This involved an extended process of public questioning during the preparation of the ES for each site through a contact group involving a range of key local interests. These contact groups were established by the developer (Hampshire Waste Services [HWS]), and were part of the contractual requirements placed on the company by the CC (Petts 2003). This approach had the potential to enable the public's views to result in reassessment of issues dealt with in the ES, and to changes in the project proposals and mitigation measures, prior to the submission of the ES to the competent authority.

The terms of reference for the Portsmouth contact group stated that it was designed (a) to allow key members of the public to develop informed decisions about waste issues and the proposal; and (b) to assist the developer (HWS) in ensuring that it understood and responded to the views of the members of the local community (Snary 2002). The arrangements for extended public participation in this case go beyond the legal requirements under the UK EIA Regulations (discussed in Chapter 6), and were the first time that such methods had been used in the UK EIA process for a waste incinerator (Snary 2002).

In the Portsmouth incinerator case, 10 members of the public were included in the contact group – they were selected by HWS to represent a range of local interests, and included a representative from the local school, the local branch of FoE, and the Portsmouth Environmental Forum, plus seven representatives from the six neighbourhood forums in closest proximity to the project site. Group members were encouraged to network with the local residents in their neighbourhood. It was made clear to participants that membership did not imply support for the proposals, and indeed almost all of the participants were opposed to the development.

The contact group met once a week over a six-week period immediately prior to the submission of the planning application and ES in August 1998. Issues covered by the contact group at these meetings included:

- waste-to-energy incinerators and EIA;
- design of the plant;
- noise and traffic assessments;
- visual and ecological issues;
- alternative sites and noise issues; and
- air quality issues and health risk assessment.

Information was provided on these issues by HWS and by its consultants at the meetings. During discussions, the participants were able to make their views known by raising questions, concerns and suggestions. Answers to questions were provided on the day and in written form at the next meeting. There was also a closing meeting to discuss the conclusions of the ES. An independent chairperson was appointed by HWS "to ensure that all participants had an equal opportunity to contribute to the meetings and that issues were fairly addressed" (Snary 2002).

9.5.5 *Evaluation of the process*

How effective were the methods of public participation employed in this case, and what were the views of the various participants in the process? Snary (2002) has assessed the success of the contact group process in the Portsmouth case, based on interviews with those involved; the process has also been evaluated by Petts (2003), drawing on observation of all three contact groups. Key findings are summarized below, focusing in particular on the limitations of the process in practice.

- *The contact group process took place too late in the EIA process.* The contact group meetings started only 6 weeks prior to the submission of the ES, and by this stage the majority of the EIA had been completed. The opportunity for the group to influence the way in which impacts were defined, assessed and evaluated was therefore very limited. This was particularly true of the health risks posed by emissions, which were discussed only at the last meeting of the group. The process would have been more effective if it had started during the scoping stage of the EIA. However, the scoping exercise was restricted to consultation with the local planning authority and statutory consultees, with no public involvement (Snary 2002).
- *Insufficient time was allowed for the process.* A number of participants commented that the meetings were attempting to cover too much information – often of a complex nature – in too short a time. Again, this suggests that the process should have been started earlier to allow the wide range of issues involved to be dealt with adequately.
- *Criticisms were made of the EIA consultants.* Some of the participants were critical of the consultants involved in the preparation of the ES. Criticisms included the view that assessments were based too much on desk studies and that the consultants lacked detailed knowledge of the locality; that the consultants did not always

provide adequate answers to questions; and that the EIA work should have been undertaken and presented by independent consultants.

- *Participants were generally better informed about the proposals.* Almost all participants stated that they felt better informed about the issues relating to the proposal as a result of attending the meetings. This is hardly surprising, but the developer's project manager also argued that the process had "informed key members of the public better than the traditional methods of public involvement could ever have done" (Snary 2002). However, doubts were expressed about the complex nature of the information provided about the health risks posed by the development. One participant commented "I am not a scientist and I found it very difficult to understand. I felt as though they were trying to blind me with figures and technical terms. The residents that I have spoken to who went to have a look at the environmental statement felt exactly the same; they didn't really understand the assessment." These criticisms are partly related to the tight timescale for the contact group process, although non-experts will always need to have a degree of trust in those providing technical information in EIA. Snary suggests that such trust could have been increased by the use of independent consultants or an independent third party to summarize and validate the information presented.
- *Limited impact on the development proposals.* The contact group meetings appear to have had only limited impacts on the scheme proposals. The project manager for the development stated that the process had resulted in changes to the architecture of the scheme (in particular the colour of the buildings) and improvements to the traffic assessment. However, apart from these relatively minor changes, many participants were sceptical about how else the views of the group had affected the proposals. These findings are not surprising, given the fact that the meetings took place at such a late stage in the planning, design and EIA work for the scheme.
- *Low levels of trust in the developer and consultants.* All participants had only low or moderate levels of trust in the developer and its consultants. Reasons for low levels of trust included a feeling that the developer was bound to be biased because its aim was to gain planning permission, a view that group members were only being provided with part of the information about health risks and concerns over the competency of the EIA consultants.
- *The process failed to resolve fundamental concerns about the proposal.* All but one of the participants still had relatively strong risk-related concerns about the proposal at the end of the contact group process. Therefore,

although the contact group was able to better inform key local stakeholders about the risks posed by emissions, it was unable to convince the majority of the group that the risks were acceptable and that waste-to-energy incineration was an appropriate waste management solution. (Snary 2002)

This is despite the fact that, as we have seen, the Portsmouth incinerator proposal emerged as part of a county-wide waste strategy that was developed through an extensive and innovative public involvement exercise. Snary attributes this to inadequacies in the earlier strategic-level consultation exercise, which had failed to reach a consensus on the appropriate role of waste-to-energy incineration in the county's waste strategy and which most of the contact group members had been unaware of prior to joining the group. It was also unclear how the views expressed in the strategic consultation had influenced the county's developing waste strategy.

In her evaluation of the Hampshire contact group process, Petts (2003) reaches broadly similar conclusions:

> While the process did open up the environmental assessment to detailed questioning by a small but representative group of the public, it arguably started too late in the limited regulatory process to allow the Contact Group members to frame and define the problems to be considered and assessed. During the author's own observation and evaluation of the process, it was evident that questions about the assessment methods were able to be raised (for example, the Portsmouth Contact Group identified deficiencies in the transport assessment based upon knowledge of cycling on the local roads). Some reassessment did take place as a result of such a public quality assurance mechanism. However, this was limited. Participants valued the opportunity provided to them to review the assessment but were suspicious that outcomes had already been decided.

9.5.6 *Conclusions*

The case study has illustrated the use of extended methods of public consultation in EIA, which go beyond the minimum legal requirements in the EU EIA Directive. These methods are not without their practical difficulties, and these have been highlighted. The main weakness in this case appears to have been that the contact group meetings started too late in the overall EIA process. Public involvement at the scoping stage of the EIA may have helped to avoid some of the problems encountered.

As a postscript, after a public inquiry was held in 2000, planning permission for the Portsmouth incinerator was finally granted in October 2001 – 10 years after the initial application for an incinerator on the site had been submitted.

9.6 Cairngorm mountain railway – mitigation in EIA

9.6.1 *Introduction*

It is appropriate that one of our case studies includes a tourism project, for tourism is the world's largest industry, it is growing apace and it contains within itself the seeds of its own destruction. That tourism can destroy tourism has become increasingly recognized over the last 30 years or so, with a focus of concern widening from initially largely economic impacts to a now wider array which includes social and biophysical impacts (see Glasson et al. 1995, Hunter & Green 1995, Mathieson & Wall 2004). Mountain areas can be particularly sensitive to tourism impacts, including from walking, skiing and associated facilities. This case study takes a particularly controversial project, the Cairngorm mountain railway, in the Highlands of Scotland, which was opening in 2001 after a long and protracted debate about its impacts and their management. This brief case study focuses on the latter aspect as an example of approaches to mitigation in EIA.

9.6.2 The project

The Cairngorm Ski Area is one of five ski areas in Scotland. It developed rapidly in the 1960s and 1970s in combination with the adjacent settlement of Aviemore. Chairlift facilities were built to take skiers to the higher slopes in winter, and also to carry walkers in other times of the year. However, the industry has been vulnerable to climate/weather trends and to the quality of the infrastructure. In 1993 the Cairngorm Chairlift Company published a Cairngorm Ski Area Development Plan designed to upgrade facilities, to give better access to reliable snow-holding in the area, to reduce vulnerability to adverse weather conditions, to improve the quality of visitor experience and to improve economic viability, whilst ensuring that all relevant environmental considerations are taken fully into account. The Cairngorm Funicular Railway was a key element in the plan.

The Cairngorm Funicular is the country's highest and fastest mountain railway. It is approximately 2 km in length and takes visitors in 8 minutes from the existing chairlift station/car park base at Coire Cas (610 m) to the Ptarmigan top station (1100 m). It comprises two carriages (or trains of carriages) running on a single-line railway track between two terminal points. The carriages, which start at opposite ends of the track, are connected by a hauling rope. As one carriage descends the track, the other travels upwards and they pass each other at a short length of double track midway. The track is carried on an elevated structure, a minimum of 1 m and a maximum of 6 m above ground level. The final 250 m runs in a "cut and cover" tunnel. The development has also included a major remodelling of the existing chairlift base station, and replacement of the existing top station with a new development, which includes catering facilities for about 250 persons, and a new interpretative centre, including various displays and an outdoor viewing terrace. The previous chairlift and towers have been removed as part of the development. It was anticipated that the railway would carry approximately 300,000 visitors p.a., with two-thirds in the non-skiing months. This would represent a three- to four-fold increase over 1990s numbers reaching the top station by the chairlift, and a doubling of numbers from the early 1970s.

9.6.3 The EIA and planning process

The original planning application for the Funicular Railway was submitted in 1994, with an ES (Land Use Consultants 1994). Revised proposals and a supplementary ES were submitted in early 1995. The scheme was very controversial, with much opposition. Particular concerns focused on the potential impact of improved visitor access to the sensitive environment of the summit plateau, which is recognized as a European candidate Special Area of Conservation and an SPA. As a condition of the planning approval, it was necessary for the developer to satisfy Scottish Natural Heritage (SNH) (the statutory body responsible for nature conservation in Scotland) that a visitor management plan (VMP) and other mitigation measures would be put in place that would avoid adverse impacts on the summit plateau.

The planning application was approved by the Highland Council (the consenting authority) in 1996. This was subject to a Section 50 (now Section 75) planning agreement to create, in partnership with the SNH and in agreement with the developer/authority, a regime for visitor and environmental monitoring and management. Amended designs for the station buildings were approved in 1999, and construction work finally began in August 1999. The railway opened in December 2001, following the approval of the proposed VMP by SNH.

9.6.4 *Visitor management and mitigation measures*

The Section 50 agreement attached to the planning approval is a legally binding agreement between the planning authority, in partnership with SNH, and the developer/operator and landowner. The agreement provides for (Highland Council 2003):

- a baseline survey of current environmental conditions and visitor usage in the wider locality;
- an implementation plan providing details of the timing and means of implementation of the development with particular reference to reinstatement following construction;
- an annual monitoring regime to identify changes and establish causes and consequences [the first monitoring report has been prepared and is under discussion];
- an annual assessment by the operator of any actions necessary to ensure acceptable impacts to the European designated conservation sites on the summit plateau;
- fall-back responsibilities in the event of default; and
- eventual site restoration if public use of the development ceases.

The Cairngorm Funicular Railway VMP was produced in the context of this agreement. The objective of the VMP is to protect the integrity of the adjacent areas which have been designated or proposed under the European Habitats and Birds Directives from the potential impacts of non-skiing visitors as a direct consequence of the funicular development. The VMP has gone through several stages and was subject to a short period of public consultation in 2000. Many issues have been raised, including the innovative or repressive (according to your perspective) "closed system", and the associated monitoring arrangements (SNH 2000).

The closed system, whereby non-skiing visitors are not allowed access to the Cairngorm plateau from the Ptarmigan top station, is a key feature of the VMP. Instead visitors must be content with a range of inside interpretative displays, access to an outside viewing terrace – plus, of course, shopping, catering and toilet facilities! This system proved very contentious, and received considerable criticism, in the public consultation on the VMP. Some saw it as violating the freedom to roam; for others, it was a cynical device for extracting economic benefit in shops and catering outlets. Others considered it unnecessary, given the recent improved pathway from the Ptarmigan top station to the summit of Cairngorm, as noted by one respondent – "For years I have been advocating stone paths. People use the paths and the ground round about recovers. Now that the path up to the summit is pretty well complete most people will be barred from using it!" (SNH 2000). Another issue has been how to allow ingress to the facilities of the top station from non-railway-using walkers on the plateau, whilst preventing egress from non-walkers. Alternatives have been suggested to the closed system including ranger-led walks and time-limited access, but the system is now in place and is part of a 25-year agreement. The guide leaflet for the funicular users includes the following:

Protecting the Mountain Environment: large areas of the fragile landscape and habitats of the Cairngorms are protected under European Law. Cairngorm

Mountain Limited is committed to ensuring that recreational activities are environmentally sustainable. For this reason the Railway cannot be used to access the high mountain plateau beyond the ski area at any time. Outwith the ski season, visitors are required to remain within the Ptarmigan building and viewing terrace, returning to the base station using the railway. Mountain walkers are welcome to walk from the car park and use the facilities at the Ptarmigan, but may not use the railway for their return journey and are asked to sign in and out of the building at the walkers' entrance.

As noted in earlier chapters of this book, monitoring can support effective mitigation measures. For this project, monitoring will cover all topics subject to baseline surveys – including visitor levels, behaviour and, habitats, birds, soils and geomorphology. It will use the Limits of Acceptable Change (LAC) method, whereby indicators and levels of acceptable change will be identified, monitored and, when levels are reached, will trigger management responses (see Glasson et al. 1995). In response to a concern about the independence of the monitoring activity, the annual monitoring reports will be presented to the SNH and the Highlands Council by an independent reporting officer jointly appointed by them.

9.6.5 Conclusions

The Cairngorm Funicular has now been operational for 3 years. Visitor numbers have been less than the predictions in the ES, but still represent a substantial increase on previous levels in the 1990s. All conditions have been complied with, the Section 50(75) agreement has been secured, and a good working partnership has been established between public authorities and the operator; there is access by all-abilities to the Ptarmigan top station, an improved footpath system and the old White Lady chairlift system has been removed. On the basis of such achievements, the Highlands Council submitted the development for a 2003 Scottish Award for Quality in Planning (Highland Council 2003).

9.7 Humber Estuary projects – assessment of cumulative effects

9.7.1 Introduction

This case study provides an example of an attempt to assess the cumulative impacts of a number of adjacent concurrent projects in the Humber Estuary, Humberside, UK, undertaken in the late 1990s. This type of cumulative effects assessment (CEA), which was undertaken collaboratively by the developers involved in the various projects, is relatively uncommon in EIA. However, a number of other examples do exist, for example, in wind energy development cases in which several wind farms have been proposed in the same area. More generally, the assessment of cumulative impacts is widely regarded as one of the weak elements in project-level EIA (see, for example, Cooper & Sheate 2002; see also Section 11.3).

Cumulative effects assessment studies of the type described here present a number of difficulties, and the case study examines how and to what extent these were overcome. The benefits derived from the CEA process are also discussed, from the viewpoint of the various stakeholders involved. This case study is based on research carried out by Jake Piper as part of her PhD studies with the IAU, Oxford Brookes University, and has previously been documented as Piper (2000). The Humber Estuary case study, together with a number of other examples of cumulative effects assessment in the UK, is also examined in Piper (2001a,b, 2002).

9.7.2 The Humber Estuary case study

This case study involved a cluster of adjacent projects, proposed at around the same time by different developers. Each of the proposed projects required EIA, and because of the variety of project types, more than one consenting authority was involved in approving the projects. The developers concerned agreed to collaborate in the preparation of a single CEA of their combined projects, which was presented to each of the consenting authorities simultaneously.

In 1996–97, five separate developments were proposed along the north bank of the Humber Estuary, within a distance of 5 km of each other. The projects included:

- a new wastewater treatment works serving the city of Hull;
- a 1200 MW gas-fired power station;
- a roll-on/roll-off sea ferry berth;
- reclamation works for a ferry terminal; and
- flood defence works.

The five proposed projects involved four separate developers and five consenting authorities. The environment in the vicinity of the projects was a sensitive one, with a European site for nature conservation – an SPA designated for its bird interest under the EU Birds Directive and EU Habitats Directive – located within a short distance of the developments. It was the presence of this site, and the almost concurrent timing of the projects, that prompted the CEA study in this case. Indeed, the CEA was designed to satisfy the requirements for an "appropriate assessment" of the effects of the proposed schemes on the SPA, under the terms of the Habitats Directive (similar to the process described in Section 9.4). It was also hoped that the CEA would help to avoid lengthy delays in securing approval for the projects, as Piper (2000) explains:

> The strategy adopted assumed that, by providing a common assessment to answer the needs of each of five competent authorities involved..., the amount of interplay and discussion required between these authorities would be reduced, avoiding lengthy delays.... The strategy means, however, that any insoluble problems associated with any one project could tie up all consent applications simultaneously.

In order to guide the CEA process, a Steering Group was established consisting initially of the developers and the two local authorities concerned. Other key statutory consultees, including the Environment Agency and EN, joined the Steering Group later, but non-governmental environmental organizations and the public were not directly involved.

A single environmental consultancy prepared the CEA, acting equally on behalf of all four developers. Draft reports were prepared in consultation with the statutory consultees and developers, with opportunities for review and comment. Close liaison with EN (the statutory body responsible for nature conservation) was an important element in the process, given the need to specifically address the potential impacts on the SPA. It was important to ensure that the document presented to the local authorities and other consenting authorities also fulfilled the requirements of this statutory consultee.

The Steering Group was involved in determining the scope of the CEA, but no public participation was arranged for this stage of the study. The scoping exercise identified those issues where there was potential for cumulative effects to occur. These included, during the construction phase, effects on bird species on the SPA site and on traffic, and during the operational phase, effects on estuary hydrodynamics, water quality and aquatic ecology. Data was made available for the study by the developers, including information from existing EIA work already undertaken; some additional modelling work was also carried out. The information provided included the probable timing of activities within the construction programmes for each project, the manpower requirements for these activities and associated traffic movements. Existing baseline data available included the range of bird species present at different times of year in the SPA, and their vulnerability to disturbance (Piper 2000). Prediction of cumulative impacts was assisted by the production of a series of tables and matrices, which brought together the levels and timing of impacts identified for each project. These included

- a combined timetable of major construction works;
- bird disturbance potential (sensitivity in each month of the year);
- timetable of construction work potentially affecting birds, and monthly sensitivity;
- potential aquatic impacts of the developments; and
- predicted traffic patterns (vehicles per day, for each month of the construction works).

In arriving at predictions, it was decided to use the developers' best estimates, rather than a worst-case scenario approach (Piper 2001a).

As a result of the cumulative impacts predicted, a number of additional mitigation measures were proposed (in addition to those measures that would have been considered had the schemes been assessed separately). Examples included the scheduling of certain noise-generating construction activities such as piling outside sensitive periods (e.g. bird roosting), and the introduction of staggered working hours to reduce peak traffic volumes. It was also proposed that the design of adjacent projects should be integrated in such a way as to minimize environmental impacts. An example was revisions to the design of the ferry berth structure to complement the design of the outfall from the water treatment works, and so enhance mixing of water in the estuary. Finally, recommendations were made for continued monitoring of the cumulative effects on birds and the aquatic environment. Responsibility for funding this work was shared amongst a sub-group of the developers involved in the proposed schemes (Piper 2000).

9.7.3 *Costs and benefits of the CEA process*

Piper (2000) has assessed the costs and benefits associated with the Humber Estuary CEA study, drawing on a series of interviews with those involved in the process, including the developers, the relevant local authorities, other consenting authorities and statutory consultees. The views of these different stakeholders are summarized below, beginning with the developers of the proposed schemes.

Views of developers

- *Greater understanding of the area and potential development impacts.* Three of the four developers felt that the CEA process had increased their understanding of the estuary and the potential impacts of the proposed developments. For example, the power station developer referred to better understanding of the impacts to the mudflats and birds and potential traffic impacts, whilst the dock developer emphasized greater understanding of the hydrodynamics and morphology of the estuary and the relationship between the schemes and the SPA.
- *Other benefits.* Other benefits identified included the development of local relationships, including closer working relationships with the other developers, LPAs and statutory consultees; the establishment of a consistent basis for mitigation and monitoring; the opportunity to share the costs of ongoing monitoring work in the estuary; and – for one of the developers – the fact that the CEA process had facilitated the rapid achievement of planning approval.
- *Financial costs of the CEA process.* The financial cost of undertaking the CEA was relatively low for all of the developers, although the majority of the cost was in fact borne by a single developer (the water utility company). The cost of the CEA to this company represented around 5 per cent of the total cost of the EIA work for its proposed scheme. Costs were much lower for the other developers.
- *Changes to the project proposals and additional mitigation.* As noted above, the CEA process resulted in some changes to the original project proposals and additional mitigation measures, which would not have occurred if the projects had been assessed separately. Examples included changes to piling operations during the power station construction to minimize noise impacts, modifications to the ferry berth construction to compensate for loss of bird habitats elsewhere in the estuary, changes to the timing of certain construction activities and staggering of working hours to minimize peak traffic flows. All the developers indicated that the additional mitigation prompted by the CEA had added relatively little to the costs of the overall development. This may reflect the ability to share the costs of mitigation measures between the developments. Without this opportunity, mitigation might have been less effective or more costly (Piper 2000).
- *Delays caused by the CEA process.* Views differed about whether the CEA process had resulted in a saving or loss of time in obtaining consent for the proposed schemes. In part, this reflected the stage in the planning approval process reached by each developer at the start of the CEA process. Delays ranged from one to two months for the water utility company to six months for the dock developer (this last delay was attributed to the late involvement of a statutory consultee, despite an earlier invitation to join the study); the power station developer felt that its timetable had not been affected. Some delay may have been caused by the fact that the CEA process began after the bulk of the initial consultation and assessment work on some of the schemes had been completed. This resulted in some duplication of effort.
- *Other issues.* One developer noted the problem of distinguishing between those changes that resulted from the CEA process and those that would have occurred anyway through the proper consideration of each scheme in isolation. A further

issue concerned the appropriate treatment of new projects that may come forward in the area after the initiation of the CEA. Should such projects be incorporated into the CEA process, implying an open-ended timescale for the process, or should a new CEA be started for the next group of schemes?

Views of local planning authorities and consenting authorities

The two local planning authorities responsible for the area in which the developments were located were supportive of the CEA study and identified a number of benefits from the process:

> The study was found to be helpful in assessing the overall impact of several major projects proposed for a relatively small geographic area. The study was very helpful in its technical assessment of impacts. The study was definitely of great value for both [councils] in understanding likely impacts.
>
> [It] was probably of equal value in demonstrating the likely impacts to the developers themselves, making them fully aware of the potential consequences of their proposals. (Comments from local authority representatives, quoted in Piper 2000).

The point was made that local planning authorities lack the technical expertise and resources to carry out detailed review of environmental assessments, and therefore rely on the integrity of ES authors and consultants to identify areas of potential concern. In this respect, "a major factor in favour of the CEA [process] is that the advisers of each scheme proponent help 'to monitor the others', thus 'producing a more balanced product'" (Piper 2000). Both authorities commented on the lack of public participation in the CEA study. One noted that, partly due to the tight timescales involved, there had been little or no public consultation, and that this represented the main weakness in the process.

Other consenting authorities included three government departments (DTI, DETR and MAFF). The DTI commented that the study had facilitated decision-making, stating that "without the CEA, the power station project would have been refused" (quoted in Piper 2000). The CEA approach would be recommended in similar cases of multiple projects elsewhere.

Views of statutory consultees

English Nature, as the statutory body responsible for nature conservation, was the principal consultee in this case and was involved in the CEA process from an early stage. It was necessary for the CEA to satisfy the requirements of EN, given its responsibilities under the Habitats Directive to ensure the protection of the SPA. These requirements were expressed in a number of planning conditions attached to the consents for the various schemes:

> The conditions covered the mitigation of construction works (via measures to reduce disturbance of birds, a code of practice for personnel and compliance with a programme of works designed to take account of other CEA-related construction projects) and the monitoring of construction. A monitoring scheme

was outlined which will last throughout construction and for 5 years subsequently and will observe the movements and ranges of population of waterfowl. Provided these stipulated conditions are met, English Nature was of the opinion that the various projects would not, individually or [in combination], adversely affect the conservation objectives of the Special Protection Area. (Piper 2000)

English Nature commented that a number of factors – some of which were unique to this case – had assisted the completion of the CEA. These included the relatively small geographical area covered by the schemes; the fact that all schemes were at an early stage at the start of the process, although some project-specific EIA work had already been completed; the absence of direct competition between the developers to be the first to obtain planning consent; and the willingness of one of the developers (the water utility) to take the initiative in getting the study underway (Piper 2000). The latter was seen as particularly important, given that responsibility for undertaking the "appropriate assessment" under the Habitats Directive properly rested with the consenting authority. As we have seen, in this case there were no fewer than five different consenting authorities. It was suggested that

> it would have been problematic to sort out exactly where responsibility lay, had the CEA strategy not been devised by the water utility and its advisers. For these reasons English Nature indicated that, whilst CEA was "an excellent solution" [in this particular case], it is not a method of immediate and general applicability but depends upon the circumstances encountered in each case. (Piper 2000)

9.7.4 Conclusions

The consideration of cumulative effects is widely regarded as one of the weak areas in EIA, both at project level (see Section 11.3) and in some SEA studies (see Section 9.8). This case study has demonstrated a novel approach to the assessment of cumulative effects, in this case associated with the impacts of a number of adjacent proposed developments. The assessment process was made possible by a number of factors, including the willingness of one of the developers to take the initiative in starting the CEA study and the fact that the developers involved were not directly in competition with each other. These circumstances may not apply in all such cases. Nevertheless, CEA studies of the type described have a number of benefits, for developers, consenting authorities and other key stakeholders, and – at least based on the evidence in this case – appear to involve relatively little additional cost.

9.8 UK offshore wind energy development – strategic environmental assessment

9.8.1 Introduction

This case study provides an example of the application of SEA in practice, and concerns the SEA of the UK government's plans for the future development (post-2004) of offshore wind energy. The SEA was carried out during 2002–03, prior to the implementation of the EU SEA Directive (2001/42/EC). Further information on the requirements of this Directive, and on SEA more generally, can be found in Chapter 12.

The context for this particular example of SEA was ambitious government targets for renewable energy generation, linked to the achievement of the UK's commitments in the Kyoto Protocol to significantly reduce CO_2 emissions. At the time of the SEA (2002–03), the UK Government was committed to supplying 10 per cent of electricity needs from renewable sources by 2010, rising to 20 per cent by 2020. Offshore wind energy was seen as a major contributor towards these targets (DTI 2003a), and the UK Government wished to see rapid development of the industry. But it was also committed to an SEA process, which was intended to influence decisions on which areas of the sea should be offered to developers (and which should be excluded), as well as to guide decisions on bids submitted by individual developers. At the time, offshore wind energy was a new industry undergoing rapid development, and there were therefore many uncertainties about environmental impacts, including potential cumulative effects. This presented difficulties for the SEA work.

9.8.2 Development of offshore wind energy in the UK

The development of offshore wind energy in the UK involves separate licensing and consent systems. The licensing system is operated by the Crown Estate, in its role as landowner of the UK sea bed. Licensing takes place under a competitive tendering process in which developers submit bids for potential wind farm sites. It is left to the developers themselves to identify potential sites, from within broad areas defined by the UK DTI. The developers submitting successful bids are then offered an option on their proposed site. Detailed technical studies, consultation and EIA work on the site is then undertaken by the developer, prior to the submission of a consent application. The necessary planning consents are granted by the DTI and the DEFRA, following consultation with the LPAs most closely affected, statutory consultees and the public. Once the necessary consents have been obtained, developers are granted a lease of 40–50 years on the site and can then begin construction of the wind farm.

In the UK, the Crown Estate's first invitation to developers for site leases (Round 1 of licensing) took place in April 2001. This resulted in 18 planned developments, each of up to 30 turbines. Most of these schemes obtained planning consent in 2002–03 and were installed from 2003 onwards. After this first round of licensing, the government published "Future Offshore", a document setting out its plans for the second licensing round (DTI 2002). This envisaged much larger developments than in the previous round, and stated that future development was to be focused in three "Strategic Areas" – the Thames Estuary, the Greater Wash and Liverpool Bay (Figure 9.4). These areas were selected as having the greatest development potential, based on the potential wind resource available, the bathymetry of the offshore area, proximity to existing grid connections and initial expressions of interest from developers (DTI 2003b); however, environmental constraints do not appear to have influenced the choice of strategic areas.

A three-month consultation period on the "Future Offshore" document started in November 2002. The SEA of the government's plans for Round 2 licensing, which is the focus of this case study, started at the same time, with the resulting SEA Environmental Report submitted in May 2003 (for a 28-day consultation period). Despite this SEA process, the government was keen to maintain the pace of development in the offshore wind energy industry, and the deadline for developers to submit expressions of interest for Round 2 site leases to the Crown Estate was the end of

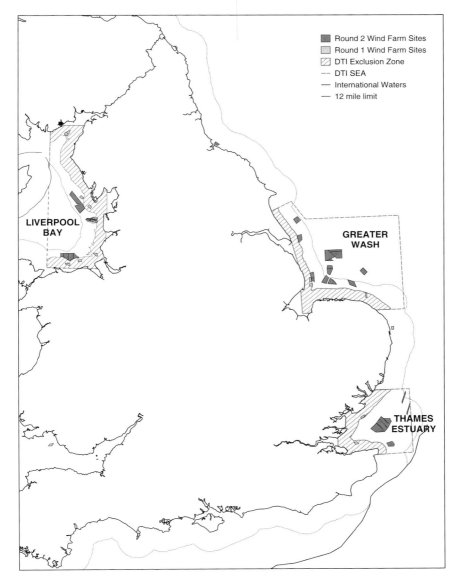

Figure 9.4 Map of the three strategic areas for offshore wind farm development (Round 2).

March 2003 – i.e. prior to the completion of the SEA Environmental Report or the receipt of consultation responses on this report.

The successful bids for Round 2 developments were finally announced in December 2003. These included 15 projects with a total capacity of between 5.4 and 7.2 GW – this compares with the 1.2 GW consented under Round 1, and so represented a step change in the development of the industry in the UK. Some of the selected sites soon proved controversial, with concerns about the potential impacts on important bird habitats raised by the RSPB (2003).

9.8.3 *The SEA approach*

The SEA in this case was of the UK Government's draft programme for the second licensing round of offshore wind energy development. The SEA was commissioned by the DTI voluntarily, in accordance with the requirements of the EU SEA Directive (although this had not yet been implemented at the time). The timescale under assessment was from the present (2003) until 2020, with separate assessments undertaken of development up to 2010 and 2020 (DTI 2003a). Two potential development scenarios ("likely" and "maximum credible") were considered and their likely impacts assessed. The "likely" scenario envisaged the development of 4.0 GW of capacity by 2010, whilst the "maximum credible" scenario envisaged 7.5 GW. By 2020, these figures increase to 10.2 and 17.5 GW respectively. A no-development option was also considered.

A Steering Group was used to guide the SEA process, with its membership drawn from specialists in coastal/marine environmental issues, wind energy development and SEA. Steering Group members included representatives from relevant government departments (DTI, DEFRA, ODPM); the Crown Estate; the British Wind Energy Association (BWEA), the body representing the UK wind energy industry; government and non-governmental environmental organizations, such as the RSPB, Joint Nature Conservancy Council (JNCC), Countryside Council for Wales, and EN; and the IEMA, the body representing the UK EIA "industry".

Consultation was undertaken on the scope and design of the SEA. A scoping workshop was held towards the end of 2002 and a scoping report was produced. Some changes to the scope of the SEA were introduced as a result of the consultation responses received, for example by including a wider range of socio-economic impacts which had been identified as important by a number of consultees (DTI 2003a). The environmental report produced at the end of the SEA process provided information on:

- the nature and extent of the technical, environmental and socio-economic constraints that may preclude or be affected by wind farm development;
- the identification of locations within the three strategic areas (the Thames Estuary, the Greater Wash and Liverpool Bay) with the lowest levels of constraint;
- the significance of the environmental and socio-economic impacts arising from different realistic scales of wind farm development in those areas with the lowest levels of constraint; and
- recommendations for managing the impacts of wind farm development in the three strategic areas.

Overall, it was concluded that:

> The likely development scenario, to 2010, is achievable for each Strategic Area without coming into significant conflict with the main significant impact risks, namely areas of high sensitivity to visual impact, concentrations of sensitive seabirds, designated and potentially designated conservation sites, MoD Practice and Exercise Areas and main marine traffic areas.
>
> [However], the 2020 likely development scenario would only be achievable subject to resolving the uncertainties concerning impacts on: physical processes, birds, elasmobranchs (shark, skate and ray species) and cetaceans.

The maximum credible scenario for all Strategic Areas, particularly the Greater Wash and Thames Estuary, for 2020, may be compromised by constraints, particularly cumulative impacts and conflict with marine traffic (commercial and recreational navigation); and large scale development could exclude fisheries from significant areas of fishing grounds, particularly if it were to coincide with severance areas associated with other offshore activities. (DTI 2003a)

In order to minimize environmental impacts, the following broad strategic approach was recommended:

- the development of fewer large wind farms, of around 1 GW (1,000 MW) or more capacity, located further offshore is generally preferable to several small-scale developments, though the latter would be preferable for development closer to the coast;
- in all strategic areas, avoid the majority of development within the zone of high visual sensitivity close to the coast;
- where development might occur close to the coast, preferentially select low constraint areas and consider small-scale development;
- pending the outcome of monitoring studies, avoid development in shallow water where birds such as common scoter and red throated diver, and other species (including marine mammals) are known to congregate (particularly in Liverpool Bay and the Greater Wash); and
- address the uncertainties of large-scale impacts, particularly cumulative effects, at a strategic level (DTI 2003a).

The environmental report was subject to a short period of public consultation (28 days). The government argued that the report and the comments received would be "a significant input to government decision-making on the nature of the second licensing round" (DTI 2003a).

9.8.4 SEA methods

It must be accepted that, in an SEA, the level of detail that can be analysed and presented, in respect of both baseline data and quantification of impacts, is less than in a project-level EIA. This was true of this particular SEA, which "focuses more on assessing constraints, sensitivities and risks instead of detailed analysis of the characteristics of specific impacts" (DTI 2003a). The methods used in the SEA included a GIS-based spatial analysis (constraint mapping exercise), followed by a risk-based analysis of the likely impacts of the selected development scenarios (including the cumulative implications). Each of these methods is described briefly below, with selected examples included to illustrate the approach used.

Spatial analysis

The spatial analysis made use of electronic overlay mapping of a variety of technical, socio-economic and environmental features to identify areas of the sea with high or low constraints within each of the three strategic areas. Examples of the main features mapped are listed below:

Technical constraints to wind farm development:

- Existing and planned licensed areas for aggregate extraction, waste disposal and military operations;
- Oil and gas structures (pipelines) and safety zones;
- Cultural heritage sites (wrecks and other sea bed obstructions);
- Cables;
- Existing shipping/navigation lanes;
- Proposed wind farm sites from the first round of licensing.

Socio-economic constraints:

- Shipping;
- Fishing effort;
- Shell-fishery areas.

Environmental constraints:

- Marine habitats of conservation interest (designated and potentially designated);
- Seascape sensitivity;
- Fish spawning areas;
- Fish nursery areas.

Because of baseline data limitations, not all relevant constraints could be mapped within the relatively tight timescale of the SEA. In particular, it was not possible to map a number of important environmental constraints, such as the distribution of certain bird and fish species and migration routes. Whether these omissions invalidate the conclusions drawn from the constraint mapping exercise is open to question (see below for a summary of consultation responses on this issue). However, those factors that could not be mapped were considered in the later risk-based analysis of impacts.

A scoring system was used in the mapping of constraints, in which each area was awarded a score between 0 and 3 for each mapped constraint (with higher scores indicating greater constraints). The scoring system used is illustrated by the following examples:

Scores for fishing effort:

- 0   None
- 1   Low (less than 500 hours per annum)
- 2   Medium (500–5,000 hours per annum)
- 3   High (over 5,000 hours per annum).

Scores for designated habitats of conservation interest:

- 0   Designated habitats are absent
- 1   Not applicable
- 2   Nationally important habitats are present (including those not yet designated)
- 3   Internationally important habitats are present (including those not yet designated).

Scores for seascape:

- 0 No sensitivity
- 1 Low sensitivity
- 2 Medium sensitivity
- 3 High sensitivity.

The scoring system allowed the identification of locations within each of the three strategic areas that had several constraints (a high total score) and those with fewer overall constraints (a lower overall score), subject to the qualification that not all relevant constraints could be mapped. Broad conclusions from the spatial analysis are summarized below, for each strategic area (DTI 2003a):

- *Liverpool Bay.* Overall, the greater amount of constraint and sensitivities occur in the southern part of this strategic area, due to the presence of bird interests, marine habitats of conservation interest, seascape, fisheries and marine traffic. Seascape constraints in the north of the area are significant.
- *Greater Wash.* The Greater Wash has the largest area of low constraint in comparison with the other strategic areas and offers the greatest potential capacity for wind farm development. Inshore areas, particularly in the southern part of the area, have the greatest amount of constraint and sensitivity, particularly with respect to visual impacts, inshore fisheries, marine mammals, birds and offshore habitats of conservation interest.
- *Thames Estuary.* Includes areas of low constraint on its eastern boundary, and has fewer environmental constraints than the other strategic areas. However, several estuaries and marshes are important bird habitats. Commercial activities (e.g. aggregate extraction) and recreational navigation are other important constraints.

Risk-based analysis of impacts

For each strategic area, the likely impacts of the two development scenarios ("likely" and "maximum credible") were assessed. This analysis incorporated factors that were mapped as part of the earlier spatial analysis, plus specific receptors that could not be mapped such as particular bird species. Impacts were quantified wherever possible, or otherwise described qualitatively, and their significance evaluated using a risk-based approach. This was based on an assessment of (a) the likelihood of the impact occurring, and (b) the expected consequences (impact on the receptor). As with the spatial analysis, a scoring system was used in the evaluation of impact significance, as shown below (DTI 2003a):

Scores for Impact Consequence:

- 5 Serious (e.g. impacts resulting in irreversible or long-term adverse change to key physical and/or ecological processes; direct loss of rare and endangered habitat or species and/or their continued persistence and viability);
- 3 Moderate (e.g. impacts resulting in medium-term (5–20 years) adverse change to physical and ecological processes; direct loss of some habitat (5–20 per cent) crucial for protected species' continued persistence and viability in the area and/or some mortality of species of conservation significance);

- 1   Minor (e.g. impacts resulting in short-term adverse change to physical and ecological processes; temporary disturbance of species; natural restoration within two years requiring minimal or no intervention);
- 0   None (e.g. impact absorbed by natural environment with no discernible effects; no restoration or intervention required);
- +   Positive (e.g. activity has net beneficial effect resulting in environmental improvement).

Scores for Impact Likelihood:

- 5   Certain (the impact will occur);
- 3   Likely (impact is likely to occur at some point during the wind farm life cycle);
- 1   Unlikely (impact is unlikely to occur, but may occur at some point during wind farm life cycle).

The impact significance scores for each impact are the product of the consequence and likelihood scores, ranging from 1 (minor consequence and unlikely) to 25 (serious consequence and certain).

9.8.5 Issues raised in consultation responses

The environmental report produced at the end of the SEA process was subject to a short period of consultation. An analysis of the consultation responses reveals a number of key issues that were raised by interested stakeholders. These include a range of concerns about the quality and effectiveness of the SEA process, and its influence on decisions for the next phase of wind energy developments. Many of these concerns arose from the tight timescale for the SEA work and the resulting practical difficulties encountered. The main points raised in consultation are high-lighted briefly below (a fuller discussion can be found in DTI (2003b)). Many of the issues raised are interlinked; for example, weaknesses in baseline data may be due to limited consultation or a tight timescale in which to complete the SEA.

- *Pre-selection of the three strategic areas, and lack of a national-level SEA.* The three strategic areas in which Round 2 development was to be focused were selected as having the greatest development potential, based on the potential wind resource available, the bathymetry of the offshore area, proximity to existing grid connections and initial expressions of interest from developers (DTI 2003b). However, the selection did not appear to take explicit account of environmental constraints, and this was a cause of concern to a number of respondents.
- *The tight timescale for the SEA and uncertainty over the influence of the SEA process on decision-making for future developments.* The timescale for the SEA was considered too tight to allow effective stakeholder engagement and consultation, or to allow additional baseline data to be collected. The fact that developer bids for Round 2 sites were invited before the completion of the SEA Environmental Report was also a source of concern.
- *Concern over the rapid development of offshore wind energy, prior to the proper consideration of potential impacts.* Some respondents argued that the development of the offshore wind energy industry was too rapid and premature; greater efforts

should be made to understand the impacts of the smaller Round 1 developments before allowing large-scale expansion of the industry.

- *The need for clearer locational guidance.* There was felt to be a need for clearer recommendations on suitable and unsuitable locations for future offshore wind energy development (including the definition of exclusion zones or "no-go" areas), and it was considered that these had not emerged sufficiently from the SEA process.
- *Concerns about the scope and methodology of the SEA.* Most respondents were supportive of the overall methodology of the SEA, including the risk-based approach to the assessment of impact significance, but there was some disagreement over the detailed scores awarded to specific receptors or geographical areas.
- *Weaknesses in the available baseline data.* A recurrent theme in the consultation responses was limitations in the baseline data available to the SEA study. This included missing data for certain important environmental constraints (which could not be mapped) and areas in which the data used in the SEA was not the most accurate or appropriate. Some respondents thought that these data limitations were sufficiently serious as to invalidate the identification of areas of high and low constraints in the SEA. Data gaps and uncertainties about impacts also led respondents to urge a precautionary approach; the need for such an approach was also strengthened by ongoing delays in the designation of offshore areas of conservation interest under the EU Habitats Directive. Other uncertainties included doubts about whether the impacts of smaller wind farms (from the first round of licensing) could necessarily be extrapolated to larger wind farms further offshore.
- *Limited consultation with certain stakeholders.* According to some respondents, the SEA had involved only limited consultation with certain stakeholders – e.g. fisheries and recreational boating interests. This lack of consultation, again partly linked to the tight timescale for the exercise, helped to explain some of the data weaknesses on certain issues in the SEA.
- *Insufficient attention to cumulative and indirect effects.* It was considered that insufficient attention was given to the impact of related onshore development in the SEA, such as transmission connections. This concern echoes the issues highlighted in the first case study in Section 9.2. More attention also needed to be devoted to cumulative impacts in the environmental report. There was no indication in the report of the carrying capacity of each of the strategic areas, and it was therefore difficult to assess the significance of the cumulative impacts arising under the two development scenarios.
- *Overlaps with project-level EIA.* There was some disagreement over the level of detail needed in the SEA, and which issues could be left to project-level EIA for individual sites. For example, bird distribution data was considered to be one area in which survey data could reasonably be collected at a more strategic level.
- *Responsibility for future SEA studies.* Some respondents requested clarification about who would be responsible for progressing further studies arising from the SEA, including additional data collection to fill existing data gaps and ongoing monitoring. Arrangements for the sharing of such data were felt to be important.

9.8.6 Conclusions

This case study of SEA was undertaken voluntarily, prior to the implementation of the EU SEA Directive. It provides an example of how SEA can be applied, within the context of a new, rapidly developing industry. The UK Government's commitment to large-scale development of offshore wind energy to meet international obligations to reduce CO_2 emissions dictated a tight timescale for this SEA. However, the resulting limitations in baseline data, and restricted timescale for stakeholder consultation and feedback, were identified as particular weaknesses in this case.

9.9 Pre-assessment of the impact of refugees in Guinea

9.9.1 Introduction

This case study of a UNEP EIA of refugee camps in Guinea, West Africa highlights several issues. First, although the UNEP calls it an "impact pre-assessment" (in this case because it is only a partial assessment, more like a scoping document), it is really a post-assessment: many impacts had already happened by the time of the assessment. The assessment was trying to prevent the problems from getting worse, and could also be used to try to prevent similar impacts in future refugee camps. Second, it is a form of SEA. It takes a broad-brush view of an issue affecting an entire country, and does not focus on site-specific problems. The solutions that it proposes are similarly large-scale. The case study also highlights the interrelationships between social and environmental impacts. This is often particularly striking in developing countries where social customs have evolved over thousands of years to help preserve environmental capacities. In such circumstances, social disruptions can lead to severe environmental repercussions and vice-versa. This case study is heavily based on UNEP et al. (2000) and ReliefWeb (2001).

9.9.2 The problem

Guinea is a western African country the size of the UK, with 7.7 million inhabitants. It is one of the world's poorest countries, with a life expectancy of under 50 years. Since 1989, Guinea has had a steady influx of refugees from the neighbouring countries of Liberia, Sierra Leone and Guinea-Bissau due to conflicts in those countries. About 400,000 refugees lived in Guinea in 2001.[1] In contrast to many other countries, which have settled refugees in large segregated camps, with corresponding alienation between citizens and refugees, in Guinea the refugees were generally allowed to settle spontaneously and peacefully around local towns and villages, leading to dispersed settlement patterns. Many Guinean households accommodated refugees, and the situation was, until recently, one of social integration and peaceful cohabitation. Many refugees lived near their country of origin, in the south-east and south-west of Guinea.

The large influx of people has presented Guinea not only with an economic burden – Guinea was able to balance its national budget before the refugee crisis but is no longer able to do so – but also with an environmental one. In rural areas the higher demand for natural resources such as arable land, wood and water caused by the greater population has led to:

- shortening of the fallow period. Traditional bush management allowed for long fallow periods to restore soil fertility, but increasing demand for arable land has shortened this, leading to soil depletion, reduced yields and generally unsustainable agricultural practices;

- conversion of swamps into agricultural areas. This has affected streams and water sources. For instance, clearing of trees for the Kaliah refugee camp resulted in the drying up of the water source for the nearby village of Berecore. It may also affect migrating birds;
- destruction of the forest cover and forest degradation because the present rate of wood taken for charcoal, firewood and construction materials does not allow the biomass to recover: it is not sustainable; and
- loss of indigenous plant and animal species due to destruction of the forest ecosystem, the practice of burning grassland and bushes to clear land for cultivation and intensified hunting to meet food demands for the larger population.

In urban areas the environmental problems include:

- large increases in population. For instance, Gueckedou's population rose from 20,000 to 150,000 in 10 years, and in the capital city Conakry there are about 100,000 refugees in a total population of 800,000–900,000;
- a rapid rise in solid wastes, unmatched by improvements in the sanitation infrastructure and waste management, leading to a serious decline in the health status of urban areas. This includes small waste dumps inside towns, polluted streams and inadequate pit latrines. Epidemics are a serious threat;
- a dramatic increase in the demand for potable water, leading to massive over-utilization of all sources of water and increased pollution; and
- depletion of the natural resources in the areas around the cities. In particular, the vegetation cover has declined due to its use to build housing and provide firewood and charcoal.

9.9.3 The "impact pre-assessment"

In 1999, at the request of the Government of Guinea, the UNEP, in close cooperation with United Nations Centre for Human Health and United Nations High Commissioner for Refugees, carried out a rapid assessment of the environmental impact of refugees in Guinea. The assessment was carried out by experts in refugee management, EIA, human settlements and environmental management. The work began with a nine-day desk study in November 1999, involving two working groups: one on substantive issues and the other on institutional issues and logistics. Directly afterwards a week-long intensive field mission took place. This involved visits to relevant government departments and NGOs (e.g. Ministry of Agriculture, World Bank), to several existing and proposed refugee camps and settlements and to affected urban areas.

The work resulted in a "pre-assessment" report (UNEP et al. 2000) that first summarized the environmental impacts of the refugees and then recommended a range of solutions. Existing initiatives to counter the refugees' impacts – both from international organizations and from the Guinean Government – were found to be sector-specific and insufficiently integrated. The report recommended the development of an urban programme to strengthen the environmental management capacities of the urban centres in southern Guinea. It also recommended that an action plan for the protection of existing natural resources should be developed, focusing on ways to integrate agricultural and development activities with sustainable natural resource management. It recommended that the UN should take the lead in developing these, and that they should be presented at a donors conference in late 2000. Finally, the report recommended that a workshop should be organized in Conakry in late 2000 to strengthen the links between all of the organizations involved in environmental management in Guinea.

This type of assessment is not the first of its kind. For instance, a BBC (1995) television programme documented a similar assessment, led by the UN, about the environmental impacts of the huge Benaco camp in Tanzania set up for Hutu refugees from Rwanda. That EIA, which took place after the camp had been in existence for several months, aimed to identify the camp's environmental impacts and recommend solutions for how to minimize them. For instance, to reduce the impact on nearby forests caused by the camp's need for firewood, key trees that should not be cut down were identified and marked; refugees were transported from the camp to nearby forests but made to carry firewood back on foot, to spread the impact of wood collection over a wider area; maize was provided as flour rather than whole grain because that required less cooking; and open fires were substituted with more energy-efficient clay stoves.

9.9.4 Since the "impact pre-assessment"

In September 2000, armed incursions into Guinea from Liberia and Sierra Leone ignited ethnic violence between local communities, forcing civilian populations to flee north and east. Between 40,000 and 100,000 Guinean civilians were displaced, as well as refugees. The displacement coincided with the harvest season, so most of the civilians returned home within weeks. However, these events caused a radical change in the Guinean authorities' attitude towards very large numbers of refugees after many years of hospitality. A national radio and television appeal by president Lansana Conte for Guineans to "crush the invaders" led to mobs attacking refugee camps, and rounding up and beating refugees in urban areas. Although a government official later announced that Guinea was still willing to host refugees, the civilian militia continued to intimidate refugees and harass aid workers.

In Tanzania, recent studies (ReliefWeb 2003) show that the high levels of environmental degradation that resulted from the highest refugee influx, between 1993 and 1996, are being reversed. A study carried out in 1997 by Tanzania Agro-Industrial Services found that, although all vegetation had been cleared within 6 km of the camps, vegetation levels were starting to increase; and that 6 million trees had been planted in and around the camps, helping to redress the former negative impacts.

9.9.5 Conclusions

The Guinean "EIA" is very different from the others in this chapter. It covers multiple activities over a wide area. It took place after many of the impacts had already occurred. It was also carried out in a remarkably short amount of time – between 18 November and 8 December 1999 – although the total number of person-days involved is probably equivalent to a more typical EIA. The techniques used – desk study, interviews, site visits – are typical of the "rapid appraisal" techniques used in similar situations worldwide. Such techniques are very effective in identifying key impacts out of a wide array of possible impacts, allowing for a rapid response to the impacts. Arguably a much more detailed assessment would still show the same key impacts.

9.10 Summary

This chapter has examined a number of case studies of EIA in practice. Most of the cases involve EIA at individual project level, although an example of SEA has also been discussed. Links with other types of assessment, such as that required under the EU Habitats Directive, have also been explored. Whilst it is not claimed that the selected case studies represent examples of best EIA practice, they do include examples of some novel and innovative approaches towards particular issues in EIA, such as new methods of public participation and the treatment of cumulative effects. But the case studies have also drawn attention to some of the practical difficulties encountered in EIA, and to some of the limitations of the process in practice.

Note

1. Even basic data on Guinea varies widely: different sources suggest population sizes of 7.4–9.1 million; per capita GDP of $400–$2,000; life expectancy of 46–51 years; and refugee numbers between 300,000 and 600,000.

References

BBC (British Broadcasting Corporation) 1995. BBC2, "Horizon: Exodus", 6 March.

Cooper, L. M. & W. R. Sheate 2002. Cumulative effects assessment – a review of UK environmental impact statements. *Environmental Impact Assessment Review* **22**(4), 415–39.

DoE/DoT 1995. *Highways Act 1980, A556(M) Improvement (M6–M56 Link)*. Inspector's Report and Secretaries of State's Decision Letter. Manchester: Government Office for the North West.

DoT 1983. *Manual of environmental appraisal*. London: HMSO.

DoT 1989. *Departmental standard HD 18/88: environmental assessment under SC Directive 85/337*. London: HMSO.

DoT 1992. *A556(M) improvement (M6–M56) environmental statement*, vols I and II. Prepared by Allott & Lomax Consulting Engineers. Sale, Greater Manchester: Allott & Lomax.

DoT 1993. *Design manual for roads and bridges*, vol. 11: *Environmental assessment*. London: HMSO.

DTI 2002. *Future Offshore – a strategic framework for the offshore wind industry*.

DTI 2003a. *Offshore wind energy generation – Phase 1 proposals and environmental report*. Report prepared by BMT Cordah Ltd for the DTI. April 2003.

DTI 2003b. *Responses to draft programme for future development of offshore windfarms and the accompanying environmental report – summary of comments, and DTI response*. June 2003.

ETSU 1996. *Energy from waste: a guide for local authorities and private sector developers of municipal solid waste combustion and related projects*. Harwell: ETSU.

Glasson, J., K. Godfrey, B. Goodey 1995. *Towards Visitor Impact Management*. Aldershot: Avebury.

Highland Council 2003. *Application for Scottish Awards for Quality in Planning 2003 – The Cairngorm Funicular Railway Project*.

House of Lords Select Committee on Science and Technology 2000. *Science and society*. Third report. London: HMSO.

Huggett, D. 2003. *Developing compensatory measures relating to port developments in European wildlife sites in the UK*.

Hunter, C. & H. Green 1995. *Tourism and the Environment: a sustainable relationship?* London: Routledge.

IWM (Institute of Waste Management) 1995. *Communicating with the public: No time to waste.* Northampton: IWM.

Land Use Consultants 1994. *Cairngorm Funicular Project – Environmental Statement (and Appendices).* Glasgow: LUC.

Mathieson, A. & G. Wall 2004. *Tourism: economic, physical and social impacts,* 2nd edn. London: Longmans.

NAO (National Audit Office) 1994. *Department of Transport: environmental factors in road planning and design.* London: HMSO.

Petts, J. 1995. Waste management strategy development: a case study of community involvement and consensus-building in Hampshire. *Journal of Environmental Planning & Management* **38**(4), 519–36.

Petts, J. 1999. Public participation and environmental impact assessment. In *Handbook of environmental impact assessment,* J. Petts (ed.), vol. 1: *Process, methods and potential.* Oxford: Blackwell Science.

Petts, J. 2003. Barriers to deliberative participation in EIA: Learning from waste policies, plans and projects. *Journal of Environmental Assessment Policy & Management* **5**(3), 269–93.

Piper, J. M. 2000. Cumulative effects assessment on the Middle Humber: Barriers overcome, benefits derived. *Journal of Environmental Planning & Management* **43**(3), 369–87.

Piper, J. M. 2001a. Assessing the cumulative effects of project clusters: a comparison of process and methods in four UK cases. *Journal of Environmental Planning & Management* **44**(3), 357–75.

Piper, J. M. 2001b. Barriers to implementation of cumulative effects assessment. *Journal of Environmental Assessment Policy & Management* **3**(4), 465–81.

Piper, J. M. 2002. CEA and sustainable development – Evidence from UK case studies. *Environmental Impact Assessment Review* **22**(1), 17–36.

ReliefWeb 2001. *USCR Country Report Guinea: Statistics on refugees and other uprooted people.* New York: UN Office for the Coordination of Human Affairs. www.reliefweb.int/w/rwb.nsf.

ReliefWeb 2003. *Tanzania: Focus on the impact of hosting refugees.* New York: UN Office for the Coordination of Human Affairs. www.reliefweb.int/w/rwb.nsf.

RSPB (Royal Society for the Protection of Birds) 2003. *Successful offshore wind farm bids raise serious concerns for birds.* RSPB press release, 18 December 2003.

Sheate, W. R. 1995. Electricity generation and transmission: a case study of problematic EIA implementation in the UK. *Environmental Policy and Practice* **5**(1), 17–25.

Sheate, W. R. & M. Sullivan 1993. *Campaigners' guide to road proposals.* London: Council for the Protection of Rural England & Transport 2000.

Snary, C. 2002. Risk communication and the waste-to-energy incinerator environmental impact assessment process: a UK case study of public involvement. *Journal of Environmental Planning & Management* **45**(2), 267–83.

SNH (Scottish Natural Heritage) 2000. *Cairngorm Funicular Railway – Visitor Management Plan.* Paper presented to the SNH Board. SNH 100/5/7.

United Nations Environment Program, United Nations Centre for Human Health and United Nations High Commissioner for Refugees 2000. *Pre-assessment of the environmental impact of refugees in Guinea.* Geneva: UNEP. www.grid.unep.ch/guinea/project.html.

Weston, J. 2003. Is there a future for EIA? Response to Benson. *Impact Assessment and Project Appraisal* **21**(4), 278–80.

Weston, J. & R. Smith 1999. The EU Habitats Directive: Making the Article 6 assessments, The case of Ballyseedy Wood. *European Planning Studies* **7**(4), 483–99.

10 Comparative practice

10.1 Introduction

Most countries in the world have EIA regulations and have had projects subject to EIA. However the regulations vary widely, as do the details of how they are implemented in practice. This is due to a range of political, economic and social factors.

Environmental impact assessment is also evolving rapidly worldwide. When the second edition of this book was being written in 1999, for instance, many African countries and countries in transition had only recently enacted EIA regulations; by now, some of these countries have had considerable experience with EIA, and are developing more detailed guidelines and regulations. Four of the seven countries discussed in the second edition – Japan, Canada, China and Poland – have gone through major changes in their EIA systems since then.

This chapter aims to illustrate the range of existing EIA systems and act as comparisons with the UK and EC systems discussed earlier. It starts with an overview of EIA practice in the various continents of the world, and some of the factors that influence the development of EIA worldwide. It then discusses the EIA systems of seven countries in six different continents, focusing in each case on specific aspects of the system:

- Benin has one of the most advanced EIA systems in Africa, with good transparency, considerable public participation, integration of environmental concerns with national planning and robust administrative and institutional tools.
- Peru's EIA system is typical of many South American countries in its sector-specific orientation, relative lack of public participation and transparency, and late timing in project planning.
- Poland resembles several other countries in transition in that its EIA system has changed dramatically since the early post-Communist days. In 2000, Poland enacted radical new regulations which brings its EIA system in line with EU requirements.
- China's EIA system is discussed because of the worldwide effect that any Chinese environmental policy is likely to have in the future. China's environmental policies are restricted by the need to harmonize them with plans for economic development.
- The Netherlands is known for its progressive and well-developed environmental policies. The Dutch EIA system incorporates a particularly high level of public consultation, and uses an independent EIA Commission to scope each EIA and subsequently review its adequacy.

- Canada is also known for its progressive environmental policies. Its federal EIA system has good procedures for mediation and public participation.
- Australia's EIA system, like Canada's, is split between the federal and state governments. The state of Western Australia provides a particularly interesting example of a good state system with many innovative features.

The chapter concludes with a discussion of the role of international institutions in developing and spreading good EIA practice for the projects and programmes they fund and support.

10.2 EIA status worldwide

Table 10.1 and Figure 2.2 show, to the best of the authors' knowledge, the status of EIA regulations worldwide at the end of 2003. Box 10.1 gives an initial list of sources of information about EIA worldwide. More than 100 countries have some form of EIA regulation.

Table 10.1 Existing EIA systems worldwide

Country	Guideline (G) or regulation (R)	Date of implementation
Europe		
Austria	R	1993
Belgium	R	1985–92
Denmark	R	1989
Finland	R	1994
France	R	1976
Germany	R	1990
Iceland	R	
Ireland	R	1989
Italy	R	1986–96
Luxembourg	R	1994
Netherlands	R	1987
Norway	R	1990
Portugal	R	1987
Spain	R	1986
Sweden	R	
Switzerland	R	1985
United Kingdom	R	1988
Africa and Middle East		
Algeria	R	1990
Bahrain	R	1991
Benin	R	1995
Burkina Faso	R	1997
Comoros	R	1994
Congo	R	1986
Cote d'Ivoire	R	1996
Egypt	R	1991
Gabon	R	1987
Gambia	R	1994
Ghana	R	1994
Guinea	R	1987

Table 10.1 (*continued*)

Country	Guideline (G) or regulation (R)	Date of implementation
Israel	R	1982
Madagascar	R	1995, 1999
Mali	R	1991
Mauritius	R	1991
Mozambique	R	1998
Namibia	R	1994
Niger	R	1998
Nigeria	R	1992
Oman	R	1993
Senegal	R	1983
Seychelles	R	1996
South Africa	R	1984
Swaziland	R	1992
Togo	R	1988
Tunisia	R	1988
Uganda	R	1994
Yemen	R	1998
Zambia	R	1990
Zimbabwe	R	1994
Americas		
Belize	R	1992
Bolivia	R	1995
Brazil	R	1986
Canada	R	1992
Caribbean	G	1991
Chile	G	1991–96
Colombia	R	1974–85
Costa Rica	R	1997
Ecuador	R	1999
El Salvador	R	1992
Guatemala	R	1990
Guyana	R	1996
Mexico	R	1988
Nicaragua	R	1994
Panama	R	1991
Paraguay	R	1993
Peru	R	1992
Uruguay	R	1994
USA	R	1969
Venezuela	R	1976
Asia		
Bangladesh	R	1995
Bhutan	G	1993
China	R	1981
Hong Kong	R	1990
India	G	1989
Indonesia	R	1987
Japan	R	1997
Korea	R	1981
Malaysia	R	1987
Nepal	G	1992

Table 10.1 (*continued*)

Country	Guideline (G) or regulation (R)	Date of implementation
Pakistan	R	1997
Papua New Guinea	R	1978
Philippines	R	1977
Sri Lanka	R	1980
Taiwan	R	1987
Thailand	R	1978
Vietnam	R	1993
Australia et al.		
Antarctica	R	1991
Australia	R	1987
New Zealand	R	1991
Central and Eastern Europe		
Albania	R	1993
Armenia	R	1995
Azerbaijan	R	1996
Belarus	R	1992
Bosnia Herzegovina	R	1996
Bulgaria	R	1993
Croatia	R	1997
Czech Republic	R	1991
Estonia	R	2001
Georgia	R	1996
Hungary	R	1993
Kazakhstan	R	1997
Kosovo	R	2000
Kyrgyzstan	R	1997
Latvia	R	1996
Lithuania	R*	1995
Macedonia	R*	1996
Moldova	R	1996
Montenegro	R*	1997
Poland	R	1990
Romania	R	1996
Russia	R	1993
Serbia	R	1992
Slovakia	R	2000
Slovenia	R	1996
Tajikistan	R	1994
Turkey	R	1997
Turkmenistan	R	1995
Ukraine	R	1995
Uzbekistan	R	pre-1995

G = guideline; R = regulation; * = partial

However, EIA practice is not even across different countries worldwide. Figure 10.1 summarizes the evolution of EIA in a typical country: it begins with an initial limited number of EIAs carried out on an *ad hoc* basis in response to public concerns, donor requirements or industries based in a country with EIA requirements carrying out EIAs

---

Box 10.1 Some references on EIA systems worldwide

General: Angelsen et al. (1994), Beanlands (1994), CEAA (2003), Goodland & Edmundson (1994), Lee & George (2000), Netherlands Commission for Environmental Impact Assessment (1999), Roe et al. (1995), Wood (2003); publications by the OECD, ODA, UNEP and the World Bank; and articles in e.g. *Environmental Impact Assessment Review* and *Impact Assessment and Project Appraisal*

North America: Clark & Richards (1999), Wood (2003)

Western Europe: CEC (2003), Wood (2003)

Australia, New Zealand: Dixon & Fookes (1995), Harvey (1998), Morgan (1995), Thomas (1998), Wood (2003)

Asia: Briffett (1999), Welles (1995)

Central and Eastern Europe: Regional Environmental Centre for Central and Eastern Europe (2003), Rzeszot (1999), World Bank (2002)

South America: Brito & Verocai (1999), Chico (1995)

Africa and Middle East: CLEIAA (2003), d'Almeida (2001), Goodland et al. (1996), Kakonge (1999), Okaru & Barannik (1996)

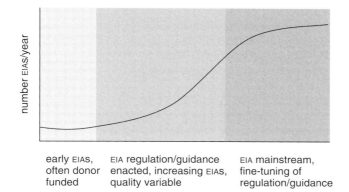

early EIAs, often donor funded

EIA regulation/guidance enacted, increasing EIAs, quality variable

EIA mainstream, fine-tuning of regulation/guidance

Figure 10.1 Evolution of EIA systems.

of their activities in the country without EIA requirements. Over time, the country institutes EIA guidelines or regulations. These may prompt a rapid surge in the number of EIAs carried out in that country, as was the case in most EU countries. However, the regulations may apply to only a limited number of projects, or may be widely ignored, leading to only a small increase. Over time, the regulations may be fine-tuned or added to, the number of EIAs carried out annually levels off and the EIA system is effectively "mature". The current status of EIA in different countries can be roughly charted on this continuum. Figure 10.2 broadly shows this status by continent, though individual countries may vary from this.

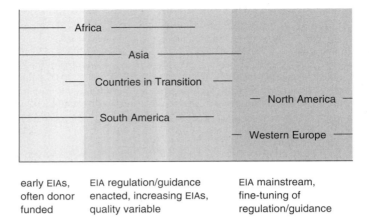

early EIAs,   EIA regulation/guidance   EIA mainstream,
often donor   enacted, increasing EIAs,   fine-tuning of
funded   quality variable   regulation/guidance

Figure 10.2 Current status of EIA systems worldwide.

The situation in *Africa* is changing rapidly, with many countries having recently instituted EIA regulations. This development has been brought about by a range of initiatives including the 1995 African Ministerial Conference on Environment that committed African environment ministers to formalize the use of EIA; an EIA stakeholders meeting in Nairobi in 1998; development of the Capacity Enhancement for EIA in Africa programme in 1999; and the establishment of the Capacity Development and Linkages for EIA in Africa (CLEIAA) in 2000, which has improved collaboration between African countries to build up EIA capacity.

On the other hand, EIA in Africa is still beset by a lack of trained personnel, cost, concern that EIA might hold back economic development and lack of political will (Kakonge 1999). Even where EIA legislation exists, this does not mean that it is put into practice, is carried out well, involves the public or is enforced (Okaru & Barannik 1996). As such, while some African countries such as Benin, the Seychelles and Tunisia have good regulations and considerable EIA practice, others such as Burundi and Guinea-Bissau are on the far left of Figure 10.2 (d'Almeida 2001).

All countries in *South and Central America* have some form of legal system for environmental protection, including at least aspects of EIA. These systems vary widely, reflecting the countries' diverse political and economic systems: average income in some South American countries is more than ten times that in others. However, in general, the development of EIA in South America has been hampered by political instability, inefficient bureaucracy, economic stagnation and external debt (Brito & Verocai 1999). This may be changing recently, but EIA in South America is often still carried out centrally, with little or no public participation, and often after a project has already been authorized (Glasson & Salvador 2000).

Environmental impact assessment in *Asia* also varies widely, from no legislation (e.g. Cambodia), through to extensive experience with robust EIA regulation set within the context of SEA (e.g. Hong Kong). EIA regulations were established in many Asian countries in the late 1980s, and EIA is practised in all countries of the region through the requirements of donor institutions. On the other hand, Briffett (1999) suggests that many Asian EIAs are of poor quality, with poor scoping and impact prediction, and limited public participation. This is due in part to the perception that EIA may retard economic growth – symbolized by the wish, in some countries, to expose large buildings and infrastructure projects to show off the country's wealth (Briffett 1999).

The *Countries in Transition* have just gone through 15 years of enormous change associated with the move from centrally planned to market systems. This has included, in many cases, a move from publicly to privately owned enterprises. Many of these countries' economies are based on heavy industry, with concomitant high use of energy and resources. Many went through an economic crisis in the 1990s.

Of the countries in transition, the Central and Eastern European countries (including Turkey) are aiming towards EU accession and are moving to harmonize their EIA legislation with European Directives 85/337 and 97/11: Poland is an example of this. The Newly Independent States of the former Soviet Union all have similar systems, based on the "state ecological expertise/review system" developed under the former Soviet Union. The countries of South East Europe – Albania, Bosnia Herzegovina, Croatia, Yugoslavia and Macedonia – have relatively undeveloped EIA systems, but are starting to harmonize their EIA legislation with that of the EU (Rzeszot 1999, World Bank 2002).

Box 10.2 summarizes some of the factors that affect the application of EIA in developing countries.

Box 10.2 Factors affecting the implementation of EIA in developing countries

- Many apply to countries in or near tropical areas, where environmental models, data requirements and standards from temperate regions may not apply.
- Socio-cultural conditions, traditions, hierarchies and social networks may be very different.
- The technologies used may be of a different scale, vintage and standard of maintenance, bringing greater risks of accidents and higher waste coefficients.
- Perceptions of the significance of various impacts may differ significantly.
- The institutional structures within which EIA is carried out may be weak and disjointed, and there may be problems of understaffing, insufficient training and know-how, low status and a poor coordination between agencies.
- EIA may take place late in the planning process and may thus have limited influence on project planning, or it may be used to justify a project.
- Development and aid agencies may finance many projects, and their EIA requirements may exert considerable influence.
- EIA reports may be confidential, and few people may be aware of their existence.
- Public participation may be weak, perhaps as a result of the government's (past) authoritarian character, and the public's role in EIA may be poorly defined.
- Decision-making may be even less open and transparent, and the involvement of funding agencies may make it quite complex.
- EIAs may be poorly integrated with the development plan.
- Implementation and regulatory compliance may be poor, and environmental monitoring limited or non-existent.

The EIA systems of *Western European* countries have already been amended at least once: Directive 85/337 was amended 12 years later by Directive 97/11, and some countries already had EIA systems in place before the original Directive of 1985. As outlined in Chapter 2, a recent review of the status of EIA in EU Member States (CEC 2003) showed that, by late 2002:

- Directive 97/11 had been legally transposed by 11 of the 15 Member States;
- the Directive had led to an increase of roughly one-quarter in the number of EIAs carried out;
- Member States varied widely in how they determined what Annex II projects require EIA: most use a "traffic lights" threshold system (green = no EIA needed, red = EIA needed, amber = EIA may be needed), but the level of the thresholds varies between countries;
- as such, the number of EIAs carried out varies considerably between Member States, from less than 20 per year in Austria to about 7,000 per year in France;
- seven of the fifteen Member States have a mandatory scoping stage;
- the consideration of alternatives and involvement of the public varies widely between Member States.

So even after two rounds of "harmonization", EIA practice still varies widely across Europe.

Environmental impact assessment procedures in *North America, Australia* and *New Zealand* are still amongst the strongest in the world, with good provisions for public participation, consideration of alternatives and consideration of cumulative impacts. All have separate procedures for federal and state/provincial projects.

The following sections discuss the EIA systems of different countries worldwide as examples of the concepts discussed above.

10.3 Benin

Benin, in West Africa, was once covered by dense tropical rain forest behind a coastal strip. This has largely been cleared and replaced by palm trees. Grasslands predominate in the drier north. Benin's main exports are cotton, crude oil, palm products and cocoa. Even compared to other nearby countries, Benin has a relatively low GDP per capita. However, Benin has had a fully functioning EIA system since 1995, and has been identified as one of three francophone African countries that is most advanced in EIA terms (d'Almeida 2001). Its EIA system is characterized by transparency, public participation, integration of environmental concerns with national planning and robust administrative and institutional tools (Baglo 2003).

Like many other African countries, Benin's early EIAs were carried out at the behest of funding institutions such as the World Bank and African Development Bank. Benin's constitution of December 1990 placed particular emphasis on environmental protection, and in 1993 the Environmental Action Plan was adopted. In 1995 the Benin Environmental Agency was created and made responsible for, *inter alia*, implementing national environmental policy, conducting and evaluating impact studies, preparing State of the Environment reports and monitoring compliance with environmental regulations. Although the Agency reports to the Ministry of the Environment, Habitat and Town Planning, it has corporate status and financial independence. The Agency subsequently developed a framework regulation on EIA, as well as a range of EIA guides (e.g. on projects that require EIA, EIA for gas pipelines and irrigation projects). Benin also has a national association of EIA professionals, established in 1998 with 16 registered members (Baglo 2003).

Benin's EIA process is led by the Benin Environmental Agency. Although this has only two officers responsible for EIA, it can also draw on a forum of experts from public and private institutions to prepare or review specific EIAs. This approach allows the Agency to operate with minimal staff, ensures ongoing cooperation with other institutions and puts into practice the principle of broad participation in decision-making (Baglo 2003).

The EIA process consists of:

- screening – this work is decentralized, with responsibility allocated to the ministries responsible for the proposed project's sector;
- scoping;
- preparation of a draft EIS, and "publication" of the report primarily through the radio and non-government organizations (NGOs);
- a public hearing to discuss the EIS, arranged by the Minister of Environment through the creation of *ad hoc* Public Hearing Commissions;
- analysis of the EIS by the Agency. Where the EIS is adequate, the Agency issues a notice of environmental compliance;
- decision about the project by the ministry responsible for the sector concerned, taking into account the notice of environmental compliance, technical feasibility study and economic feasibility study; and
- follow-up assessment and audit.

Examples of projects subject to EIA in Benin since 1998 include several livestock development projects, road improvement projects, urban and industrial development projects and the West African Gas Pipeline. These have been paid for through a mixture of donor, private property and national government funds.

Environmental impact assessment practice in Benin, as in many other African countries, is limited by poor collaboration between some key ministries, a low level of public environmental awareness, illiteracy and poverty. Early problems with lack of indigenous expertise have been reduced through a range of EIA training courses run by the Benin Environmental Agency, the activities of its national association of EIA professionals and its participation in pan-African capacity building activities (e.g. CLEIAA).

10.4 Peru

Peru, the third largest country in South America, includes a thin dry strip of land along the coast, the fertile sierra of the Andean foothills and uplands, and the Amazon basin. Fishing, agriculture and mining are the main industries. A change in government in 1990 led to the reconstruction of the country after years of economic difficulties, and an extensive privatization programme – including the privatization of many of the state-owned mines – encouraged dramatic increases in foreign investment.

In September 1990, the Peruvian government enacted Decree 613–90, the Code of Environment and Natural Resources. This established EIA as a mandatory requirement for any major development project, but did not specify the EIA contents or legal procedures. The Ministry of Energy and Mining was the first ministry to put this decree into practice. In 1992, it approved the General Law of Mining by Supreme Decree no. 014–92-EM; this, in turn, contained regulations for environmental protection in mining activities, which were approved in 1993 by Supreme Decree 016–93-EM. The ministries for fishing, agriculture, transport and communication, housing and building all established similar but separate requirements in 1994 and 1995. The remainder of this section focuses on the mining sector.

According to Supreme Decree 016–93-EM, any developer who plans to exploit minerals, or to expand existing exploitation by 50 per cent or more, must carry out an EIA. The EIA has to be carried out by an institution authorized and registered with the Ministry of Energy and Mining. The decree lists the EIA contents in two parts. The first part is mandatory. If, after reviewing the EIA, the Ministry considers that the project will have significant environmental impacts, it can also require the EIA to address some of the aspects in the second part:

Part I EIA contents

1.   executive summary;
2.   antecedents (e.g. applicable legislation);
3.   introduction (project description and estimated cost);
4.   project area description (general information about location, geological and biotic components, etc.);
5.   project description (e.g. estimated volume of water used, wastewater and wastes produced, energy demand, employment, etc.);
6.   predicted impacts (human health, flora and fauna, ecosystems, etc.);
7.   control and mitigation (e.g. measures to control noise and dispose of wastes);
8.   cost–benefit analysis.

Part II Additional information

1.   project alternatives (description, justification of chosen alternative);
2.   affected environments (detailed studies of continental and marine waters, etc.).

Every EIA must include an environmental management plan which lists the activities and programmes to be implemented before and during the project to guarantee the fulfilment of existing environmental standards and practices. It must also include a programme to incorporate new technologies and alternative measures into mining activities so as to reduce emissions or discharges.

The resulting EIS is reviewed by the Ministry's environment directorate. In theory, this review should be carried out within 45 days, or else automatic consent is granted. In practice, since the ministry has 45 days "after receiving the EIS and/or additional information", and since the relevant directorate only has eight officers, who also have to deal with other matters, the officers often request additional information so that they have longer time to review the EIS. Until 1996, only the EIS's non-technical summary was publicly available; the main body of the EIS was felt to be commercially sensitive and thus confidential. However, Ministerial Resolution 335–96-MEM/SG requires that a public inquiry should be held before a decision is made.

Once a project is approved, the developer must carry out programmes of management, control and monitoring throughout the operations to ensure that the environmental management plan is adhered to. The developer has to contract a registered auditing consultancy to inspect its activities twice a year. The consultancy must prepare a report on activities at the site and submit it to the Ministry, which can apply sanctions for non-compliance.

Peru's EIA system is typical of that of many South American countries in its relative lack of public participation or sectoral integration, and frequent preparation of the EIS after the project construction has begun (Brito & Verocai 1999). A survey of environmental consultants showed that almost two-thirds of EISs are begun after the project planning is "more than 50 per cent completed", with the sections on control, mitigation and cost–benefit appraisal being particularly difficult to carry out. Early indications are that EIS quality is quite high, but the discussions of mitigation measures are weak (Iglesias 1996).

10.5 China

Since 1978, China has been undergoing a rapid shift towards economic growth, decentralization of power, industrialization, private enterprise and urbanization. This has engendered many development projects with significant environmental impacts. China's EIA system has struggled to keep up, and to date has not managed to prevent some serious environmental harm (Mao & Hills 2002).

Environmental impact assessment in China was officially introduced through the Provisional Environmental Protection Law of 1979, which was revised and finalized in 1989. Shortly after the law's introduction, several guidelines for its implementation were prepared, of which the central ones were the "Management Rules on Environmental Protection of Capital Construction Projects" of 1981. These were revised and formalized in 1986, with details on timing, funding, preparation, review and approval of EISs. Further, stronger ordinances were enacted in 1990 and December 1998. The 1998 "Ordinance of Environmental Management for the Construction Project" requires EIA for regional development programmes as well as individual projects, added requirements for public participation, and strengthened legal liabilities and punishment for violation of EIA requirements. A further law requiring SEA for plans and programmes was enacted in October 2002, and came into force in September 2003.

The State Environmental Protection Agency (SEPA) administers EIA for projects of national economic or strategic significance; the provincial environmental protection bureaus (EPB) administer EIA for projects of regional importance; and so on down to the village level.

Figure 10.3 summarizes China's EIA procedures. The process begins when a developer asks a competent authority to determine whether or not a proposed action requires full EIA. Most projects require only the preparation of an environmental impact form, which describes the project and briefly states its environmental impacts. Large projects with significant impacts and smaller projects in inappropriate locations require full EIAs. The competent authority personnel, sometimes assisted by outside experts, conduct a preliminary study and then make a ruling. If an EIA is needed, the competent authority identifies those factors most likely to affect the environment and prepares a brief. The EIA's management is then entrusted to licensed, state-approved experts, who work to the brief.

The expert analyses the relevant impacts in greater detail and compares them with relevant environmental quality standards. Baseline environmental assessments are

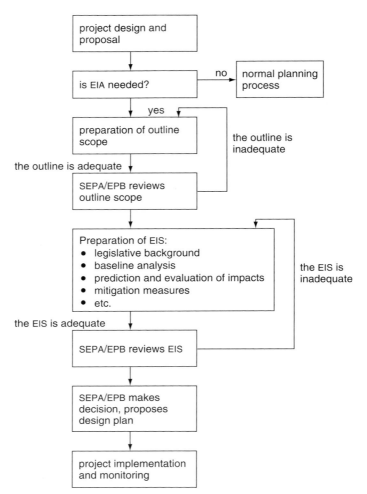

Figure 10.3 China's EIA procedures. (Based on Wang et al. 2003.)

carried out if the project is proposed for an area of low industrial activity where environmental standards are high. The impacts are then predicted, and mitigation measures proposed. An EIS is then produced, which, according to the guidelines, needs to discuss:

- the general legislative background;
- the proposed project, including materials consumed and produced;
- the baseline environmental conditions and the surrounding area;
- the short-term and long-term environmental impacts of the project;
- impact significance and acceptability;
- a cost–benefit analysis of the environmental impacts;
- proposals for monitoring; and
- conclusions.

The EIS or environmental impact form is submitted to the competent authority, which checks the proposal against environmental standards and makes a decision, after considering the comments of relevant experts. For a controversial project, and for a project that crosses provincial boundaries, the document is submitted to the higher authority for examination and approval. If the project is approved, conditions for environmental protection may be included, such as monitoring and verification procedures. The competent authority must submit a report that states how the project will be carried out and how any required environmental protection measures will be implemented. Once this has been approved by the provincial authorities, a certificate of approval is issued.

Ortolano (1996) estimates that tens of thousands of EISs or environmental impact forms are prepared in China every year. The effectiveness of the new, post-1998 EIA system is not yet clear, but the pre-1998 system seemed to be inadequate in controlling the effects of economic growth in China. Whereas EISs were prepared for 76 per cent of large and medium-scale projects approved between 1981 and 1985, and 90 per cent between 1986 and 1990, this compliance rate plummeted to just over half in the early 1990s, although it was again close to 90 per cent by the late 1990s (Wang et al. 2003). It was particularly low for foreign investment projects and for the emerging development zones. Many small local projects that should require EIA are not tracked, so the compliance rate for those EIAs is not known (Mao & Hills 2002). The quality of EISs has also been variable, and many were prepared post-construction (China 1999).

This is due in part to the speed of the economic transition and the decentralization of power. Although local governments have gained much fiscal and administrative authority, they were – and are – also faced with budget constraints: as such they compete with each other in trying to attract business, and are unwilling to impose constraints in the form of EIA (Mao & Hills 2002). Competent authorities are often unwilling to antagonize other government departments or local leaders who strongly favour the proposed projects. In some cases, even where an EIA is prepared and environmental protection measures are agreed, "the mayor . . . steps in and asks [the competent authority] . . . whether less money couldn't be spent on [pollution control] equipment" (Jahiel 1994), effectively cancelling out the project's environmental protection features. China's EIA process has also been criticized for its complexity; narrow historic focus on air, water and soil pollution; lack of effective public participation; and lack of consideration of alternatives (Wang et al. 2003). The changes of 2003 may redress some of these problems.

10.6 Poland

Poland, like the other countries in transition, is undergoing a rapid evolution in environmental policy, with three clearly distinguishable phases in the last 25 years.

In the latter days of Communism, the government's policy of borrowing from the West was backfiring, the Communist regime was providing only the most basic services and environmental issues were being virtually ignored (Fisher 1992). Several areas of Poland were subject to severe pollution, causing widespread concern. In response to this, the government enacted the Environmental Protection Act (EPA) in 1980. This was subsequently strengthened through various amendments and by the T & CP Act of 1984. Under this system, an EIA began when a developer asked a local environmental authority whether an EIA was needed. If it was, the authority drew up a list of suitable consultants to carry out the work. Once the chosen consultant had completed the EIS, consultation with the public might be carried out but was not mandatory. If the EIS was accepted by the environmental authority, then the local planning authority could issue a "location indication" which listed alternative locations for the project. Based on this, the developer chose a site and continued to design the project, and the environmental consultants prepared a final EIS. The developer delivered the EIS along with a planning application to the local planning authority, which again consulted with the local environmental authority before making a decision about the project. Construction consent required yet a third EIS to accompany the technical design of the project.

This system was criticized for lacking screening criteria, being cumbersome and redundant with multiple EISs prepared for each project, having minimal procedures for public participation and for resource constraints on the commission that reviewed the EISs (Jendroska & Sommer 1994). It also could not deal with the huge social and economic changes that Poland went through after the overthrow of the Communist regime in late 1989:

> There are no more economic plans and central planners, the currency is convertible and the best technology accessible, and the whole economy is being privatised. Moreover, administrative arrangements have been redesigned in order to create a strong central agency as an environmental watchdog... [but] old industry is still operating. The observed improvement of environmental records since 1989 is only a side effect of the recession ... EIA law in Poland still reflects two characteristic features of the Communist regime: an aversion to getting the general public involved in decision-making, and a reluctance to developing procedural rules for dispute settlement. This means that this legislation not only is not efficient enough from the "environmental" point of view, but also does not match the political and economic transformation towards an open and democratic society and a free market. (Jendroska & Sommer 1994)

In 1994, Poland became an associate partner of the EU. This requires Poland to incrementally change its laws and statutes to bring them in line with those of the EU. Since then, two key regulations have been enacted: the Act of 9 November 2000 on Access to Information on the Environment and its Protection and on EIA, and the Environmental Protection Law (EPL) Act of 27 April 2001. These include new requirements for screening and scoping, expand the requirements for public consultation in EIA and shift the emphasis of EIA from the preparation of a report to the process as a whole (Wiszniewska et al. 2002).

The new system is much more like that of the EU, but includes some interesting variants:

- EIA is required for two types of projects: group I projects that require EIA in all cases, and group II projects that may require EIA. The former are linked to the permit requirements of other acts, and include mineral extraction, groundwater abstraction and flood control, restructuring of rural land holdings, and motorways. The latter are judged on e.g. whether the project requires the establishment of a restricted use area, whether it may cause transboundary impacts and whether relevant environmental bodies feel that it should require EIA.

- The EIS must include a description of alternatives to the proposed project, including the "do nothing" option and the option that would be the most advantageous for the environment; an indication of whether it is necessary to establish a restricted-use area; an analysis of probable social conflicts related to the project and proposals for impact monitoring.
- The competent authority must review the quality of the EIS to ensure that it complies with its interim decision on the scope of the report and the requirements of the EPL Act, and that it provides a sufficient basis for issuing the decision and informing society about the project. The *Handbook on Environmental Impact Assessment Procedures in Poland* (Wiszniewska et al. 2002) gives criteria for how this can be done.
- An EIA may still be required during the procedure of granting a building consent, but the *Handbook* suggests ways of avoiding duplication of EIAs.

It is still too early to determine the effectiveness of these changes, but clearly Poland has made great efforts to improve its EIA system. On the other hand, the OECD (2003) *Environmental Review of Poland* suggests that the country still faces major challenges before it can achieve convergence with EU Member States on environmental standards. Priority issues include pollution prevention, wastewater treatment, waste management, biodiversity and landscape protection. The report suggests that Poland needs to invest up to 2.7 per cent of GDP annually for 10 years simply to comply with EU accession rules, and recommends the wider use of economic instruments, implementation of the "polluter pays" principle and stronger enforcement of environmental regulations.

10.7 The Netherlands

The Netherlands, with its small area of densely populated and highly industrialized land, has developed a worldwide reputation for powerful and progressive environmental legislation. EIA in the Netherlands, as in the UK, is required by EC Directive 85/337 amended by Directive 97/11. It is implemented through the Environmental Management Act and the EIA Decree 1994 as amended by the Decree of 7 May 1999. EIA is required for all projects in Annex I of the Directives, and on a case-by-case basis for Annex II projects that exceed specified exclusion thresholds: these thresholds are generally quite high compared with those of other Member States (CEC 2003). In addition, SEA has been required for sectoral plans on waste management, the supply of drinking water, energy and electricity and some land-use plans since 1987.

Figure 10.4 summarizes the Dutch EIA procedures. Once a (public or private) developer decides to carry out an activity that requires EIA, they prepare a "notification of intent" with a brief description of the proposed project, they inform the competent authority and the competent authority makes the "notification of intent" public. The public and advisers – notably the EIA Commission[1] – have four weeks to comment to the competent authority on what should be in the EIS guidelines: this is the scoping stage. Within 13 weeks of the "notification of intent", the competent authority must produce formal EIS guidelines for the action, specifying the alternatives and the main environmental impacts that the EIS must address.

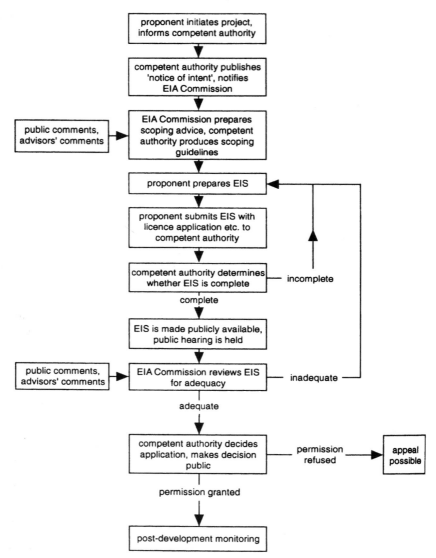

Figure 10.4 The Netherlands' EIA procedures.

The developer is responsible for preparing the EIS, which must include:

- a statement of the purpose and reason for the activity;
- a description of the activity and "reasonable" alternatives (including that least harmful to the environment and the do-nothing option);
- an overview of the specific decision(s) for which the EIS is being prepared and of decisions already taken regarding that activity;
- a description of the existing environment, and the expected future state of the environment if the activity is not carried out;

- a description of the environmental impacts of the proposed activity and alternatives, and the methods used to determine these impacts;
- a comparison of the activity with each alternative;
- gaps in knowledge; and
- a non-technical summary.

The developer submits the EIS to the competent authority, which has six weeks to decide whether it meets the criteria of the guidelines or whether any corrections or amendments are required. The findings of this inspection are made public. Once it has been accepted, the EIS is made publicly available for at least four weeks. During this time, a public hearing must be held, and the public may comment on the EIS. At this time other environmental bodies may also provide advice to the competent authority concerning the contents of the EIS. A record of the public review and other advice is then passed to the EIA Commission.

The EIA Commission then reviews the EIS against current legislation and the EIS guidelines. It also considers the advice and public review. The Commission's review is generally guided by two issues: whether the EIS can assist in decision-making, and, if so, whether it is complete and accurate. The review concerns the adequacy of the EIS, not the environmental acceptability of the activity. Within two months of receiving the EIS, the Commission sends the results of this review to the competent authority, which makes the final decision. About 40 per cent of the EISs reviewed by the Commission between 1999 and 2001 were inadequate. In such cases, the developer may be required to provide further environmental information.

The competent authority makes a decision based on the EIS, the advisers' comments, the Commission's review and the results of the public hearing. It makes the results of this decision known, including how a balance was struck between environmental and other interests, and how alternatives were considered. The competent authority must subsequently monitor the project, based on information provided by the developer, and make the monitoring information publicly available. If actual impacts exceed those predicted, the competent authority must take measures to reduce or mitigate them.

About 70 EIAs are carried out in the Netherlands each year: this has not changed much since Directive 97/11 came into force. EIA in the Netherlands does appear to influence decisions on projects. An evaluation of the EIA system carried out in 1996 suggested that 79 per cent of respondent competent authorities and project proponents felt that EIA had brought up new relevant information, leading in 52 per cent of cases to changes in the project that made it more environment-friendly (including withdrawal of project proposals). In 71 per cent of cases, EIA led to a more receptive attitude towards environmental issues that might affect future development decisions; and in 70 per cent of cases competent authorities and project proponents felt that the benefits of EIA outweighed its costs (Scholten 2003).

The Dutch EIA Commission also promotes good EIA and SEA practice internationally, both by funding international studies on EIA and by publishing many of its documents (in English as well as Dutch) on its website.

10.8 Canada

Canada has also set up a powerful and evolving system of environmental legislation, but under conditions different from those in the Netherlands. Its wealth of natural resources, which were originally plundered indiscriminately by the giant "trusts" in coal, steel, oil and railroads; its lack of strong planning and land-use legislation; and the conflicting needs of its powerful provincial governments all prompted the development of a mechanism by which widespread environmental harm could be prevented. Canada's EIA system is characterized by a split between national and provincial procedures; quite complex routeing of different types of projects through different types of EIA processes and innovative approaches to mediation and public participation in EIA.

The responsibility for EIA in Canada is shared between the federal and the provincial governments. The *federal* procedures apply to projects for which the Government of Canada has decision-making authority. Early federal EIA guidelines (FEARO 1977) were progressively strengthened throughout the 1970s and 1980s, and made legally binding in 1989. However concern over the limitations of this "Environmental Assessment and Review Process" caused it, in turn, to be replaced in 1995 by the Canadian Environmental Assessment Act. This was further strengthened by Bill C-9, which came into force on 30 October 2003. SEA of policy has been required since 1993, and SEA requirements were strengthened in 1999. Gibson (2002) gives a useful review of the development of Canada's federal EIA system upto 2002.

The Canadian Environmental Assessment Agency (CEAA) administers the Canadian Environmental Assessment Act. An initial self-assessment by the responsible agency proposing the action determines whether the action requires EIA under the Act, i.e. whether it

- is a "project" as defined by the Act;
- is not excluded by the Act's Exclusion List regulation;
- involves a federal authority; and
- triggers the need for an EIA under the Act.

The Exclusion List Regulation identifies projects for which EIAs are not required because their adverse environmental effects are not regarded as significant (e.g. simple renovation projects).

Once an EIA is determined to be required, a decision is made as to which of four EIA tracks to follow: screening, comprehensive study, mediation or review panel. Most projects require a "screening" involving documentation of the project's environmental effects and recommended mitigation measures. "Class screening" may be used to assess projects with known effects that can easily be mitigated. "Model class screenings" provide a generic assessment of all projects within a class: the responsible authority uses a model report as a template, accounting for location- and project-specific information. "Replacement class screenings" apply to projects for which no location- or project-specific information is needed.

A small number of projects will require a fuller "comprehensive study". These are listed in the Comprehensive Study List Regulations and include, for instance, nuclear power plants, large oil and gas developments and industrial plants.

If a screening requires further review, it is referred to a mediator or review panel. Similarly, early in a comprehensive study, the Minister of the Environment must decide whether the project should continue to be assessed as a comprehensive study, or whether it should be referred to a mediator or review panel. Projects are normally referred to a mediator or review panel – essentially changing them from self-assessment by the responsible agency to independent, outside assessment – where the significance of their impacts is uncertain, where the project is likely to cause significant adverse environmental effects and there is uncertainty about whether these are justified, or where public concern warrants it (CEAA 2003).

The mediation option, new since 1995, is a voluntary process in which an independent mediator appointed by the Minister helps the interested parties to resolve their issues through a non-adversarial, collaborative approach to problem-solving. A review panel is a group of experts approved by the Minister of the Environment which reviews and assesses a project with likely adverse environmental impacts. Any project requiring comprehensive study, mediation or a review panel must consider alternative means of carrying out the project, the project's purpose and its effects on the sustainability of renewable resources; and must include a follow-up programme. The responsible authority must take the results of the comprehensive study, or the mediator's or review panel's recommendations into account when making a decision on the project (CEAA 2003).

Public comments must be considered at various stages of the EIA process, though it is more restricted for screenings. A participant-funding programme allows stakeholders to participate in comprehensive studies, panel reviews and mediation. A specific consultation process for Aboriginal people was established by Bill C-9 (Sinclair & Fitzpatrick 2002).

The Canadian Environmental Assessment Agency publishes many of its reports on the Web. Between 1995 and 1999, more than 25,000 federal EIAs had been carried out: 99 per cent were screenings, but 62 comprehensive studies (37 completed) and 11 panel reports (9 completed) had also been carried out (Gibson 2002).

Most of Canada's *provinces* have quite widely varying EIA regulations for projects under their own jurisdictions. These include Ontario's EA Act of 1976, very advanced at the time, but subsequently weakened in 1997; and Manitoba's sustainable development code of practice of 2001 which requires public officials to promote consideration of sustainability impacts in EIA. In early 1998, federal and provincial environment ministers signed an accord on EIA harmonization which promotes cooperative use of existing processes to reduce duplication and inefficiency (Gibson 2002).

10.9 Australia and Western Australia

Like Canada, Australia also has a federal (Commonwealth) system with powerful individual states. Its environmental policies, including those on EIA, have some interesting features but are generally not as powerful as those of Canada or the Netherlands. The Commonwealth EIA system was established as early as 1974 under the Environmental Protection (Impact of Proposals) Act. It applied only to federal activities. During the life of the Act (1974–2000) about 4,000 proposals were referred for consideration, but on average less than 10 formal assessments were carried out each year (Wood 2003). As such, the states put in place their own legislation or procedures to extend the scope of EIA to their own activities, and many of these state systems have become stronger and more effective than the national system. Over time there has been concern about the variation in EIA procedures, and their implementation, between states in Australia and there have been attempts to increase harmonization (Australian and New Zealand Environment and Conservation Council – ANZECC 1991, 1996, 1997) (see also Harvey 1998, Thomas 1998). In addition, a major review of Commonwealth EIA processes was undertaken in 1994 producing a set of very useful reports on cumulative impact and strategic assessment; social impact assessment; public participation; the public inquiry process; EIA practices in Australia and overseas comparative EIA practice (CEPA 1994). The review highlighted, among other issues, the need to reform EIA at the Commonwealth level – including a better consideration of cumulative impacts, social and health impacts, SEA, public participation and monitoring.

Following Government changes and a further review of federal/state roles in environmental protection, Australia repealed its Commonwealth EIA legislation, and several other environmental statutes, to create the Environmental Protection and Biodiversity Conservation Act (EPBCA) in 1999. The EPBCA provides a lot more procedural detail than the original EIA legislation, and a range of documents has been produced to explain the processes (Environment Australia 2000). EIA is undertaken for matters of national environmental significance, defined as World Heritage properties, Ramsar wetlands, threatened and migratory species, the Commonwealth marine environment and nuclear actions. The Act promotes ecologically sustainable development; it also provides for SEA (IEMA 2002). In its first year, the EPBCA did not appear to increase the rate of Commonwealth EIA activity (Wood 2003). See Padgett & Kriwoken (2001), Scanlon & Dyson (2001), and Marsden & Dovers (2002) for further commentary on the Act and on recent developments in EIA and SEA in Australia.

10.9.1 EIA in Western Australia (WA)

The Western Australian (WA) EIA system provides an interesting example of a good state system that includes many innovative features. Central to the success of the Western Australian system is the role of the EPA (Wood & Bailey 1994). The Environmental Protection Authority (EPA) was established by the WA Parliament as an Authority with the broad objective of protecting the State's environment and it is the independent environmental adviser that recommends to the WA government whether projects are acceptable. It is independent of political direction. The EPA determines the form, content, timing and procedures of assessment and can call for all relevant information; the advice it provides to the Minister for the Environment must be published. The EPA overrides virtually all other legislation, and the environmental decision (with conditions) is central to the authorization of new proposals. Other permits must await the environmental approval, based on the EIA.

Proposals may be referred to the EPA by any decision-making authority, the proponent, the Minister for the Environment, the EPA or any member of the public. The EPA determines the level of assessment, normally adopting one of the five levels illustrated in Table 10.2a. This sets the general form, content, timing and procedure for the EIA process, the most comprehensive being the Public Environmental Review (PER) and the Environmental Review and Management Programme (ERMP). Guidance is provided on scoping and on the content of the assessment (the environmental review) (Table 10.2b,c). The environmental review document is produced by the proponent, and it is subject to public review. The guidance on scoping for a PER or ERMP contains interesting features, especially in relation to peer review and public consultation. The assessment (environmental review) pays particular attention to the regional setting, and seeks to highlight potential "fatal flaws". Waldeck et al. (2003) found that such EIA guidance influenced the practice of consultants and was perceived as effective in enhancing the outcomes of the EIA process – including

Table 10.2 Some features of the Western Australian EIA system

(a) **Five levels of assessment**

- *Assessment of Referral Information* (ARI): where proposal raises one or a small number of significant environmental factors which can be readily managed, but where environmental conditions are required to ensure the proposal is implemented and managed in an environmentally acceptable manner.
- *Proposal unlikely to be Environmentally Acceptable* (PUEA): where proposal clearly cannot meet EPA's environmental objectives, and which cannot be reasonably modified, or is proposed in special environmental area. In such cases, the Chair of EPA may have discussions with the proponent to try to achieve a better location and/or design, prior to deciding on level of assessment.
- *Environmental Protection Statement* (EPS): this level of assessment typically applies to proposals of local interest that raise a number of significant environmental factors which can be readily managed, and where formal public review may be unnecessary because the proponent has adequately consulted with stakeholders.
- *Public Environmental Review* (PER): applies to proposals of local or regional significance that raise a number of significant environmental factors, some of which are considered complex and require detailed assessments to determine whether, and if so, how they can be managed. PERs are subject to a formal public review period of between four and eight weeks.
- *Environmental Review and Management Programme* (ERMP): applies to proposals of statewide interest that raise a number of significant environmental issues – many of which are considered to be complex or of a strategic nature, and require substantial assessment to determine whether, and if so how, they can be managed in an acceptable way. Such proposals are subject to extensive public review.

(b) **A formal Environmental Scoping stage for PER and ERMP**

- *An Environmental Scoping document shall include*:
  - (i) a summary description of the project;
  - (ii) a summary description of the environment which places the proposal in a regional biophysical and social context;
  - (iii) a preliminary impact assessment with identification of the environmental issues/factors arising from the project and their relative significance. In doing this, the proponent should have preliminary discussions with decision-making authorities (DMAs) and any other relevant agencies, regarding issues/factors they consider should be addressed, and any specific requirements they may have;
  - (iv) a Scope of W310orks setting out the proposed environmental surveys/ investigations to be carried out as part of the EIA for preparation of the PER/ ERMP. The surveys/investigations should be clearly linked to the environmental issues/factors identified from the preliminary impact assessment, and be aimed at demonstrating that:
    - (a) best practicable measures have been taken in planning and designing the proposal to avoid, and where this is not possible, to minimize impacts; and
    - (b) unavoidable impacts should be found to be environmentally acceptable, taking into account cumulative impacts which have already occurred in the region.
  - (v) a list of the people, if any, proposed to provide peer review of findings and conclusions of the environment surveys/investigations;

Table 10.2 (*continued*)

> (vi) a planned programme of consultation with the public, key stakeholders and relevant government agencies, as appropriate; and
> (vii) a proposed timetable for undertaking the environmental surveys/investigations and submission of the draft PER/ERMP.

- The Authority will maintain a publicly available database of generic environmental factors, and associated broad EPA environmental objectives, as a guide to proponents for preparing their Environmental Scoping document and PER/ERMP.

(c) **Guidance on the assessment content (environmental review document) for PER and ERMP**

- The proponent shall ensure that the environmental review document is written and is presented in a style that is readily understandable, accurate and concise. The proponent should ensure that an environmental review focuses on addressing the more significant environmental issues/factors and should include but not be limited to:

  (a) a description of the proposal and any alternatives considered, including alternative locations with a view to minimizing environmental impacts;
  (b) a description of the receiving environment and key ecosystem processes, and discussion of their significance in a regional setting. This should focus on those elements of the environment that may affect or be affected by the proposal;
  (c) placing the proposal in a regional setting in relation to existing biophysical impacts and potential for future cumulative impacts;
  (d) identification of the key issues (and list the environmental factors associated with these issues) and their relative significance. The environmental review should concentrate on the more significant issues/factors, including potential "fatal flaws";
  (e) discussion of the impacts of the proposals, both of the footprint and in the context of the regional setting, and description of the management arrangements and commitments to ameliorate those impacts to the most practical extent possible;
  (f) provision of information as to the extent that best practice will be adopted in the pursuit of the proposal and discussion of how the principles of sustainability have been incorporated, where appropriate;
  (g) details of public and government agency consultation and how comments received have been responded to;
  (h) a synthesis of the environmental costs and benefits of the proposal with the aim of achieving an overall net environmental benefit; and
  (i) presenting a case as to why the EPA should find the proposal environmentally acceptable, and demonstration that the proposal would be implemented adopting best practicable measures to minimize the impacts on the environment.

- The environmental review shall also include an audit table with environmental management commitments that will form part of the conditions of approval and will become legally binding and be audited if the proposal is implemented.

(*Source*: WA EPA 2002.)

increased certainty of outcome of the EIA process, and better design of proposals to meet environmental objectives from the outset.

The EPA then assesses the environmental acceptability of the proposals on the basis of the review document, public submissions, proponents' response, expert advice and its own investigations. The resulting EPA report to the Minister for the

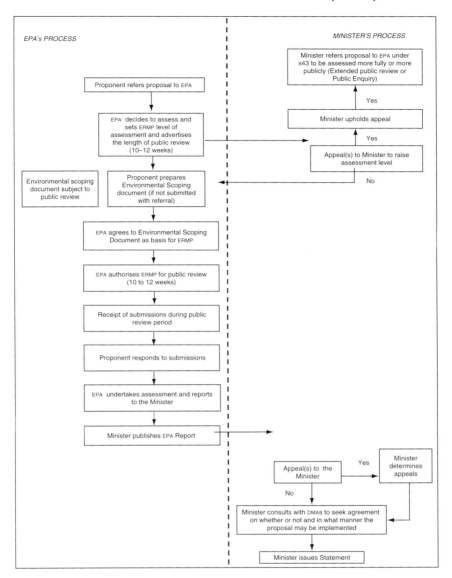

Figure 10.5 Outline of Procedure for ERMP Assessment in Western Australia. (*Source:* WA EPA 2002.)

Environment pronounces on the environmental acceptability or otherwise of the proposal and on any recommended conditions to be applied to ministerial approval. Figure 10.5 provides an outline of the full procedure for ERMP assessment. The centrality of the EPA's review of the relevant environmental information to the Minister's decision, which itself has predominance, is the most remarkable aspect of the WA system, and one which highlights the significance of the EIA impact on decisions. The WA system also has a high level of public participation, especially in controversial EIAs. The central role of the EPA also ensures consistency. However, the limited integration of the EIA and planning procedures is a weaker feature of the WA procedures. Amendments in the mid-1990s were designed to secure better integration, improving the EIA of land-use schemes, but they did also reflect a shift of control away from the EPA to the Ministry of Planning. This is symptomatic of challenges faced by an effective system. In 1993, WA lost its pioneering Social Impact Unit, which had provided expert advice on social impacts, and there is a strong development lobby, in a state highly dependent on major mineral projects, to further "soften green laws".

10.10 International bodies

Several of the major international funding institutions and other international organizations have established EIA procedures. In several cases these have evolved over time, with some handbooks or guidance manuals now on their second edition or with multiple updates.

The *European Bank for Reconstruction and Development*'s EIA requirements are typical of those of other lending institutions. The Bank's *Environmental Procedures* (EBRD 2003) aim to ensure that the environmental implications of potential bank-financed investment and technical cooperation projects are identified and assessed early in the bank's planning and decision-making process, and that environmental considerations – including potential benefits – are incorporated into the preparation, approval and implementation of projects.

Initial screening discussions between the bank and the project sponsor sort the proposed project into one of several categories:

- Category A projects likely to cause significant adverse impacts which, at the time of screening cannot readily be identified or assessed. A full EIA is required for these projects. Some types of projects automatically come under Category A; others are put into Category A on a case-by-case basis.
- Category B projects likely to have less severe impacts than Category A projects. These require a less stringent environmental analysis. The scope of the environmental analysis is determined on a case-by-case basis.
- Category C projects likely to result in minimal or no adverse impacts and that do not need analysis.

Where it is not possible to categorize the project during early screening, an initial environmental examination is carried out to determine the appropriate screening category. The screening process also determines whether an environmental audit is required to identify the impact of past or current operations of existing facilities.

Public disclosure and consultation is required throughout the planning and assessment/analysis processes of Categories A and B projects. The EBRD sets environmental standards for each project. It may also specify an environmental action plan and/or monitoring to be carried out by the sponsor as a condition of investment. Prior to making a final decision about whether to lend money for the project, Bank officials review the environmental due diligence information available, ensure that proposed mitigation measures are agreed with the project sponsor, highlight opportunities for environmental improvements, identify environmental monitoring requirements and any technical/environmental cooperation initiatives that should be undertaken and advise on whether the project complies with the Bank's environmental policy and procedures.

The *Asian Development Bank's Environmental Assessment Guidelines* (ADB 2003) have similar screening and assessment requirements. The guidelines give more detailed information about assessment of projects vs. programmes, a range of different sectors, equity investments, etc.; they also stress consultation.

The *World Bank* perceives EIA as one of its key environmental and social safeguard policies. Its Operational Policy/Bank Procedures 4.01 of 1991 require EIA for relevant lending operations, and its *Environmental Assessment Sourcebook* (World Bank 1991) and various updates explain the EIA process. EIA involves screening of the project into assessment categories by Bank staff; preparation of an environmental assessment report by the proponent; review of the report by Bank staff; an appraisal mission in which Bank staff discuss and resolve environmental issues with the proponent; documentation of the findings; and supervision and evaluation during project implementation (World Bank 1999). Between 1989 and 1995, the Bank screened over a thousand projects for their potential environmental impacts: of these 10 per cent were in Category A (primarily energy, agriculture and transport projects), 41 per cent in Category B, and 49 per cent in Category C (World Bank 1997). Category A projects are those expected to have "adverse impacts that may be sensitive, irreversible and diverse" (World Bank 1999), and they require a full EIA. For Category B projects, where impacts are "less significant – not as sensitive, numerous, major or diverse", a full EIA is not required, but some environmental analysis is necessary. Category C projects have negligible or minimal direct disturbance on the physical setting, and neither EIA nor environmental analysis is required. Typical category C projects focus on education, family planning, health and human resource development.

Notable features of the World Bank process include a holistic environment definition, including physical, biological and socio-economic aspects, a high profile for public consultation and considerable focus on project implementation. A report (World Bank 1995) identified five main challenges ahead: moving EIA "upstream" (into project design stages and at sectoral and regional levels); more effective public consultation; better integration of EIA into the project work programme (including mitigation, monitoring and management plans); learning from implementation (the "feedback loop"); and engaging the private sector (especially financiers and project sponsors) to ensure that projects are subject to EIA of acceptable quality. Mercier (2003) reinforces the emphasis now placed on implementation of the mitigation, prevention and compensation measures contained in the EIA. Also, because many of the client countries now have their own EIA requirements and their own EIA staff and review mechanisms, the Bank is increasingly involved in enhancing that capacity upfront during project preparation.

The *African Development Bank* and *European Investment Bank* (EIB 2002) have less comprehensive EIA guidance, but both require EIA to be carried out, promote public participation in the EIA process, and take account of these when deciding on whether to fund a project. Social issues are considered as part of the EIA where appropriate.

Other organizations have also published EIA guidance. For instance UNEP's very useful *Environmental Impact Assessment Training Resources Manual*, now in its second edition (UNEP 2002), includes case studies (primarily from developing countries), transparencies and detailed chapters on various stages of EIA. The UK Department for International Development has produced a similar *Environmental Guide* (DFID 1999).

10.11 Summary

Gibson (2002) suggests that EIA worldwide has been moving, over the last 30 years, towards being:

- earlier in planning (beginning with purposes and broad alternatives);
- more open and participative (not just proponents, government officials and technical experts);
- more comprehensive (not just biophysical environment, local effects, capital projects, single undertakings);
- more mandatory (gradual conversion of policy-based to law-based processes);
- more closely monitored (by the courts, informed civil society bodies and government auditors);
- more widely applied (through law at various levels, but also in land-use planning, through voluntary corporate initiatives, and so on);
- more integrative (considering systemic effects rather than just individual impacts);
- more ambitious (overall sustainability rather than just individually "acceptable" undertakings); and
- more humble (recognizing and addressing uncertainties, applying precaution).

However, Goodland & Mercier (1999) suggest that, worldwide, EIA is still constrained by lack of political will, insufficient budget to implement proposed mitigation measures and lack of institutional capacity.

Chapters 11 and 12 draw on some of the ideas discussed here, and elsewhere, to identify possibilities for the future, focusing primarily on the UK system, but set in the wider EU and global context.

Note

1. The EIA Commission is an independent body which carries out research on the EIA system, and which advises on the scope and adequacy of each EIA. The core of the Commission is composed of a chairman, who is appointed by the Council of Ministers, two vice-chairmen and a full-time staff of about 25. In addition, about 200 members who are experts in EIA-related fields assist on a case-by-case basis.

References

ADB (Asian Development Bank) 2003. ADB *Environmental Assessment Guidelines*, www.adb.org.

Angelsen, A., F. Odd-Helge, U. R. Sumaila 1994. *Project appraisal and sustainability in less developed countries*, R. 1994:1. Fantoft-Bergen, Norway: Chr. Michelsen Institute.

ANZECC (Australia and New Zealand Environment and Conservation Council):

1991   *A National Approach to EIA in Australia*

1996   *Guidelines and Criteria for Determining the Need for and Level of EIA in Australia*

1997   *Basis for a National Agreement on EIA*
        Canberra: ANZECC.

Baglo, M. A. 2003 Benin's experience with national and international EIA processes, www.ceaa.gc.ca/017/0005/0002/2b_e.htm.

Beanlands, G. 1994. *Summary of country policies and procedures – coherence of environmental assessment for international bilateral aid.* Paris: OECD/DAC Working Party on Development Assistance and Environment.

Briffett, C. 1999. Environmental impact assessment in East Asia. In *Handbook of Environmental Impact Assessment,* J. Petts (ed.), vol. 2, 143–67. Oxford: Blackwell Science Ltd.

Brito, E. & I. Verocai 1999. Environmental impact assessment in South and Central America. In *Handbook of Environmental Impact Assessment,* J. Petts (ed.), vol. 2, Chapter 10, 183–202. Oxford: Blackwell Science Ltd.

CEAA (Canadian Environmental Assessment Agency) 2003. *Canadian Environmental Assessment Act: An Overview.* Ottawa: CEAA. www.ceaa.gc.ca.

CEC (Commission of European Communities) 2003. *Application and Effectiveness of the EIA Directive (Directive 85/337/EEC as amended by Directive 97/11/EC),* report to the European Parliament and the Council. CEC: Brussels.

CEPA 1994. *Review of Commonwealth environmental impact assessment.* Canberra: CEPA.

Chico, I. 1995. EIA in Latin America. *Environmental Assessment* 3(2), 69–71.

China 1999 (in Chinese). *A summary of the environmental protection and management work of construction projects,* www.China-eia.com/chegxu/hpcx_main0.htm.

Clark, R. & D. Richards 1999. Environmental impact assessment in North America. In *Handbook of Environmental Impact Assessment,* J. Petts (ed.), vol. 2, 203–22. Oxford: Blackwell Science Ltd.

CLEIAA (Capacity Development and Linkages for Environmental Impact Assessment in Africa) 2003. www.cleiaa.org.

d'Almeida, K. 2001. *Cadre institutional législatif et réglementaire de l'évaluation environnementale dans les pays francophones d'Afrique et de l'Océan Indien.* Montréal, Canada: EIPF et Secrétariat francophone de l'AiEi/IAIA.

DFID (Department for International Development) 1999. *Environmental guide.* London: DFID. www.dfid.gov.uk.

Dixon, J. & T. Fookes 1995. Environmental assessment in New Zealand: prospects and practical realities. *Australian Journal of Environmental Management* 2(2), 104–11.

EBRD (European Bank for Reconstruction and Development) 2003. *Environmental Procedures,* www.erdb.org.

EIB (European Investment Bank) 2002. *Environmental procedures.* Luxembourg: European Investment Bank.

Environment Australia 2000. *EPBC Act: Various Documents, Including Environmental Assessment Processes; Administrative Guidelines on Significance; and Frequently Asked Questions.* Canberra: Department of Environment and Heritage.

FEARO (Federal Environmental Assessment Review Office) 1977. *A guide to the federal environmental assessment and review process.* Ottawa: FEARO.

Fisher, D. 1992. *Paradise deferred: environmental policymaking in Central and Eastern Europe.* London: Royal Institute of International Affairs.

Gibson, R. 2002. From Wreck Cove to Voisey's Bay: the evolution of federal environmental assessment in Canada. *Impact Assessment and Project Appraisal* 20(3), 151–60.

Glasson, J. & N. N. B. Salvador 2000. EIA in Brazil: a procedures-practice gap. A comparative study with reference to EU, and especially the UK. *Environmental Impact Assessment Review* 20, 191–225.

Goodland, R. & V. Edmundson 1994. *Environmental assessment and development.* Washington, DC: World Bank.

Goodland, R. & J. R. Mercier 1999. *The Evolution of Environmental Assessment in the World Bank: From "Approval" to Results,* World Bank Environment Department Paper No. 67, Washington DC: World Bank.

Goodland, R., J. R. Mercier, S. Muntemba 1996. *Environmental assessment (EA) in Africa: a World Bank commitment.* Washington, DC: World Bank.

Harvey, N. 1998. *EIA: procedures and prospects in Australia.* Melbourne: Oxford University Press.

IEMA (Institute of Environmental Management and Assessment with EIA Centre (University of Manchester)) 2002. *Environmental Assessment Yearbook 2002*. Manchester: EIA Centre, University of Manchester.

Iglesias, S. 1996. *The role of EIA in mining activities: the Peruvian case* (MSc dissertation, Oxford Brookes University, Oxford).

Jahiel, A. R. 1994. *Policy implementation through organisational learning: the use of water pollution control in China's reforming socialist system* (PhD dissertation, University of Michigan, Ann Arbor).

Jendroska, J. & J. Sommer 1994. Environmental impact assessment in Polish law: the concept, development, and perspectives. *Environmental Impact Assessment Review* 14(2/3), 169–94.

Kakonge, J. O. 1999. Environmental impact assessment in Africa. In *Handbook of Environmental Impact Assessment*, J. Petts (ed.), vol. 2, 168–82. Oxford: Blackwell Science Ltd.

Lee, N. & C. George 2000. *Environmental assessment in developing and transitional countries: principles, methods and practice*. Chichester: Wiley.

Mao, W. & P. Hills 2002. Impacts of the economic-political reform on environmental impact assessment implementation in China. *Impact Assessment and Project Appraisal* 29(2), 101–11.

Marsden, S. & S. Dovers 2002. *Strategic Environmental Assessment in Australasia*. Annadale, NSW: Federation Press.

Mercier, J. R. 2003. Environmental Assessment in a Changing World at a Changing World Bank, in IEMA/EIA Centre 2003. *Environmental Assessment Outlook*, Manchester: EIA Centre, University of Manchester.

Morgan, R. 1995. Progress with implementing the environmental assessment requirements of the Resource Management Act in New Zealand. *Journal of Environmental Planning and Management* 38(3), 333–48.

Netherlands Commission for Environmental Impact Assessment 1999. *Country status reports on EIA*. The Hague: Netherlands Commission for EIA, www.eia.nl/os/publications/cprofiles.htm.

OECD 2003. *Environmental Review of Poland*, www.oecd.org.

Okaru, V. & A. Barannik 1996. Harmonization of environmental assessment procedures between the World Bank and borrower nations. In *Environmental Assessment (EA) in Africa*, R. Goodland et al. (eds), 35–63. Washington, DC: World Bank.

Ortolano, L. 1996. Influence of institutional arrangements on EIA effectiveness in China. *Proceedings of the 16th Annual Conference of the International Association for Impact Assessment*, 901–05. Estoril: IAIA.

Padgett, R. & L. K. Kriwoken 2001. The Australian Environmental Protection and Biodiversity Conservation Act 1999: what role for the Commonwealth in environmental impact assessment? *Australian Journal of Environmental Management* 8, 25–36.

Regional Environmental Centre for Central and Eastern Europe 2003. www.rec.org.

Roe, D., B. Dalal-Clayton, R. Hughes 1995. *A directory of impact assessment guidelines*. London: International Institute for Environment and Development.

Rzeszot, U. A. 1999. Environmental impact assessment in Central and Eastern Europe. In *Handbook of Environmental Impact Assessment*, J. Petts (ed.), vol. 2, Chapter 7, 123–42. Oxford: Blackwell Science Ltd.

Scanlon, J. & M. Dyson 2001. Will practice hinder principle? Implementing the EPBC Act. *Environment and Planning Law Journal* 18, 14–22.

Scholten, J. 2003 EIA in the Netherlands: The paradox of successful application and mediocre reputation, www.ceaa.gc.ca/017/0005/0002/2g_e.htm.

Sinclair, A. J. & P. Fitzpatrick 2002. Provisions for more meaningful public participation still elusive in proposed Canadian EA Bill. *Impact Assessment and Project Appraisal* 20(3), 161–76.

Thomas, I. 1998. *EIA in Australia*, 2nd edn. New South Wales: Federation Press.

UNEP (United Nations Environment Programme) 2002. *UNEP Environmental Impact Assessment Training Resource Manual*, 2nd edn, www.unep.ch/etu/publications/EIAMan_2edition_toc.htm.

Waldeck, S., A. Morrison-Saunders, D. Annadale 2003. Effectiveness of non-legal EIA guidance from the perspective of consultants in Western Australia. *Impact Assessment and Project Appraisal* **21**(3), 251–56.

Wang, Y., R. Morgan, M. Cashmore 2003. Environmental impact assessment of projects in the People's Republic of China: new law, old problem. *Environmental Impact Assessment Review* **23**, 543–79.

Welles, H. 1995. EIA capacity-strengthening in Asia: the USAID/WRI model. *Environmental Professional* **17**, 103–16.

WA EPA (Western Australian Environmental Protection Authority) 2002. *Environmental Impacts Assessment (Part IV Division 1) Administrative Procedures*. Perth: EPA.

Wiszniewska, B., J. Farr, J. Jendrośka 2002. *Handbook on Environmental Impact Assessment Procedures in Poland*. Warsaw: Ministry of Environment.

Wood, C. 2003. *Environmental impact assessment: a comparative review*, 2nd edn. Prentice Hall.

Wood, C. & J. Bailey 1994. Predominance and independence in EIA: the Western Australian model. *Environmental Impact Assessment Review* **14**(1), 37–59.

World Bank 1991. *Environmental Assessment Sourcebook*. Washington, DC: World Bank, www.worldbank.org.

World Bank 1995. *Environmental assessment: challenges and good practice*. Washington, DC: World Bank.

World Bank 1997. *The Impact of Environmental Assessment: A Review of World Bank Experience*. World Bank Technical Paper no. 363, Washington, DC: World Bank.

World Bank 1999. *Environmental Assessment*, BP 4.01, Washington, DC: World Bank.

World Bank 2002. Environmental impact assessment systems in Europe and Central Asia Countries, www.worldbank.org/eca/environment.

Part 4

Prospects

11 Improving the effectiveness of project assessment

11.1 Introduction

Overall, the experience of EIA to date can be summed up as being like the proverbial curate's egg: good in parts. Current issues in the EIA process were briefly noted in Section 1.6: they include scoping, EIA methodology, the roles of the participants in the process, EIS quality, monitoring and the extension of EIA to more strategic levels of decision-making. The various chapters on steps in the process have sought to identify best practice, and Chapter 8 provides an overview of the quantity and quality of UK practice to date. Detailed case studies of good practice and comparative international experience provide further ideas for possible future developments. The evolving, but still in some cases limited, experience in EIA among the main participants in the process – consultants, local authorities, central government, developers and affected parties – explains some of the current issues.

However, 15 years after the implementation of EC Directive 85/337, there is less scepticism in most quarters and a general acceptance of the value of EIA. There are still some major shortcomings, and there is considerable scope for improving quality, but practice and the underpinning knowledge and understanding are quickly developing; EIA is on a steep learning curve. The procedures, process and practice of EIA will undoubtedly evolve further, as evidenced by the comparative studies of other countries. The EU countries can learn from such experience and from their own experience since 1988.

This chapter focuses on the prospects for project-based EIA. The following section briefly considers the array of perspectives on change from the various participants in the EIA process. This is followed by a consideration of possible developments in some important areas of the EIA process and in the nature of EISs. The chapter concludes with a discussion of the parallel and complementary development of environmental management systems and audits. The nature and types of system and audit are explained, and their important relationships to EIA are discussed. Chapter 12 closes the book by widening the scope of EIA to SEA, from projects to programmes, plans and policies.

11.2 Perspectives on change

An underlying theme in any discussion of EIA is change. This has surfaced several times in the various chapters of this book. EIA systems and procedures are changing in many countries. Indeed, as O'Riordan (1990) noted (see Section 1.4), we should expect EIA to change in the face of shifting environmental values, politics and managerial capabilities. This is not to devalue the achievements of EIA to date. As the World Bank (1995) notes, "Over the past decade, EIA has moved from the fringes of development planning to become a widely recognised tool for sound project decision making."

The practice of EIA under the existing systems established in the EU Member States has also improved rapidly (see Chapter 8). This change can be expected to continue in the future, as the provisions of the amended Directive 97/11 work through, and further amendments are introduced. Changes in EIA procedures, like the initial introduction of EIA regulations, can generate considerable conflict between levels of government: between federal and state levels, between national and local levels and, in the particular case of Europe, between the EU and its Member States. They also generate conflict between the other participants in the process: the developers, the affected parties and the facilitators (see Figure 3.1).

The *Commission of the European Communities* (CEC) is generally seen as positive and proactive with regard to EIA. The CEC welcomed the introduction of common legislation as reflected in Directive 85/337, the provision of information on projects and the general spread of good practice, but was concerned about the lack of compatibility of EIA systems across frontiers, the opaque processes employed, the limited access to the public and lack of continuity in the process. It pressed hard for amendments to the Directive, and achieved some of its objectives in the amended Directive. The CEC is committed to reviewing and updating EIA procedures, which may involve further changes. An SEA Directive will be implemented from 2004 (see Chapter 12). Other areas of attention include, for example, cumulative assessment, public participation, economic valuation and EIA procedures for development aid projects. In contrast with the CEC, Member States tend to be more defensive and reactive. They are generally concerned in maintaining "subsidiarity" with regard to activities involving the EU; this has been an issue with EIA, as reflected in the exchanges between the EC Commissioner for the Environment and the UK Government in 1991–92 (see Section 6.6). Governments are also sensitive to increasing controls on economic development in an increasingly competitive and global economy.

For example, within the UK *Government*, the DoE (now ODPM) has been concerned to tidy up ambiguities in the project-based procedures, and to improve guidance and informal procedures for example, but is wary of new regulations. However, it has commissioned and produced research reports, for example on an EIA good-practice guide, on the evaluation and review of environmental information and on mitigation in EIA. Its response to many of the proposals in the amended Directive also reflect an acceptance of the value of EIA. *Local government* in the UK has begun to come to terms with EIA, and there is evidence that those authorities with considerable experience (e.g. Essex, Kent, Cheshire) learn fast, apply the regulations and guidance in user-friendly "customized" formats to help developers and affected parties in their areas, and are pushing up the standards expected from project proponents (see Essex Planning Officers' Association 2001).

Pressure groups – exemplified in the UK by the Campaign to Protect Rural England (CPRE), the Royal Society for the Protection of Birds (RSPB) and Friends of the Earth (FoE) – and those parties affected by development proposals view project EIA as a very useful tool for increasing access to information on projects, and for advancing the protection of the physical environment in particular. They have been keen to develop EIA processes and procedures; see, for example, the reports by CPRE (1991,

1992). Many *developers* are less enthusiastic about changes in the regulations, but would welcome clarification on ambiguities – especially on whether EIAs are needed in the first place for their particular projects. For *facilitators* (consultants, lawyers, etc.), EIA has been a welcome boon; their interest in longer and wider procedures, involving more of their services, is clear.

Other participants in the process in the UK, such as the IEMA, the Association of Environmental Consultants, academics and some environmental consultancies, are carrying out ground-breaking studies into topics such as best-practice guidelines, the use of monetary valuation in EIA and approaches to types of impact study. In addition, the production of over 600 EISs a year in the UK is generating a considerable body of expertise, innovative approaches and comparative studies. EISs are also becoming increasingly reviewed, and it is hoped that bad practice will be exposed and reduced. Training in EIA skills is also developing.

11.3 Possible changes in the EIA process: the future agenda

11.3.1 An overview of possible changes

In the important *International study of the effectiveness of environmental assessment* for the IAIA, Sadler (1996) provided a summary of "best case" and "worst case" EA performance (Box 11.1). He also provides a five-part agenda for action:

- "Going back to basics" involves building on well-established procedures, by providing, for example, more good-practice guidance, explicit periods for the process and the removal of duplication.
- "Upgrading EIA processes and activities" involves, in particular, better quality control, public involvement and addressing the issue of cumulative effects.
- "Extending SEA as an integral part of policy making" includes the development of methods, and extended applications.
- "Sharpening EA as a sustainability instrument" includes incorporating relevant sustainability indicators, the consideration of capacities, dealing with risks and uncertainty and linking EIA with other forms of assessment and other policy instruments, such as environmental accounting.
- "New opportunities and challenges" covers issues such as the transboundary management of common property resources (e.g. the Antarctic), global change and the decommissioning or replacement of major infrastructure items.

A pragmatic approach to change could subdivide the future agenda into proposals to *improve* EIA procedures, usually sooner and maybe more easily than proposals to *widen* the scope of EIA, which are likely to come later and will probably be more difficult to implement.

Improvements to project EIA cover some of the changes introduced by the amended EC Directive, including developments in approaches to screening, the mandatory consideration of alternatives and a strong encouragement to undertake scoping at an early stage in the project development cycle. There is also more support for transparent procedures, and encouragement for consultation, for the explanation and publication of decisions and for the inclusion of cumulative impacts and risk assessment. There may be a case for further changes in the legal basis of project EIA, especially in the

---

Box 11.1 Summary of international best- and worst-case EA *performances*

Best-case performance

The EA process:

- facilitates informed decision-making by providing clear, well-structured, dispassionate analysis of the effects and consequences of proposed actions;
- assists the selection of alternatives, including the selection of the best practicable or most environmentally friendly option;
- influences both project selection and policy design by screening out environmentally unsound proposals, as well as modifying feasible action;
- encompasses all relevant issues and factors, including cumulative effects, social impacts and health risks;
- directs (not dictates) formal approvals, including the establishment of terms and conditions of implementation and follow-up;
- results in the satisfactory prediction of the adverse effects of proposed actions and their mitigation using conventional and customized techniques; and
- serves as an adaptive, organizational learning process in which the lessons experienced are fed back into policy, institutional and project designs.

Worst-case performance

The EA process:

- is inconsistently applied to development proposals with many sectors and classes of activity omitted;
- operates as a "stand alone" process, poorly related to the project cycle and approval process and consequently is of marginal influence;
- has a non-existent or weak follow-up process, lacking surveillance and enforcement of terms and conditions, effects monitoring, etc.;
- does not consider cumulative effects or social, health and risk factors;
- makes little or no reference to the public, or consultation is perfunctory, substandard and takes no account of the specific requirements of affected groups;
- results in EA reports that are voluminous, poorly organized and descriptive technical documents;
- provides information that is unhelpful or irrelevant to decision-making;
- is inefficient, time consuming and costly in relation to the benefits delivered; and
- understates and insufficiently mitigates environmental impacts and loses credibility.

(*Source*: Sadler 1996.)

UK, where the wide array of regulations can cause the fragmentation of the elements of a project-linked EIA activity, as we revealed in Chapter 9. The methods of assessment could also benefit from further attention. Uncertainty about the unknown may mean the EIA process starts too late and results in a lack of integration with the management of a project's life cycle. The EIA process and the resulting EISs may lack balance, focus on the more straightforward process of describing the project and its baseline environment and consider much less the identification, prediction and evaluation of impacts. The forecasting methods used in EIA are not explained in most cases (see Tables 8.3 and 8.7). It is to be hoped that there will be advances in the application of concepts and techniques in operational practice, in the areas of predicting the magnitude of impacts and determining their importance (including the array of multi-criteria and monetary evaluation techniques). A good "method statement", explaining how a study has been conducted – in terms of techniques, consultation, the relative roles of experts and others – should be a basic element of any EIS.

Widening the scope of EIA includes, in particular, the development of tiered assessment through the introduction of SEA (discussed in the next chapter). Another important extension of the scope of EIA includes "completing the circle" through the more widespread use of monitoring and auditing. Unfortunately, this vital step in the EIA process is still not mandatory under the amended EC Directive. More wide-ranging possibilities include the move to a "whole of environment" approach, with a more balanced consideration of both biophysical and socio-economic impacts. Such widening of scope should lead to more integrated EA. There may also be a trend towards what might be termed "environmental impact design", with the use of EIA to identify environmental constraints before the design process is begun.

The following sections discuss possibilities for some of these short- and long-term proposals, including allowing for cumulative impacts, building in better procedures for public participation, widening the scope to include socio-economic impacts, embracing the growing area of health impact assessment, developing integrated EA and moving towards environmental impact design.

11.3.2 Cumulative impacts

Many projects are individually minor, but collectively may impose a significant impact on the environment. Activities such as residential development, farming and household behaviour normally fall outside the scope of conventional EIA. The ecological response to the collective impact of such activities may be delayed until a threshold is crossed, when the impact may come to light in sudden and dramatic form (e.g. flooding). Odum (1982) refers to the "tyranny of small decisions" and the consequences arising from the continual growth of small developments. While there is no particular consensus on what constitutes cumulative impacts, the categorization by the Canadian Environmental Assessment Research Council (CEARC) (Peterson et al. 1987) is widely quoted, and includes:

- time-crowded perturbations – which occur because perturbations are so close in time that the effects of one are not dissipated before the next one occurs;
- space-crowded perturbations – when perturbations are so close in space that their effects overlap;
- synergisms – where different types of perturbation occurring in the same area may interact to produce qualitatively and quantitatively different responses by the receiving ecological communities;
- indirect effects – those produced at some time or distance from the initial perturbation, or by a complex pathway; and
- nibbling – which can include the incremental erosion of a resource until there is a significant change/it is all used up.

"Cumulative impact assessment is predicting and assessing all other likely existing, past and reasonably foreseeable future effects on the environment arising from perturbations which are time-crowded; space-crowded; synergisms; indirect; or, constitute nibbling" (CEPA 1994). The need to include cumulative impact assessment in EIA has been long recognized. In the CEQA of 1970, significant impacts are considered to exist if "the possible effects of a project are individually limited but cumulatively considerable". Subsequent legislative reference is found in the 1991 Resource Management Act of New Zealand, which makes explicit reference to cumulative effects, and now also in the amended Directive 97/11/EC, which refers to the need to consider the characteristics of projects having regard to "the cumulation with other projects". In Canada, which has been at the forefront in the development of "cumulative effects assessment" (CEA) in recent years, the consideration of cumulative effects is now explicit and mandatory in legislation both federally and in several provinces.

However, it is in the practical implementation of the consideration of cumulative impacts that the problems and deficiencies become clear, and cases of good practice and useful methodologies have been limited until recently. In Australia, assessments have largely been carried out by regulatory authorities, rather than by project proponents, and have focused on regional air quality and the quality and salinity of water in catchment areas (CEPA 1994). Figure 11.1 provides an example of a simple perturbation impact model developed by Lane and Associates (1988). It is basically an "impact tree" which links (a) the principal causes driving a development with, (b) the main perturbations induced with, (c) the primary biophysical and socio-economic impacts and (d) the secondary impacts. The figure shows some of the potential cumulative impacts associated with a number of area-related tourism developments. In the US, the CEQ produced a practice guide "Considering Cumulative Effects" (CEQ 1997), based on numerous case studies. The guide consists of 11 steps for CEA, in three main stages.

In Canada, a "Cumulative Effects Assessment Practitioners' Guide" (CEAA 1999) provides a very useful overview and clarification of terms and fundamentals, of practical approaches to completing CEAs, and case studies of approaches used by project proponents. The guide provides some clear and simple definitions – "Cumulative effects are changes to the environment that are caused by an action in combination with other past, present and future human actions. A CEA is an assessment of those effects." Further Canadian work has sought to improve the practice of CEA (see Baxter et al. 2001). In the EU, there has also been an attempt to support practice in the area through the development of "Guidance for the assessment of indirect and cumulative impacts" (Hyder Consulting 1999). Piper (2000, 2001a,b) provides valuable evidence on the state of UK practice in CEA drawing on research on a number of case studies (see Chapter 9, Section 9.7).

Between them, these guides and assessments of practice highlight some of the key process and organizational issues in considering cumulative impacts/effects. Process issues include, for example: establishing the geographic scope of the analysis (how wide should the impacts region be), establishing the time frame for the analysis (including not only present projects, but also those in the non-immediate time frame – past and reasonably foreseeable future) and determining the magnitude and significance of the effects. A key organizational issue in the UK (see Piper 2001b) is "Which organization has the responsibility to require or commission the CEA work?" This is complicated when, as is often the case in CEA, there is more than one competent

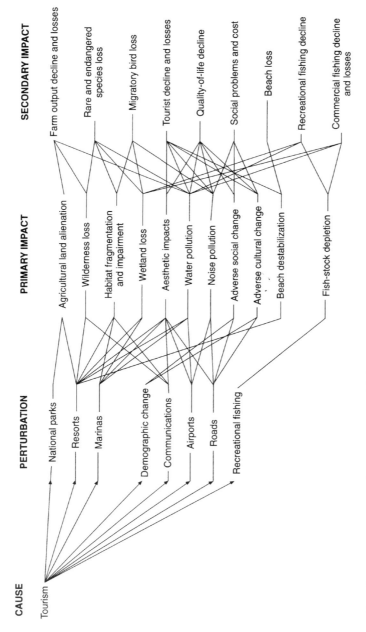

CAUSE   **PERTURBATION**   **PRIMARY IMPACT**   **SECONDARY IMPACT**

Figure 11.1 Cumulative impacts – perturbation impact model. (*Source:* Lane and associates 1988.)

authority involved. Piper's case studies show a range of initiators, including developers, statutory consultees and local planning authorities.

11.3.3 Public participation

The lack of effective public participation in EIA is a major weakness in the UK, in most of Europe and in many other countries. It tends to occur late, if at all, and is often tokenistic and limited to minimum requirements and to the lowest rungs on Arnstein's ladder (see Section 6.2). There is an unequal balance of participants between the "impactors" and the "impactees". We hear much of "expert speak" but often very little of "people speak". Yet the public have much to contribute: they may offer a superior knowledge of local conditions; they bring their own values as stakeholders and they contribute a non-scientific discourse to a process which is often *too* scientific. Lack of effective public participation is not only inequitable and inefficient; it may also be very counter-productive, as frustrated and unequal participants resort to other means, including direct action.

More effective public participation needs both the will and the methods. In Europe, there are signs from the EU, and more widely, that the will is strengthening, as evidenced by the following declaration from the Third Conference of European Environment Ministers (Sofia, Bulgaria, October 1995):

> We believe it is essential that, in accordance with Principle 10 of the Rio Declaration, States should give the public the opportunity to participate at all levels in decision-making processes relating to the environment, and we recognize that much remains to be done in this respect. We call upon countries in the region to ensure that they have a legal framework and effective and appropriate mechanisms to secure public access to environmental information, to facilitate and encourage public participation, inter alia through environmental impact assessment procedures, and to provide effective public access to judicial and administrative remedies for environmental harm.
>
> We invite countries to ensure that in relevant legislation effective public participation as a foundation for successful environmental policies is being introduced.

As noted in Section 2.7, public participation in EIA in the EU is still problematic even after the 1997 EIA Directive amendments; however, there is some hope that the transposition of the UNECE Aarhus Convention (UNECE 1998) into EIA legislation may produce a catalyst for change (Dagg et al. 2003). The Convention has three main elements relating to public participation: access to environmental information and its collection and dissemination (Articles 4 and 5); public participation measures (Articles 6–8) and access to justice (Article 9). An EC Directive (CEC 2003) on public participation is a result of the Convention; it seeks to increase the rights of public involvement within the EIA Directive with the aim of producing better quality decisions as a result.

New methods are needed to empower people in EIA to participate genuinely and constructively. These could include deliberative techniques, such as focus groups, Delphi panels and consultative committees, and appropriate resourcing, perhaps through intervenor funding. Petts (2003) highlights some of the possibilities and problems of deliberative participation, or communication through dialogue; she also stresses the need for such participation to be integral to the EIA process rather than an "add-on". Balram et al. (2003) provide an interesting development of the Delphi approach, Collaborative Spatial Delphi, using a GIS-based approach. There may be considerable potential for using spatial technology in participation in EIA. There is also the potential of the rapidly evolving Internet. As discussed in the Hong Kong case study in Section 7.3, the Internet can be used to facilitate participation at several stages in the EIA process.

11.3.4 Socio-economic impacts

Widening the scope of EIA to include socio-economic impacts in a much better way is seen as a particularly important item for the agenda. While there are varying interpretations of the scope of socio-economic or social impacts, a number of recent reports have highlighted the importance of this area (see, for example, CEPA 1994, IAIA 1994, Vanclay 2003). SIA has been defined by Burdge (1999) as "the systematic analysis, in advance, of the likely impacts a proposed action will have on the life of individuals and communities". Most development decisions involve trade-offs between biophysical and socio-economic impacts. Also, development projects affect various groups differently; there are invariably winners and losers. Yet the consideration of socio-economic impacts is very variable in practice, and often very weak. There is useful practice and associated legislative impetus for SIA in some countries, for example the USA, Canada and some states of Australia. Some of the procedures of international funding institutions also give a high profile to such impacts. But in Europe the profile is lower, and the consideration of socio-economic impacts has continued to be the poor relation (Chadwick 2002, Glasson 2001). The uncertain status of such impacts, plus the lack of best-practice guidance on their assessment, has resulted in a partial approach in practice. When socio-economic impacts are included, there tends to be a focus on the more measurable direct employment impacts; the consideration of the social–cultural impacts (such as severance, alienation, social polarization, crime and health) are often very marginal.

Yet although most of the environmental receptors listed in EC and UK regulations are biophysical in nature, the inclusion of "human beings" as one of the receptors to be considered in EIA would appear to imply a wider definition of "the environment", encompassing its human (i.e. social, economic and cultural) dimensions. The inclusion of socio-economic impacts may help to better identify all of the potential biophysical impacts of a project, because socio-economic and biophysical impacts are interrelated (Newton 1995). Early inclusion of socio-economic considerations in the EIA can also provide an opportunity to modify project design or implementation, to minimize adverse socio-economic effects and to maximize beneficial effects (Chadwick 2002). Inclusion of socio-economic impacts in the EIS also allows a more complete picture of a project's impacts, in a consistent format, in a publicly available document. Failure to include such impacts can lead to delays in the EIA process, since the competent authority may request further information on such matters.

The fuller and better consideration of socio-economic impacts raises issues and challenges, for example about the types of impact, their measurement, the role of public participation and their position in EIA. One categorization of socio-economic impacts is into: (a) quantitatively measurable impacts, such as population changes, and the effects on employment opportunities or on local financial implications of a proposed project, and (b) non-quantitatively measurable impacts, such as effects on social relationships, psychological attitudes, community cohesion, cultural life or social structures (CEPA 1994). Such impacts are wide-ranging; many are not easily measured, and direct communication with people about their perceptions of socio-economic impacts is often the only method of documenting such impacts. There is an important symbiotic relationship between developing public participation approaches and the fuller inclusion of socio-economic impacts. SIA can establish the baseline of groups which can provide the framework for public participation to further identify issues associated with a development proposal. Such issues may be more local, subjective, informal and judgemental than those normally covered in EIA, but they cannot be ignored. Perceptions of the impacts of a project and the distribution of those impacts often largely determine the positions taken by various groups on a given project and any associated controversy.

11.3.5 *Health impact assessment (HIA)*

Health impact assessment (HIA) is a major growth area in the field of impact assessment, as evidenced by the popularity in recent years of the HIA 'track' in the annual conference of the influential International Association for Impact Assessment (IAIA). There has been a surge of academic papers, reviews, guidelines and websites relating to HIA (Ahmad 2004), but what is HIA, where is it best practised, how is it practised and how does it relate to EIA as discussed in this book?

Health includes social, economic, cultural and psychological well-being – and the ability to adapt to the stress of daily life (Health Canada 1999). Health impact refers to a change in the existing health status of a population within a defined geographical area over a specified period of time. HIA is a combination of procedures and methods by which a policy, plan, programme or project (PPPP) may be judged as to the effects it may have on the health of a population. It provides a useful, flexible approach to helping those developing and delivering proposals to consider their potential (and actual) impacts on people's health, and on health inequalities, and to improve and enhance a proposal (Taylor and Quigley 2002, Taylor and Blair-Stevens 2002, WHO Regional Office for Europe 2003, Douglas 2003).

HIA is well advanced in a number of developed countries, particularly Canada, the Netherlands, in parts of Scandinavia, and more recently in the UK. Some developing countries are also finding it very relevant to their needs (see Phoolchareon et al. 2003, for Thailand). Policy drivers can be found at various levels of government. In the UK for example, see Secretary for State for Health (1999). In the EU, the Directive on SEA specifically refers to the impact of plans and programmes on human health (see Section 12.3.4).

The main stages in the HIA process are similar to those used in EIA, including: screening, scoping, profiling (identifying the current health status of people within the defined spatial boundaries of the project using existing health indicators and population data), assessment (HIA stresses the importance of consultation with community groups to identify potential impacts), implementation and decision-making, and monitoring and continual review (Douglas 2003). Of particular use to practitioners are the procedures and methods, known as the 'Merseyside guidelines for HIA' (Scott-Samuel et al. 2001).

The overlap between HIA and EIA in terms of process, and in terms of many categories of baseline data, does raise questions as to why HIA and EIA are not better integrated. Ahmad (2004) suggests an interesting list of reasons as to why health has been overlooked in EIA. These include, for example: difficulty of establishing causality between population health and multiple pollutants; limitations on resources to carry out such assessments within the often tight timeframes of EIA; confidentiality of some health data; lack of mandatory legal framework requiring HIA; and bias amongst EIA professionals towards engineering and ecology backgrounds. However, he also concludes that there are many benefits to be gained from closer integration, in terms of shared experience, procedures, data and values. With regard to the last, HIA can bring to EIA 'values such as equity, transparent use of evidence and the consideration of differential impacts of the policy or project on various population subgroups' (Ahmad 2004). The SEA Directive 2001/42/EC provides an important milestone on the desirable path to a more integrated approach – a concept which is developed a little further in the following section.

11.3.6 Integrated environmental assessment

Hopefully such widening of scope will lead to IEA, with decisions based on the extent to which various biophysical, social and economic impacts can be traded (Figure 11.2). For example, decision-makers might be unwilling to trade critical biophysical assets (e.g. a main river system and the quality of water supply) for jobs or lifestyle, but willing to trade less critical biophysical assets. Integrated environmental assessment or IEA (Bailey et al. 1996, Davis 1996) differs from traditional EIA in that it is consciously multi-disciplinary, does not take citizens' participation or the ultimate users of EIA for granted and recognizes the critical role of complexity and uncertainty in most decisions about the environment. Hence it tolerates a much broader array of methods and perspectives (quantitative and qualitative; economic and sociological; computer modelling and oral testimony) for evaluating and judging alternative courses of action. However, integration is not without its problems, including limitations on the transferability of assessment methods (see *Project*

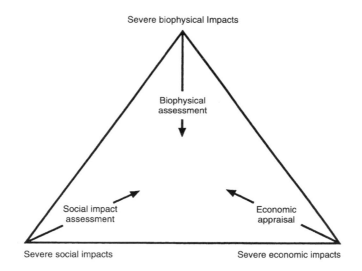

Figure 11.2 Integrated environmental assessment.

Appraisal 1996). The Integrated Assessment Workshop at the IAIA 2002 Conference highlighted the continuing problems of including social processes in integrated assessment (IAIA 2002).

Another equally important perspective is of the integration of relevant planning, environmental protection and pollution procedures. At the one extreme the UK still has over 20 regulations for EIA, grafting the procedures into an array of relevant planning and other legislation; there is also parallel environmental protection and pollution legislation. At the other, there is the New Zealand "one-stop shop" *Resource Management Act*. A better integration of relevant procedures represents another challenge for most EIA systems.

11.3.7 *Extending EIA to project design: towards environmental impact design*

An important and positive trend in EIA has been its application at increasingly early stages of project planning. For instance, whilst the DoT's 1983 *Manual of environmental appraisal* applied only to detailed route options, its 1993 *Design manual for roads and bridges* requires a three-stage approach covering, in turn, broadly defined route corridors, route options and the chosen route. British Gas also now uses three levels of environmental analysis for its pipelines, from broad feasibility studies to detailed design (Parkinson 1996). This application of EIA to the early stages of project planning helps to improve project design and to avoid the delayed and costly identification of environmental constraints that comes from carrying out EIA once the project design is completed.

McDonald & Brown (1995) suggest that the project designer must be made part of the EIA team:

> Currently, most formal administrative and reporting requirements for EIA are based on its original role as a stand alone report carried out distinct from, but in parallel with the project design ... We can redress [EIA limitations] by transferring much of the philosophy, the insights and techniques which we currently use in environmental assessments, directly into planning and design activities.

A further evolution of this concept is to use EIA to identify basic environmental constraints before the design process is begun, but then allow designers freedom to design innovative and attractive structures as long as they meet those constraints. Holstein (1996) calls this postmodern approach "environmental impact design" (EID),[1] and distinguishes it from EIA's traditionally conservative, conservation-based focus. The following paragraphs explain Holstein's view of EID.

> EIA as presently practised deconstructs a site: it takes an environment apart to highlight the different interacting components within it (e.g. soil, water, flora). EIA suggests that the site has another (environmental) function other than that for which it is being developed. Yet this relationship to deconstruction is only superficial because EIA is conservation based; it makes little challenge to the fixed hierarchies of modernism that underpin it, such as development-induced growth and technological subservience. Environmental design within

EIA is too often merely a by-product of assessment or is even handed back to the developer to have another shot at the design themselves. It makes little use of artistic-based metaphors to provide any re-enchantment or return to human landscape values, it makes no attempt to rip apart environmental function and form, and creates no demand for the kind of relative individualism needed to reflect cultural sustainability to an uninterested-unless-aroused population (all characteristics of postmodernism). Through this passivity of EIA, time, space, communication, leadership – all the key elements of good flowing design are lost.

This said, initially it might be argued that true postmodernism is simply beyond the remit of an EIA which exists for objective assessment rather than artistic purposes. The above description should be called EID. EID emphasizes the artistic contribution to EIA; it requires a different set of approaches (and probably personnel) than pure EIA, as well as creativity and elements of cultural vision. To an extent some of the principles of EID are already being undertaken in EIA, in the mitigation sections of EISs, and especially within environmental divisions of the larger developers (e.g. the utilities) who often seem to see the formal EIA process as merely a lateral extension to their own design policies. Even so, rarely is it recognized as an artistic activity.

The key difference between EIA and EID lies in the concept of "unmodifiable design". Traditionally, EIAs are carried out on projects in which most of the structural elements have already been finalized. In more EID-oriented approaches, there is less unmodifiable design and thus more scope for introducing environmentally sound design as mitigation measures. An even more radical path would be a postmodern EIA which aims to begin with so few unmodifiable design ideas that the EIA essentially becomes the leading player in design (adapted from Holstein 1996).

11.3.8 Complementary changes: enhancing skills and knowledge

The previous discussions indicate that EIA practitioners need to develop further their substantive knowledge of the wider environment. There is an important role for "State of the Environment reports" and the development of "carrying capacity and sustainability indicators" – if not interpreted too narrowly. For example, carrying capacity is multi-dimensional and multi-perspective (see Figure 11.3 for an example for tourism impact assessment). Carrying capacity is also an elastic concept, and the capacity can be increased through good management.

Practitioners also need to develop both "technical" and "participatory" approaches, such as the focus group, the Delphi approach and the mediation approaches noted earlier. EIA has been too long dominated by the "clinical expert" with the detached quantitative analysis. Notwithstanding, there is still a place for the sensible use of the rapidly developing technology – including expert systems, GIS, participatory techniques and text-oriented analysis (e.g. non-numerical unstructured data – NUD*IST – pulling out issues from focus group transcripts) (Rodriguez-Bachiller with Glasson 2003). There is also a need for more capacity building of EIA expertise, plus relevant research, including, for example, more comparative studies and longitudinal studies (following impacts over a longer life cycle – moving towards adaptive EIA).

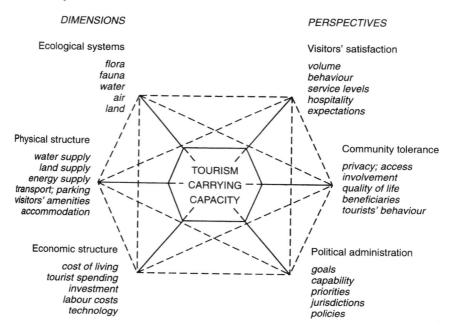

DIMENSIONS

Ecological systems

flora
fauna
water
air
land

Physical structure

water supply
land supply
energy supply
transport; parking
visitors' amenities
accommodation

Economic structure

cost of living
tourist spending
investment
labour costs
technology

TOURISM
CARRYING
CAPACITY

PERSPECTIVES

Visitors' satisfaction

volume
behaviour
service levels
hospitality
expectations

Community tolerance

privacy; access
involvement
quality of life
beneficiaries
tourists' behaviour

Political administration

goals
capability
priorities
jurisdictions
policies

Figure 11.3 Carrying capacity – a tourism example. (*Source*: Glasson et al. 1995.)

11.4 Extending EIA to project operations: environmental management systems and environmental audits

An environmental management system (EMS), like EIA, is a tool which helps organizations to take more responsibility for their actions, by determining their aims, putting them into practice and monitoring whether they are being achieved. However, in contrast with the orientation of EIA to future development actions, EMS involves the review, assessment and incremental improvement of an *existing organization*'s environmental effects. EMS can thus be seen as a continuation of EIA principles into the operational stage of a project.

EMS has evolved from environmental audits, which were first carried out in the 1970s by private firms in the USA for financial and legal reasons as an extension of financial audits. Auditing later spread to private firms in Europe as well and, in the late 1980s, to local authorities in response to public pressure to be "green". In the early 1990s environmental auditing was strengthened and expanded to encompass a total quality approach to organizations' operations through EMS. EMS is now seen as good practice and has mostly subsumed environmental auditing.

This section reviews existing standards on EMS, briefly discusses the application of EMS and environmental auditing by both private companies and local authorities and concludes by considering the links between EMS and EIA.

11.4.1 Standards and regulations on EMS

Two EMS standards apply in the UK: the EC's Eco-Management and Audit Scheme (EMAS) of 1993, which was revised in 2001, and the International Organization for Standardization's ISO series 14000. A third early standard, British Standard 7750, was withdrawn in 1997. EMAS and ISO 14001 are compatible with one another, but differ slightly in their requirements.

The *EC's EMAS scheme* was adopted by EC Regulation 1836/93 in July 1993 (EC 1993), and became operational in April 1995. It was originally restricted to companies in industrial sectors, but since the 1993 regulations were replaced in 2001 by Regulation 761/2001 (EC 2001a) it has been open to all economic sectors, including public and private services. It is a voluntary scheme and can apply on a site-by-site basis. To receive EMAS registration, an organization must:

- Establish an environmental policy agreed by top management, which includes provisions for compliance with environmental regulation, and a commitment to continual improvement of environmental performance;
- Conduct an environmental review that considers the environmental impacts of the organization's activities, products and services; its framework of environmental legislation; and its existing environmental management practices;
- Establish an EMS in the light of the results of the environmental review which aims to achieve the environmental policy. This must include an explanation of responsibilities, objectives, means, operational procedures, training needs, monitoring and communication systems;
- Carry out an environmental audit that assesses the EMS in place, conformity with the organization's policy and programme and compliance with relevant environmental legislation; and
- Provide a statement of its environmental performance which details the results achieved against the environmental objectives, and steps proposed to continuously improve the organization's environmental performance (EC 2001b).

The environmental review, EMS, audit procedure and environmental statement (ES) must be approved by an accredited eco management and audit scheme (EMAS) verifier. The validated statement must be sent to the EMAS competent body for registration and made publicly available before an organization can use the EMAS logo. In the UK the competent body is the IEMA (EC 2001b). Although EMAS was originally oriented towards larger private organizations, it can also apply to local authorities and smaller companies. The DoE/Welsh Office's Circular 2/95 discusses EMAS for local authorities, and the Small Company Environmental and Energy Management Assistance Scheme was established in November 1995 to help companies with fewer than 250 employees to carry out EMAS.

The *International Organization for Standardization's ISO 14000* series was first discussed in 1991, and a comprehensive set of EMS standards was published in September 1996. These include ISO 14001 on EMS specifications (IOS 1996a), ISO 14004 on general EMS guidance (IOS 1996b) and ISO 14010–14014 which give guidance on environmental auditing and review. EMAS and ISO 14001 are compatible, but have some differences. These are shown in Table 11.1.

Table 11.1 Differences between EMAS and ISO14001

	EMAS	ISO 14001
Preliminary environmental review	Verified initial review	No review
External communication and verification	Environmental policy, objectives, EMS and details of organization's performance made public	Environmental policy made public
Audits	Frequency and methodology of audits of the EMS and of environmental performance	Audits of the EMS (frequency of methodology not specified)
Contractors and suppliers	Required influence over contractors and suppliers	Relevant procedures are communicated to contractors and suppliers
Commitments and requirements	Employee involvement, continuous improvement of environmental performance and compliance with environmental legislation	Commitment of continual improvement of the EMS rather than a demonstration of continual improvement of environmental performance

(*Source:* EC 2001b.)

11.4.2 Implementation of EMS and environmental auditing

By 1998, over 600 UK sites had received ISO 14001 certification, and 57 UK sites – including 19 local authority sites – were EMAS registered (INEM 1998). Organizations perceive EMS as a way to reduce their costs through good management practices such as waste reduction and energy efficiency. They also see EMS as good publicity and, less directly, as a way of boosting employees' morale. However, private companies still have problems implementing EMS due to commercial confidentiality, legal liability, cost and lack of commitment. Smaller companies are especially affected by the cost implications of establishing EMS systems, and have been slower than the larger companies in applying it to their operations.[2] The use of EMS by local authorities has been limited by cutbacks in central government funding, government reorganization and growing public concerns about economic rather than environmental issues.

11.4.3 Links between EMS and EIA

The growth in EMS is important to EIA for several reasons. First, EMS of both private and public sector organizations will increasingly generate environmental information that will also be useful when carrying out EIAs. Local authorities' state of the environment reports provide data on environmental conditions in the area that can be used in EIA baseline studies. Generally, a state of the environment report will contain information on such topics as local air and water quality, noise, land use, landscape, wildlife habitats and transport. Unfortunately, unless this information is regularly (and expensively) updated, it quickly becomes outdated. It is also often collected only on a large-scale (e.g. countywide) basis, and so may not be suitable for any specific site. However, state of the environment reports do generally identify sources for environmental data that can be contacted for the most up-to-date information. Similarly, the reports may be useful when determining suitable locations for new developments, by identifying sites that are particularly environmentally sensitive and should clearly be avoided, or those that are environmentally robust and more suitable for development, for input to SEA (see Chapter 12).

Private companies' environmental audit findings have traditionally been kept confidential: it is noticeable that many more companies have ISO accreditation – which requires only limited disclosure of information – than EMAS accreditation. Thus a private company's EMS is likely to be useful for EIA only if that company intends to open a similar facility elsewhere. However, environmental auditing information about levels of wastes and emissions produced by different types of industrial processes, the types of pollution abatement equipment and operating procedures used to minimize these by-products, and the effectiveness of the equipment and operating procedures will be useful for determining the impact of similar future developments and mitigation measures. Some of these audits are also likely to provide models of "best practice", which other firms can aspire to in their existing and future facilities. Most interestingly, however, project EIAs are increasingly used as a starting point for their projects' EMSs. For instance, emission limits stated in an EIA can be used as objectives in the company's EMS, once it is operational. The EMS can also test whether the mitigation measures discussed in the EIA have been installed and whether they work effectively in practice.

Overall, EMS is likely to increase the level of environmental monitoring, environmental awareness and the availability of environmental data. All of this can only be of help in EIA.

11.5 Summary

As in a number of other countries discussed in Chapter 10, the practice of EIA for projects in the UK, set in the wider context of the EU, has progressed rapidly up the learning curve. Understandably however, practice has highlighted problems as well as successes. The resolution of problems and future prospects are determined by the interaction between the various parties involved. In the EU the introduction of the amended EIA Directive in 1999 has helped to improve some steps in the EIA process, including screening, scoping, the consideration of alternatives and consultation. However, some key issues remain unresolved, including the lack of support for mandatory monitoring. This chapter has identified an agenda for other possible changes, including cumulative impacts, public participation, socio-economic impacts, IEA and EID. Some of these will be easier to achieve than others, and there will no doubt be other emerging issues and developments in a dynamic area, and systems and procedures will continue to evolve in response to the environmental agenda and to our managerial and methodological capabilities.

There is an urgent need to "close the loop", to learn from experience. While the practice of mandatory monitoring is still patchy, there is some notable progress in the development of environmental management and auditing systems. Assessment can be aided by the recent development of environmental auditing for existing organizations, be they private-sector firms or local authorities. The information from such auditing could provide a significant change in the quality and quantity of baseline data for EIA.

As EIA activity spreads, more groups will become involved. Capacity building and training is vital both in the EIA process, which may have some commonality across countries, and in procedures that may be more closely tailored to particular national contexts. EIA practitioners also need to develop their substantive knowledge of the wider environment and to improve both their technical and participatory approaches in the EIA process.

Notes

1. This term was originally coined for a slightly different context by Turner (1995).
2. To help small firms in particular the UK Government introduced a new initiative in 2003.

> A new British Standard – BS 8555 (Guide to the implementation of an environmental management system including environmental performance evaluation) – has been developed to assist organisations, in particular small and medium sized enterprises, to implement on environmental management system and subsequently achieve EMAS registration. The standard includes guidance on how to develop indicators, so right from the start it is possible to know whether environmental impacts have been successfully reduced.

For further information, see DEFRA (2003), *An Introductory Guide to EMAS*, London: DEFRA.

References

Ahmad, B. 2004. Integrating health into impact assessment: challenges and opportunities. *Impact Assessment and Project Appraisal* 22(1), 2–4.

Bailey, P. et al. 1996. *Methods of integrated environmental assessment: research directions for the European Union*. Stockholm: Stockholm Environmental Institute.

Balram, S., S. Dragicevic, T. Meredith 2003. Achieving effectiveness in stakeholder participation using the GIS-based Collaborative Spatial Delphi methodology. *Journal of Environmental Assessment Policy and Management* 5(3), 365–94.

Baxter, W., W. Ross, H. Spaling 2001. Improving the practice of cumulative effects assessment in Canada. *Impact Assessment and Project Appraisal* 19(4), 253–62.

Burdge, R. 1999. The practice of social impact assessment – background. *Impact Assessment and Project Appraisal* 2(2), 84–88.

Canadian Environmental Assessment Agency (CEAA) 1999. *Cumulative Effects Assessment: Practitioners' Guide*. Quebec, Canada: CEAA (www.ceaa.gc.ca).

CEC 2003. Directive 2003/35/EC of the European Parliament and of the Council of May 26th 2003 on public participation in respect of the drawing up of certain plans and programmes relating to the environmental and amending with regard to public participation and access to justice. Council Directive 85/337/EEC and 96/61/EC, *Official Journal of the European Communities*, No. L156, 0017–0025.

CEPA (Commonwealth Environmental Protection Agency) 1994. *Review of Commonwealth environmental impact assessment*. Canberra: CEPA.

CEQ (Council on Environmental Quality, USA) 1997. *Considering Cumulative Effects under the National Environmental Policy Act*. Washington, DC: Office of the President (available at: http://ceq.eh.doe.gov/nepa/nepanet/htm).

Chadwick, A. 2002. Socio-economic impacts: are they still the poor relations in UK Environmental Statements? *Journal of Environmental Planning and Management* 45(1), 3–24.

CPRE (Council for the Protection of Rural England) 1991. *The environmental assessment directive – five years on.* London: Council for Protection of Rural England.

CPRE 1992. *Mock directive.* London: Council for Protection of Rural England.

Dagg, S., R. Aschemann, J. Palerm 2003. The changing process of public and stakeholder participation in response to diverse and dynamic contexts. *Journal of Environmental Assessment Policy and Management* 5(3), v–ix.

Davis, S. 1996. *Public involvement in environmental decision making: some reflections on West European experience.* Washington, DC: World Bank.

Douglas, C. H. 2003. Developing health impact assessment for sustainable futures. *Journal of Environmental Assessment Policy and Management* 5(4), 477–502.

EC (European Commission) 1993. *Regulation No. 1836/93 allowing voluntary participation by companies in the industrial sector in an eco-management and audit scheme.* Brussels: EC.

EC 2001a. *Regulation No. 761/2001 of the European Parliament and of the Council of 19 March 2001 allowing voluntary participation by organisations in a Community eco-management and audit scheme (EMAs).* Brussels: EC.

EC 2001b. EMAS and ISO/EN ISO 14001: differences and complementarities, http://europa.eu.int/comm/environment/emas/pdf/factsheets.

Essex Planning Officers' Association 2001. *The Essex Guide to Environmental Impact Assessment.* Chelmsford: Essex County Council.

Glasson, J., K. Godfrey, B. Goodey 1995. *Towards visitor impact management.* Aldershot: Avebury.

Glasson, J. 2001. Socio-economic impacts. In *Methods of Environmental Impact Assessment* P. Morris & R. Therivel (eds), Chapter 2. London: Spon.

Health Canada 1999. *Canadian Handbook of Health Impact Assessment,* vol 1: *The Basics.* Ottawa: Health Canada.

Holstein, T. 1996. Reflective essay: postmodern EIA or how to learn to love incinerators (written as part of MSc course in Environmental Assessment and Management, Oxford Brookes University, Oxford).

Hyder Consulting 1999. *Guidelines for the assessment of indirect and cumulative impacts as well as impact interactions.* Brussels: CEC-DGXI (available at: http://europa.eu.int/comm/environment/eia/eia_support.htm).

IAIA (International Association for Impact Assessment) 1994. Guidelines and principles for social impact assessment. *Impact Assessment* 12, Summer.

IAIA 2002. Workshop on Integrated Assessment, in IAIA *Annual Conference.* The Hague: IAIA.

INEM (International Network for Environmental Management) 1998. Website: www.inem.org.

IOS (International Organization for Standardization) 1996a. *ISO 14001 Environmental Management Systems – Specification with Guidance for Use.*

IOS 1996b. *ISO 14004 Environmental Management Systems – General guidelines on principles, systems and supporting techniques.*

Lane, P. and Associates 1988. *A reference guide to cumulative effects assessment in Canada,* vol. 1. Halifax: P Lane and Associates/CEARC.

McDonald, G. T. & L. Brown 1995. Going beyond environmental impact assessment: environmental input to planning and design. *Environmental Impact Assessment Review* 15, 483–95.

Newton, J. 1995. *The integration of socio-economic impacts in EIA and project appraisal.* MSc dissertation, University of Manchester Institute of Science and Technology.

Odum, W. 1982. Environmental degradation and the tyranny of small decisions. *Bio Science* 32, 728–29.

O'Riordan, T. 1990. EIA from the environmentalist's perspective. VIA 4, March 13.

Parkinson, P. 1996. *Best practice for the environmental assessment of pipelines* (paper presented at Conference on Advances in Environmental Impact Assessment), 9 July. London: IBC UK Conferences.

Peterson, E. et al. 1987. *Cumulative effects assessment in Canada: an agenda for action and research.* Quebec Canadian Environmental Assessment Research Council (CEARC).

Petts, J. 2003. Barriers to deliberative participation in EIA: learning from waste policies, plans and projects. *Journal of Environmental Assessment Policy and Management* 5(3), 269–93.

Phoolchareon, W., D. Sukkumnoed & P. Kessomboon 2003. Development of health impact assessment in Thailand: recent experiences and challenges. *Bulletin of the World Health Organisation* 81(6), 465–67.

Piper, J. 2000. Cumulative effects assessment on the Middle Humber; barriers overcome, benefits derived. *Journal of Environmental Planning and Management* 43(3), 369–87.

Piper, J. 2001a. Assessing the cumulative effects of project clusters: a comparison of process and methods in four UK cases. *Journal of Environmental Planning and Management* 44(3), 357–75.

Piper, J. 2001b. Barriers to implementation of cumulative effects assessment. *Journal of Environmental Assessment Policy and Management* 3(4), 465–81.

Project Appraisal 1996. Special edition: Environmental Assessment and Socio-economic Appraisal in Development 11(4).

Rodriguez-Bachiller, A. with J. Glasson 2003. *Expert Systems and Geographical Information Systems for Impact Assessment.* London: Taylor and Francis.

Sadler, B. 1996 *International study in the effectiveness of environmental assessment.* Ottawa: Canadian Environmental Assessment Agency.

Scott-Samuel, A., M. H. Birley & K. Ardem, 2001. *The Merseyside guidelines for health impact assessment.* International Health Impact Assessment Consortium, University of Liverpool, 2nd edition. Available at http://www.ohn.gov.uk/.

Secretary of State for Health 1999. *Saving Lives: Our Healthier Nation.* Cm 4386, London: HMSO. Available at http://www/ohn.gov.uk/.

Taylor, L. & Blair-Stevens 2002. *Introducing Health Impact Assessment (HIA): Informing the Decision-Making Process.* London: Health Development Agency.

Taylor, L. & R. Quigley 2002. *Health Impact Assessment: A Review of Reviews.* London: Health Development Agency.

Turner, T. 1995. *City as landscape: a post-postmodernist view of design and planning.* London: E&FN Spon.

UNECE (United Nations Economic Commission for Europe) 1998. *Convention on Access to Information, Public Participation in Decision-Making and Access to Justice in Environmental Matters.* (ECE/CEP/43). Aarhus, Denmark.

Vanclay, F. 2003. International principles for social impact assessment. *Impact Assessment and Project Appraisal* 21(1), 5–12.

WHO Regional Office for Europe 2003. *Health Impact Assessment Methods and Strategies.* Available at http://www.euro.who.int/eprise/main/WHO/progs/HMS/home.

World Bank 1995. *Environmental assessment: challenges and good practice.* Washington, DC: World Bank.

12 Widening the scope: strategic environmental assessment

12.1 Introduction

One of the most recent trends in EIA is its application at the level of policies, plans and programmes (PPPs). In the USA, since the enactment of the NEPA, this so-called SEA has been carried out as an extension of project EIA in a relatively low-key manner. However, in the EU it has increasingly come to be viewed as a valuable technique for achieving sustainable development, and a Europe-wide SEA Directive became operational in July 2004. SEA is also a strong growth area in other parts of the world.

This chapter discusses the need for SEA and some of its limitations. It reviews the status of SEA in the USA, Canada, New Zealand, European Union and UNECE. It then discusses in more detail how the European SEA Directive is being implemented in the UK. It concludes with a discussion of techniques for carrying out SEA in accordance with the European SEA Directive. By necessity this chapter must radically simplify many aspects of SEA. The reader is referred to Partidario & Clark (2000), Sadler & Verheem (1996), Therivel (2004) and Therivel & Partidario (1996) for a more in-depth discussion. Reference should also be made to Chapter 9 for two SEA case studies.

12.2 Strategic environmental assessment (SEA)

12.2.1 Definitions

Strategic environmental assessment can be defined as

> a systematic process for evaluating the environmental consequences of proposed policy, plan or programme initiatives in order to ensure they are fully included and appropriately addressed at the earliest appropriate stage of decision making on par with economic and social considerations. (Sadler & Verheem 1996)

It is, in other words, a form of EIA for PPPs, keeping in mind that evaluating environmental impacts at a strategic level is not necessarily the same as evaluating them at a project level.

Several things are important in Sadler and Verheem's definition. First, SEA is a process, not a snapshot at the end of a process. It should take place in parallel with the plan-making process, providing environmental information at all relevant stages.

Strategic decision-making process   **Environmental input**

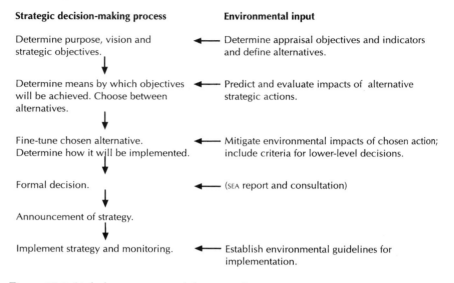

Determine purpose, vision and   ◄——— Determine appraisal objectives and indicators
strategic objectives.   and define alternatives.

Determine means by which objectives ◄——— Predict and evaluate impacts of alternative
will be achieved. Choose between   strategic actions.
alternatives.

Fine-tune chosen alternative.   ◄——— Mitigate environmental impacts of chosen action;
Determine how it will be implemented.   include criteria for lower-level decisions.

Formal decision.   ◄——— (SEA report and consultation)

Announcement of strategy.

Implement strategy and monitoring.   ◄——— Establish environmental guidelines for
                                                                implementation.

Figure 12.1 Links between SEA and the PPP-making process.

The definition also emphasizes the importance of integrating SEA in decision-making. Figure 12.1 shows the links between PPP-making and SEA.

The definition distinguishes between policies, plans and programmes (PPPs). Although they are often lumped together in the SEA literature, PPPs are not the same things, and may require quite different forms of SEA. A policy is generally defined as an inspiration and guidance for action (e.g. "to supply electricity to meet the nation's demands"), a plan as a set of co-ordinated and timed objectives for the implementation of the policy (e.g. "to build X megawatts of new electricity generating capacity by 2012"), and a programme as a set of projects in a particular area (e.g. "to build four new Combined Cycle Gas Turbine power stations in region Y by 2012") (Wood 1991). In this chapter, policies, plans and programmes will jointly be referred to as PPPs. PPPs can relate to specific sectors (e.g. transport, mineral extraction) or to all activities in a given area (e.g. land use, development or territorial plans).

In theory PPPs are tiered: a policy provides a framework for the establishment of plans, plans provide frameworks for programmes, programmes lead to projects. In practice, these tiers are amorphous and fluid, without clear boundaries. SEAs for these different PPP tiers can themselves be tiered, as we show in Figure 12.2, so that issues considered at higher tiers need not be reconsidered at the lower tiers. PPPs can also result in activities that have environmental impacts but are not development projects, for instance privatization or different forms of land management.

12.2.2 The need for SEA

Various arguments have been put forward for a more strategic form of EIA, most of which relate to *problems with the existing system of project EIA*. Project EIAs react to development proposals rather than anticipating them, so they cannot steer development towards environmentally robust areas or away from environmentally sensitive sites.

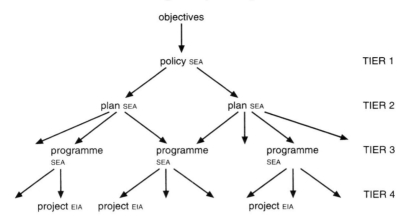

Figure 12.2 Tiers of SEA and EIA.

Project EIAs do not adequately consider the cumulative impacts[1] caused by several projects, or even by one project's subcomponents or ancillary developments. For instance, small individual mineral extraction operations may not need an EIA, but the total impact of several of these projects may well be significant. Section 9.7 provides another example. At present in most countries there is no legal requirement to prepare comprehensive cumulative impact statements for projects of these types.

Project EIAs cannot consider the impacts of potentially damaging actions that are not regulated through the approval of specific projects. Examples of such actions include farm management practices, privatization and new technologies such as genetically modified organisms. Project EIAs cannot fully address alternative types and locations for developments, or the full range of possible mitigation measures, because these alternatives will often be limited by choices made at an earlier, more strategic level. In many cases a project will already have been planned quite specifically, and irreversible decisions taken at the strategic level, by the time an EIA is carried out.

Project EIAs often have to be carried out very quickly because of financial constraints and the timing of planning applications. This limits the amount of baseline data that can be collected and the quality of analysis that can be undertaken. For instance, the planning periods of many projects have required their ecological impact assessments to be carried out in the winter months, when it is difficult to identify plants and when many animals either are dormant or have migrated. The amount and type of public consultation undertaken in project EIA may be similarly limited.

By being carried out early in the decision-making process and encompassing all the projects or actions of a certain type or in a certain area, SEA can ensure that alternatives are better assessed, cumulative impacts are considered, the public is better consulted, and decisions about individual projects are made in a proactive rather than reactive manner.

Strategic environmental assessment can also help to *promote sustainable development*. In the UK, for instance, SEA is often expanded or integrated into sustainability assessment/appraisal.[2] This involves not only broadening the scope of assessment to also consider social and economic issues, but also setting sustainability objectives and testing whether the PPP will help to achieve them. In other words, sustainability assessment tests whether the PPP helps to promote a sustainability vision, whilst SEA tends to focus on preventing environmental problems.

12.2.3 Problems with SEA

In the early days of SEA, lack of experience and appropriate techniques limited the quality of SEAs. As SEA practice has evolved, these problems have eased but others have emerged.

First, many PPPs are nebulous, and they evolve in an incremental and unclear fashion, so there is no clear time when their environmental impacts can be best assessed: "the dynamic nature of the policy process means issues are likely to be redefined throughout the process, and it may be that a series of actions, even if not formally sanctioned by a decision, constitute policy" (Therivel et al. 1992). Second, where SEA is required only for programmes and/or plans but not policies, an environmentally unfriendly policy can lead to environmentally unfriendly plans: in such a case, the plan-level SEA can at best mitigate the plan's negative impacts, not consider more sustainable policy level alternatives. Third, multiple PPPs often affect a single area or resource. For instance, energy and transport PPPs – and many others – affect climate change. Waste and minerals PPPs are often integrally interconnected, as are land-use and transport PPPs. As such, it is often difficult to assess a PPP on its own.

There has also been considerable uncertainty about whether SEA should be broadened out to also cover social and economic issues. Considering environmental issues separately from social and economic issues may give them an additional "weight" in decision-making and helps to keep the integrity of the environmental assessment. On the other hand, sustainability appraisal more closely reflects actual decision-making, and is legally required for many UK PPPs anyway, so integrating the two procedures makes sense in terms of efficiency.

Finally, and most importantly, policy making is a political process. Decision-makers will weigh up the implications of a PPP's environmental impacts in the wider context of their own interests and those of their constituents. SEA does not make the final decision: it merely (sometimes maddeningly so) informs it.

12.3 SEA worldwide

Despite these problems, SEA has been increasingly carried out worldwide. For instance, the USA, European Union Member States and New Zealand have all established SEA regulations; Canada requires SEA by Cabinet decision; South Africa has guidance on SEA; and SEAs are regularly carried out in Hong Kong and elsewhere. Here the SEA systems of the USA, Canada, New Zealand, the EU and UNECE are discussed because they are well developed and demonstrate a range of possible approaches. They differ in terms of whether they require or just encourage the preparation of SEAs; the types of strategic actions that require SEA; whether the SEAs consider only environmental issues or the full range of sustainability considerations; and the level of detail that they go into.

12.3.1 The USA

The USA has no separate SEA regulations. Instead, the National Environmental Policy Act of 1969 (NEPA) requires that

> all agencies of the Federal Government shall include in every recommendation or report on proposals for legislation and other major Federal *actions* significantly affecting the quality of the human environment, a detailed statement by the responsible official on the environmental impact of the proposed action... [42 USC §4332]

The term "actions" has been interpreted through the courts and through Council on Environmental Quality regulations (CEQ 1978) as including a range of PPPs. Federal agencies must prepare "programmatic environmental impact statements" (PEISs) for the following actions, if these are likely to significantly affect the quality of the human environment: agency proposals for legislation; the adoption of rules, regulations, treaties, conventions or formal policy documents; the adoption of formal plans that guide the use of federal resources; the adoption of groups of connected actions that implement a policy (40 CFR 1508.18[b]).

Several hundred PEISs have been prepared to date under the NEPA, although these form only a small percentage of all the assessments carried out in the USA. For instance, PEISs carried out in 2003/4 include statements for mountaintop mining and valley fills in Appalachia (Army Corps of Engineers and others), licensing launches of horizontally launched vehicles and re-entry vehicles (Federal Aviation Administration), and carbon sequestration (DoEn).

The main problem with the USA's system of SEA seems to be its basis in EIA, and the detailed – arguably severely over-detailed – approach it fosters:

> Unfortunately, NEPA's role as a strategic planning tool has not been fully realized...Congress envisioned that federal agencies would use NEPA as a planning tool to integrate environmental, social, and economic concerns directly into projects and programs. However... application has focused on decisions related to site-specific construction, development, or resource extraction projects. NEPA is virtually ignored in formulating specific policies and often is skirted in developing programs, usually because agencies believe that NEPA cannot be applied within the time available or without a detailed proposal. Instead, agencies tend to examine project-level environmental effects in microscopic detail. The reluctance to apply NEPA analysis to programs and policies reflects the fear that microscopic detail will be expected. (CEQ 1997)

Only a few of the USA's 50 states have SEA regulations. Of these, the SEA system established by the California Environmental Quality Act of 1986 (State of California 1986) is the most well developed. "Program environmental impact reports" (PEIRs) are required for series of actions that can be characterized as one large project and are related geographically, as logical parts in a chain of contemplated actions, in connection with the issuance of rules or regulations, or as individual activities carried out under the same authority and having generally similar environmental effects (CEQA 15168). Like project EIAs, PEIRs must include a description of the action, a description of the baseline environment, an evaluation of the action's impacts, a reference to alternatives, an indication of why some impacts were not evaluated, the organizations consulted, the responses of these organizations to the EIS and the agency's response to the responses.

12.3.2 Canada

A two-tier system, with somewhat overlapping requirements, operates at the federal level in Canada. On the one hand, the Canadian Environmental Assessment Act of 1995 requires project EIAs. One type of EIA, for projects with known effects that can be easily mitigated, is a "class screening": a generic assessment of all projects within a class. These class screenings, which are discussed further in Section 10.8, are arguably a form of programme SEAs. Only two class screenings had been carried out by the end of 2003: for routine projects in the town of Banff and for the importation of certified honeybees to Canada.

On the other hand, a recent Cabinet Directive requires SEA of PPPs. The 1999 "Cabinet Directive on the Environmental Assessment of Policy, Plan and Program Proposals" (CEAA 1999) is an evolution and strengthening of a 1990 Cabinet Directive which required federal departments and agencies to carry out SEA for policy and programme proposals submitted to Cabinet. In 1993 the Federal Environmental Assessment and Review Office (FEARO) published procedural guidance for these SEAs (FEARO 1993). The 1999 Directive expects federal departments and agencies to conduct an SEA of policy, plan or programme proposals where (1) the proposal is submitted to a minister or Cabinet for approval and (2) implementation of the proposal may result in important environmental effects. The Directive does not propose a single SEA methodology, but notes that an SEA should:

- consider the scope and nature of the likely environmental effects;
- consider the need for mitigation to reduce or eliminate adverse effects;
- consider the likely importance of any post-mitigation environmental effects;
- involve the public as appropriate; and
- be documented in a public statement.

The SEA should

> contribute to the development of policies, plans and programs on an equal basis with economic or social analysis... The environmental considerations should be full integrated into the analysis of each of the options developed for consideration, and the decision should incorporate the results of the [SEA]. (CEAA 1999)

12.3.3 New Zealand

In New Zealand, SEA is seen as a tool for achieving sustainability as part of an integrated planning and assessment process. The basis of SEA in New Zealand is the Resource Management Act (RMA) of 1991, which reworked over 70 statutes, regulations, laws and guidance to promote a sustainable process of development.

> (1) The purpose of [the RMA] is to promote the sustainable management of natural and physical resources.

(2) In this act "sustainable management" means managing the use, development and protection of natural and physical resources in a way, or at a rate, which enables people and communities to provide for their social, economic and cultural well-being and for their health and safety while –

  (a) Sustaining the potential of natural and physical resources (excluding minerals) to meet the reasonably foreseeable needs of future generations; and

  (b) Safeguarding the life-supporting capacity of air, water, soil and ecosystems; and

  (c) Avoiding, remedying or mitigating any adverse effects of activities on the environment. (MftE 1991, Section 5)

To help achieve this purpose, the Act requires all policies, rules, etc. to undergo an environmental assessment, and for the policy-makers to be "satisfied that any such [PPP] (i) Is necessary in achieving the purpose of [the] Act; and (ii) Is the most appropriate means of exercising this function" (MftE 1991, Article 32(c)).

Guidance on how the RMA should be applied is set out in the MftE's (2000) *Guide to Using Section 32 of the Resource Management Act*. This guide was partly written in response to the limited way in which individual authorities were applying the RMA's requirements. It clarifies that authorities must:

• have regard to the extent to which the policy/rule/etc. is needed at all, other possible means apart from the policy/rule/etc., and possible reasons for and against the proposed method and the principal alternative means;

• evaluate the benefits and costs of the proposed option and of the principal alternative means;

• satisfy themselves that the proposed means is necessary to achieve the purpose of the RMA and is the most appropriate in terms of effectiveness and efficiency.

It proposes 15 stages for doing this, as shown in Box 12.1. A unique feature of this approach is its emphasis on, and explanation of, efficiency, which is a key factor in deciding on whether a policy/rule/etc. should go ahead or not:

The extent to which the purpose of the Act is achieved is calculated by subtracting environmental costs from environmental benefits. How much is foregone is worked out by subtracting social and economic benefits from social and economic costs. Efficiency is then determined by comparing the first value with the second. Obviously, if social and economic benefits outweigh costs then nothing would be foregone as a result of implementing the policy or method. But if there are net social and economic costs, efficiency has to be measured by the degree to which the environmental benefits outweigh these costs. (MftE 2000)

Marsden & Dovers (2002) give more information about SEA in Australasia.

12.3.4 European Union

It was initially intended that one European Directive would cover projects and PPPs, but by the time that Directive 85/337 was approved in 1985, its application was

Box 12.1 *Sequence of steps in a New Zealand Resource Management Act Section 32 analysis*

1. Identify and specify the resource management issues
2. Identify objective(s)
3. Determine whether the objective is necessary
4. If not, review the objective and/or the issue
5. Identify the range of policies that realistically could achieve the objective
6. For each policy, identify the range of methods that could be used to implement it
7. Assess the effectiveness of each method, i.e. how successful it is in achieving the objective
8. Reject methods that are insufficiently effective
9. Identify and evaluate the benefits and costs of each method not rejected at 8.
10. Assess the relative efficiency of each method
11. Determine whether all methods are very inefficient (i.e. costs are large in relation to benefits)
12. If so, review the methods, the policy and/or the objective and consider alternatives
13. Decide which of the means are the most appropriate
14. Go back to assess the benefits and costs, effectiveness and efficiency of objectives and policies
15. Once the plan has been adopted and implemented, assess whether the plan provisions are achieving the anticipated environmental results and, if not, review the objectives, policies and methods

(*Source:* MftE 2000.)

restricted to projects only (Wood 1988). In the absence of a Europe-wide SEA requirement, several European Member States established SEA systems starting in the late 1980s: these are discussed in Therivel & Partidario (1996) and Kleinschmidt & Wagner (1998).

After 25 years of discussion and negotiations between the European Member States, the European Commission finally agreed on Directive 2001/42/EC "on the assessment of the effects of certain plans and programmes on the environment" (EC 2001) on 21 July 2001. The full text of the Directive is given in Appendix 2. The Directive became operational on 21 July 2004. Like the EIA Directive, the SEA Directive does not have a direct effect in individual European Member States, but instead needs to be interpreted into regulations in each Member State. Section 12.4 discusses how this has been done in England.

Directive 2001/42/EC requires SEA for plans and programmes (not policies) that:

1. are subject to preparation and/or adoption by an authority *and*
2. are required by legislative, regulatory or administrative provisions *and*
3. are likely to have significant environmental effects *and*
4. (a) are prepared for agriculture, forestry, fisheries, energy, industry, transport, waste management, water management, telecommunications, tourism, T&CP or land use *and* set the framework for development consent of projects listed in the EIA *or*

(b) in view of the likely effect on sites, require an appropriate assessment under the Habitats Directive *or*

(c) are other plans and programmes determined by Member States to set the framework for future development consent of projects *and*

5.   are begun after 21 July 2004 or are completed after 21 July 2006.

Just what plans and programmes are affected by this complex definition varies from Member State to Member State, and will probably only become clear in time through precedent and lawsuits.

Box 12.2 summarizes the SEA Directive's requirements. Draft plans and programmes must be accompanied by an "environmental report" that discusses the current baseline, the likely effects of the plan or programme and alternative options, how the negative effects have been minimized and proposed monitoring arrangements. The public must be consulted on the proposed plan or programme together with the environmental report, and the authority preparing the plan or programme has to show how the information in the report and the comments of consultees have been taken on board. European guidance (EC 2003) gives more details on some aspects of the Directive.

Box 12.2 Requirements of the EU SEA Directive

Preparing an environmental report in which the likely significant effects on the environment of implementing the plan, and reasonable alternatives taking into account the objectives and geographical scope of the plan, are identified, described and evaluated. The information to be given is (Article 5 and Annex I):

(a)   An outline of the contents, main objectives of the plan, and relationship with other relevant plans and programmes;

(b)   The relevant aspects of the current state of the environment and the likely evolution thereof without implementation of the plan;

(c)   The environmental characteristics of areas likely to be significantly affected;

(d)   Any existing environmental problems which are relevant to the plan including, in particular, those relating to any areas of a particular environmental importance, such as areas designated pursuant to Directives 79/409/EEC and 92/43/EEC;

(e)   The environmental protection objectives, established at international, Community or national level, which are relevant to the plan and the way those objectives and any environmental considerations have been taken into account during its preparation;

(f)   The likely significant effects on the environment, including on issues such as biodiversity, population, human health, fauna, flora, soil, water, air, climatic factors, material assets, cultural heritage including architectural and archaeological heritage, landscape and the interrelationship between the above factors. (These effects should include secondary, cumulative, synergistic, short, medium and long-term permanent and temporary, positive and negative effects);

(g)   The measures envisaged to prevent, reduce and as fully as possible offset any significant adverse effects on the environment of implementing the plan;

(h)   An outline of the reasons for selecting the alternatives dealt with, and a description of how the assessment was undertaken including any difficulties (such as technical deficiencies or lack of know-how) encountered in compiling the required information;

(i)   A description of measures envisaged concerning monitoring in accordance with Article 10;

(j)   A non-technical summary of the information provided under the above headings.

Box 12.2 (continued)

The report must include the information that may reasonably be required taking into account current knowledge and methods of assessment, the contents and level of detail in the plan, its stage in the decision-making process and the extent to which certain matters are more appropriately assessed at different levels in that process to avoid duplication of the assessment (Article 5.2).

Consulting:

- authorities with environmental responsibilities, when deciding on the scope and level of detail of the information which must be included in the environmental report (Article 5.4)
- authorities with environmental responsibilities and the public, to give them an early and effective opportunity within appropriate time frames to express their opinion on the draft plan and the accompanying environmental report before the adoption of the plan (Articles 6.1, 6.2)
- other EU Member States, where the implementation of the plan is likely to have significant effects on the environment in these countries (Article 7)

Taking the environmental report and the results of the consultations into account in decision-making (Article 8)

Providing information on the decision:

When the plan is adopted, the public and any countries consulted under Article 7 must be informed and the following made available to those so informed:

- the plan as adopted
- a statement summarising how environmental considerations have been integrated into the plan and how the environmental report of Article 5, the opinions expressed pursuant to Article 6 and the results of consultations entered into pursuant to Article 7 have been taken into account in accordance with Article 8, and the reasons for choosing the plan as adopted, in the light of the other reasonable alternatives dealt with; and
- the measures decided concerning monitoring (Article 9)

Monitoring the significant environmental effects of the plan's implementation (Article 10)

(*Source:* ODPM 2003.)

12.3.5 United Nations

In May 2003, the United Nations Economic Commission for Europe (UNECE) adopted an SEA Protocol similar to the European SEA Directive as a supplement to its 1991 Convention on EIA in a Transboundary Context (the Espoo Convention). States have been able to ratify the Protocol as of 1 January 2004. It will come into force 90 days after the sixteenth State has ratified, though it may be several years before this occurs. Although negotiated under the UNECE (which covers Europe, US, Canada, the Caucasus and Central Asia), the Protocol is open to all UN members.

The Protocol's requirements are broadly similar to, and compatible with, those of the EU Directive. Broadly the same types of plans and programmes require SEA under the Protocol; the environmental report required by the Protocol is similar to that required by the Directive, and the consultation requirements are similar. The Protocol is more focused on health impacts; makes more references to public participation; and addresses policies and legislation, although it only requires SEA of plans and programmes.

12.4 SEA in the UK

The SEA Directive has had a huge influence on SEA practice in Europe, and, indirectly through the UNECE Protocol, worldwide. In the UK, prior to the Directive, an abbreviated form of SEA – "environmental appraisal" or "sustainability appraisal" – was widely carried out. Post-Directive, SEA practice will be much more robust, but will also require more resources and expertise.

12.4.1 Environmental and sustainability appraisal

The UK government's White Paper on the Environment of September 1990 (DoE 1990) promised that it would carry out "a review of the way in which the costs and benefits of environmental issues are assessed within the Government". A year later the DoE published a guidebook entitled *Policy appraisal and the environment* and distributed copies to central government mid-level managers (DoE 1991); it has been updated recently (DEFRA 2003). The guidebook's procedures are not mandatory, but they aim to help civil servants consider the environmental repercussions of their decisions and to promote a "cultural change" in how civil servants formulate policies. *Policy appraisal and the environment* suggests that the department or agency from which a PPP originates should carry out the policy appraisal. A policy appraisal is needed whenever a policy is likely to have a significant effect on the environment, particularly if that impact would be irreversible. According to the latest guidebook, policy appraisal should aim to answer the following questions:

- What does the policy or programme aim to achieve? (Include possible trade-offs, conflicts and constraints.)
- What are the options for achieving your objectives?
- What impacts will these have on the environment at home and abroad? (Consider both direct and indirect costs and benefits; possible mitigation measures; and need for risk assessment.)
- How significant are the impacts? How large are they in relation to the other costs and benefits of the policy concerned?
- How far can the costs and benefits be quantified without disproportionate effort?
- What method will be used to value the costs and benefits? (Options include using monetary values, calculating physical quantities and systematic ranking or listing of impacts.)
- What is the preferred option and why?
- What arrangements are in place for effective monitoring and evaluation? (What data will be needed and when?)
- How will the appraisal be publicized? (DEFRA 2003)

At the central government level, these guidelines have been of limited effectiveness in promoting SEA. A DoE (1994) publication entitled *Environmental appraisal in government departments*, which summarizes central government studies carried out in response to *Policy appraisal and the environment*, does discuss a range of studies, but these were mainly cost benefit analyses (CBAs), not SEAs, and the booklet's publication was widely seen as *a pro forma* exercise. The Sustainable Development Unit's (2003) *Integrated Policy Appraisal* guidance aims to deal with some of these problems.

In contrast, a real "SEA-change" in the UK was begun by the PPG 12, *Development plans and regional planning guidance* (DoE 1992), which required local authorities to conduct an "environmental appraisal" of development plan policies, and referred them to *Policy appraisal and the environment* for guidance on this appraisal process. In response to this guidance, some local authorities began to carry out environmental appraisals, albeit using much simpler techniques than those advocated by the government. In November 1993, on the basis of the results of these early appraisals, the DoE published *Environmental Appraisal of Development Plans: A Good Practice Guidance* (DoE 1993). The 1993 guidance, which was widely used in the following years by many local authorities, recommended that environmental appraisal should involve:

1. identifying environmental components (e.g. air quality, urban "liveability") that could be affected by the plan;
2. ensuring that the plan is in accordance with government environmental and planning advice; and
3. determining whether the plan's objectives/policies are internally consistent, using a policy compatibility matrix, which resembles a triangular road mileage chart; and assessing the policies' likely environmental effects, using a "policy impact" matrix with the plan policies on one axis and the environmental components identified in Step 1 on the other axis.

The second edition of this book gave an example of an appraisal based on this guidance.

PPG12 was revised in 1999. It advised authorities to consider a range of social and economic effects as well as environmental ones in a broader "sustainability appraisal". A similar approach was taken for Regional Planning Guidance. By October 2001 over 90 per cent of English and Welsh local authorities and all regional authorities had had some experience with appraisal. About half of the appraisals were "environmental" and the other half "sustainability" (Therivel & Minas 2002).

Research on the effectiveness of these appraisals (Therivel 1995, 1996; Therivel & Minas 2002) showed that they were being increasingly carried out during (rather than after) plan making, were leading to more changes in the plans (Figure 12.3), and were providing side-benefits such as increased transparency in plan making, and a better understanding of the plan and the environment by plan makers. The appraisals were most effective in improving the plan when they were carried out by several people, including the person who wrote the plan and someone external; during rather than after the plan-making process; and with enough resources (Figures 12.4 and 12.5).

12.4.2 *Implementing the Directive*

In the UK, the SEA Directive is being implemented through different regulations in each devolved administration,[3] and through several additional guidance documents

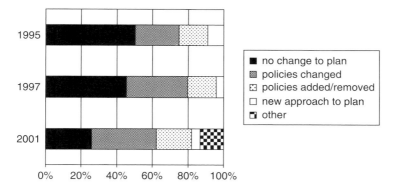

Figure 12.3 Changes to plan as result of appraisal. (*Source*: Therivel and Minas 2002.)

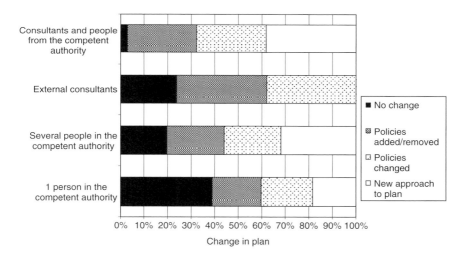

Figure 12.4 Who carries out appraisal vs. changes to plan. (*Source*: Therivel and Minas 2002.)

prepared by different government agencies for their specific plans. Authorities are still trying to determine what plans and programmes are covered by the Directive's complex definition, but it is clear that local and regional development plans and Local Transport Plans will require SEA, as will many Community Strategies. SEA guidance has already been prepared for English local and regional land-use plans (ODPM 2003), Scottish development plans (Scottish Executive 2003) and Local Transport Plans (DfT 2004). Other guidance is being developed for water-management plans (for the Environment Agency), SEA and biodiversity (English Nature et al. 2004) and SEA and climate change (CCW et al. 2004). Box 12.3 shows the table of contents of the ODPM (2003) guidance on SEA for land-use plans; Box 12.4 gives more detail on the five main stages of SEA, which are also used in the guidance on Local Transport Plans (DfT 2004).

Much of the discussion about how to implement the SEA Directive in the UK has been about how SEA should relate to the existing system of sustainability appraisal. Under changes to the planning system proposed in the Planning and Compulsory

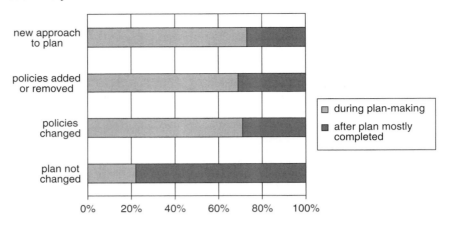

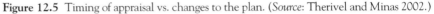

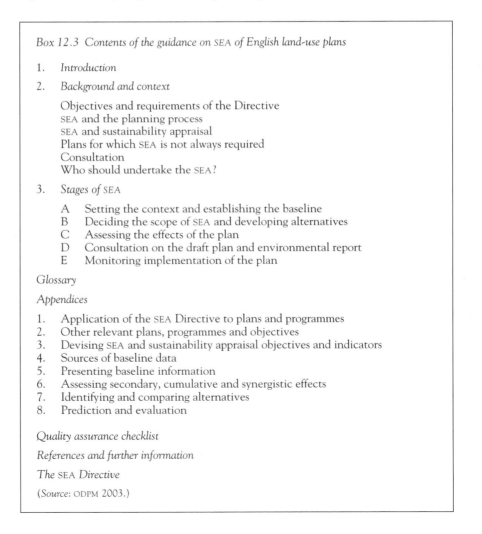

Figure 12.5 Timing of appraisal vs. changes to the plan. (*Source*: Therivel and Minas 2002.)

Box 12.3 Contents of the guidance on SEA of English land-use plans

1. *Introduction*

2. *Background and context*

    Objectives and requirements of the Directive
    SEA and the planning process
    SEA and sustainability appraisal
    Plans for which SEA is not always required
    Consultation
    Who should undertake the SEA?

3. *Stages of SEA*

    A   Setting the context and establishing the baseline
    B   Deciding the scope of SEA and developing alternatives
    C   Assessing the effects of the plan
    D   Consultation on the draft plan and environmental report
    E   Monitoring implementation of the plan

Glossary

Appendices

1. Application of the SEA Directive to plans and programmes
2. Other relevant plans, programmes and objectives
3. Devising SEA and sustainability appraisal objectives and indicators
4. Sources of baseline data
5. Presenting baseline information
6. Assessing secondary, cumulative and synergistic effects
7. Identifying and comparing alternatives
8. Prediction and evaluation

Quality assurance checklist

References and further information

The SEA Directive

(*Source*: ODPM 2003.)

Box 12.4 SEA *steps for development plans/frameworks*

A. *Setting the context and establishing the baseline*

- Identify other relevant plans and programmes
- Identify environmental protection objectives, and state their relation to the plan
- Propose SEA and sustainability appraisal objectives
- Propose indicators
- Collect baseline data, including data on likely future trends
- Identify environmental and sustainability problems

B. *Deciding the scope of* SEA *and developing alternatives*

- Identify alternatives
- Choose preferred alternatives
- Consult authorities with environmental responsibilities and other bodies concerned with aspects of sustainability

C. *Assessing the effects of the plan*

- Predict the effects of the plan
- Evaluate the plan's effects
- Propose measures to prevent, reduce or offset adverse environmental effects

D. *Consultation on the draft plan and the Environmental Report*

- Present the results of the SEA up to this point
- Seek inputs from the public and authorities with environmental responsibilities
- Take consultation results into account
- Show how the results of the Environmental report were taken into account in finalising the plan

E. *Monitor the significant effects of implementing the plan on the environment*

(*Source:* ODPM 2003.)

Purchase Bill 2003, sustainability appraisal would become mandatory for Regional Spatial Strategies and Local Development Documents. Sustainability appraisal and SEA both involve identifying and evaluating likely effects of the plan (and options), and reducing significant negative effects where possible. Both aim to help make a plan more sustainable. The key additional requirements of SEA are a description of the environment, more transparent consideration of alternatives, more formal consideration of mitigation measures for any significant negative impacts, monitoring and public consultation. The SEA Directive and guidance also focus more clearly on the environment than sustainability appraisal does. Some socio-economic factors such as (human) population, human health, material assets and cultural heritage are included, whilst others such as equity and jobs are not. This is in contrast to sustainability appraisal which covers the full range of social and economic issues. In practice, this complicated wording has made it difficult to identify just where SEA ends and sustainability appraisal begins (Therivel 2004).

The Directive also differs from sustainability appraisal in its "baseline-led" approach (Smith & Sheate 2001). Essentially SEA, according to the Directive, involves describing the baseline environment, identifying environmental problems from this, and possibly identifying different ways of resolving them (or at least not aggravating them). Sustainability appraisal, instead, is "objectives-led": it identifies sustainability objectives that the plan should aim to achieve, and possibly different ways of achieving them. SEA thus tries to solve today's problems; sustainability appraisal to achieve tomorrow's vision.

The ODPM's (2003) guidance notes that SEA objectives *can* be set, and proposes some objectives, but does not *require* them; notes that the assessment process can apply to the full gamut of social and economic as well as environmental topics; and suggests that alternatives can focus on problems *or* visions. Further guidance on integrated sustainability appraisal and SEA is expected in 2004.

12.5 Carrying out SEA under the SEA Directive

At the time of writing (early 2004), no SEAs had yet been completed in England under the SEA Directive. However, nine full and partial pilot SEAs were carried out as part of the development of the ODPM guidance (ODPM 2004). Table 12.1 lists the authorities involved and the key themes of these studies. Other early partial SEAs (e.g. from Chapter 9) had also been carried out on an *ad hoc* basis. These give an indication of the stages and techniques that one might expect of future SEAs. This section discusses these stages and techniques. It is based heavily on Levett-Therivel (2004).

A. Setting the context and establishing the baseline

The early, context-setting stages of SEA – identification of SEA objectives, collection of environmental baseline data, description of links to other plans and identification of environmental problems – are best carried out in parallel, with feedback from one to the other. They provide a basis for early scoping discussions with statutory authorities about how the rest of the SEA should be carried out.

The SEA Directive requires a description of the *baseline environment* but does not specify how this should be done. UK experience with sustainability appraisal suggests that environmental objectives provide a good framework for assessing the plan, and indicators are useful for describing and monitoring the baseline environment. Box 12.5 lists possible SEA objectives organized by Directive topic, but this would need to be updated and adapted for the plan in question.

Using the SEA objectives as a framework, the current environmental/sustainability baseline, likely future environmental trends and any significant gaps in the data should be described. Ideally, for each indicator, a description should be made of:

- The current environmental/sustainability status of the area covered by the plan.
- Comparators and targets: how does the authority compare with other similar authorities or with agreed targets?
- Trends: are things getting better or worse? If time series are available, any trends can be noted. If they are not, planners' "expert judgement" can be used.

Table 12.1 SEA pilot studies carried out as part of the development of the ODPM (2003) guidance on SEA

Authority	Plan	Key issues considered in the case study
*Full pilots**		
Vale of White Horse District Council	Local Plan 2011 First Deposit Draft	• linkages between the different stages of SEA • feasibility of carrying out entire SEA in 50–100 person days
Oxfordshire County Council	Minerals and waste local plan review	• scoping stages of SEA: links to other plans and programmes; baseline • monitoring
Partial pilots		
Cotswold District Council	Local Plan First Review 2001–11	• the level of detail feasible in the prediction of effects of plan policies • how to deal with uncertainty
East of England Local Government Conference	Regional Planning Guidance (RPG14) for the East of England 2021	• SEA objectives, in particular their relationship with the Regional Sustainable Development Framework and SA objectives • options generation and appraisal at a regional level
Hampshire County Council/ Portsmouth City Council/ Southampton City Council	Hampshire, Portsmouth and Southampton Minerals and Waste Local Plan Review	• identifying relevant policies, plans and programmes and how they relate to the plan • appraising policy choices
Luton Borough Council	Local Plan (Borough of Luton)	• "tiering": the relationship between the appraisal objectives used at different levels of the planning hierarchy, and links between SEA and EIA • stakeholder involvement
Newcastle City Council	Unitary Development Plan	• identifying relevant policies, plans and programmes and how they relate to the plan • identifying (sustainable) options for dealing with issues • assessing the effects of plan options on the SEA objectives and targets
Swale Borough Council	Swale Borough Local Plan	• option generation at a local level • constraints to options
Somerset County Council and Taunton Deane Borough Council	Taunton Urban Extension	• description of the baseline environment, including future trends • identification of environmental constraints

* These were carried out by students from Oxford Brookes University's MSc course in Environmental Assessment and Management.
(Based on ODPM 2004.)

Box 12.5 Possible SEA *objectives*

- Biodiversity, flora and fauna: conserve and enhance biodiversity at ecosystem, species and genetic levels
- Population and human health: promote the health of all residents
- Soil: maintain and enhance soil quality
- Water: maintain and enhance the quality of ground and surface waters
- Air: maintain and improve air quality, including greenhouse gases
- Climate change: respond to the effects of climate change
- Material assets: improve the accessibility of key services to local communities
- Cultural heritage: safeguard the cultural heritage, including the architectural and archaeological heritage
- Landscape – safeguard and enhance the townscape and rural landscape

(*Source*: Levett-Therivel 2004.)

Table 12.2 Table for describing the baseline (illustrative only)

Indicator	Quantified data	Comparators and targets	Trend	Problems/ constraints	Data sources
Conserve and enhance biodiversity at the ecosystem, species and genetic levels					
% SSSIs in good condition	42 % in authority A	58 % nationally; target 75% in 2012	getting worse: 45 % five years ago	Problem: below target, worse than average and getting worse	...
Area of semi-natural woodland	16 % of authority A	8 % England-wide	getting better: 13 % ten years ago	Not a problem: better than average and improving	
...					

(Based on Levett-Therivel 2004.)

- Problems/constraints: the foregoing information may help to identify problems, for instance where the authority is doing worse than other similar authorities and trends are getting worse.
- Data sources.

The baseline can be presented in many different ways: as text, maps (e.g. of designated areas, areas of deprivation, tranquil areas) and matrices (e.g. Table 12.2).

Other relevant PPPs *and environmental objectives* should be analysed to identify constraints on the plan in question, and requirements that the plan must achieve. Table 12.3 shows how the relationship between the plan in question and the other plans could be documented: the final column can help to clarify any clashes or inconsistencies between the other policies/etc. or between the plan being assessed and the other policies/etc.

Table 12.3 Links to other policies, plans, programmes or objectives

Relevant policy, plan, programme or objective	Requirement of the policy/etc.	How the requirement is (or may be) integrated into the plan	Any gaps or problems?
International			
EU Habitats Directive	Art. 2 – [which says . . .]	Plan provides (or is expected to provide) for this by: . . .	
	Art. 3.1 – [. . .]		
. . .			
National/regional			
PPS 1	Relevant policies are:	Plan provides for this by: . . .	
	. . .		
. . .	. . .		
Local			
. . .			

(*Source*: Levett-Therivel 2004.)

Finally, this stage of SEA should identify any *environmental or sustainability problems* that the plan should take into consideration. This could be done by identifying where the authority is not achieving environmental targets/thresholds, doing worse than other similar authorities, and/or trends are getting worse (an example is shown in Table 12.2); analysing Community Strategies to identify local residents' perceptions of key environmental problems; identifying where, under current trends, the plan is not expected to meet the requirements of other policies/etc.; and/or discussions with statutory consultees and other environmental experts. Ideally several of these approaches would be used, and the draft list of problems should be tested against planners' expert judgement.

B. Deciding the scope of SEA and developing alternatives

Next the SEA should describe the plan options or sub-components being put forward, and what issues were considered when putting them forward. The options put forward should be informed by:

- the environmental problems identified in A: ideally the plan should help to solve the problems, but at minimum it should not make them worse;
- any options suggested by statutory consultees;
- the sustainable "hierarchy of alternatives" – demand reduction, different ways of meeting demand, location of projects/infrastructure – recommended by the ODPM (2003).

Options should also meet the plan objectives and be realistic.

At this point, the statutory authorities should be consulted about the scope and level of detail of the information which must be included in the environmental report. The statutory consultees in England are the Countryside Agency, EN, English Heritage and the Environment Agency.

C. Assessing the effects of the plan

Several aspects of a plan may require impact assessment and mitigation at different stages during the development of the plan:

- broad strategic option(s) (e.g. whether housing should be at the edge of existing towns, scattered throughout an authority or in one large new town). These may need to be evaluated and compared early in the plan-making process before preferred option(s) can be agreed on;
- more detailed sub-components of the plan (e.g. plan policies on housing density and design). These may need to be evaluated and fine-tuned once the plan is closer to completion;
- proposed locations for development (e.g. specific housing sites). These may need to be evaluated and fine-tuned at a level of detail close to that of project EIA.

As such, this section may need to be repeated several times at different levels of detail and at different stages in the plan-making process.

Impact *assessment* involves testing how well each plan option or sub-component fulfils the SEA objectives/indicators identified in A. Where plan options are proposed, it should also compare, and inform the choice of, these options. Table 12.4 shows how plan options can be assessed and compared. Table 12.5 shows how the

Table 12.4 Assessment and comparison of options (illustrative only)

	Option			
	A.	B.	C.	D.
	Maintain the Green Belt in area X	Remove the Green Belt in area X	Have strategic wedges in the Green Belt in area X	Make the Green Belt narrower in area X
SEA *objective* (Objectives from Box 12.5 used as example)				
Conserve and enhance biodiversity	0	−	?	...
Promote the health of all residents	+	−	I	
Maintain and enhance soil quality	...		...	
...	...			

Preferred option and why: C is preferred because ...
Modifications to the preferred option (mitigation) suggested by this assessment:

Key:
+ positive impact
− negative impact
0 neutral impact
? uncertain impact
I depends on how option is implemented
(Based on Levett-Therivel 2004.)

Table 12.5 Assessment and mitigation of plan sub-components (illustrative only)

Plan sub-component	Conserve and enhance biodiversity at ecosystem, species, and genetic levels	Promote the health of all residents	Maintain and enhance soil quality	*Comments and overall assessment* (e.g. assumptions made, further studies needed, how implementation might make impact negative or positive)	*Proposed changes to the plan:*
1. Provide 2000 new houses in area X	–	+	–	… Impact on biodiversity depends on where houses are sited	• where text is not clear, possible changes to clarify it • where impact is negative (–), possible changes to reduce or reverse impact • where impact is positive (+), possible changes to further enhance impact • where impact depends on how the plan is implemented (I), measures needed to ensure that implementation is positive
2. Maintain the Green Belt in area X*	0	+	…	…	Change 1. to require no net loss to biodiversity. Impact on soil impossible to mitigate; social benefits outweigh this environmental cost.
…					

(Based on Levett-Therivel 2004.)
* Preferred option from previous, more strategic level of assessment, see Table 12.4.

impacts of plan sub-components can be assessed and mitigated. In each case, the table cells should be filled in, option by option or sub-component by sub-component, noting whether the option/sub-component:

- is clearly written: if not, it might be possible to rewrite it to make it clearer;
- has a negative impact (−): if so, this impact might be mitigated, for instance by rewriting the sub-component, adding a different sub-component, etc;
- has a positive impact (+): if so, it might be possible to rewrite it to make it even more positive;
- has an uncertain impact (?): if so, it may be necessary to collect further information before the assessment can be completed, and the plan finalized;
- has an impact which depends on how the plan is implemented (I): if so, it may be possible to rewrite the plan to ensure that it is implemented positively.

The focus of the assessment should not be on the symbol, but rather on making appropriate changes to the plan: these are the *mitigation measures* required by the SEA Directive. For some plans, or plan sub-components, it may be appropriate to carry out more detailed, quantitative prediction of environmental impacts. Before a final list of preferred options and/or plan sub-components is put forward, it may be useful to check whether they are internally compatible. This is not strictly part of SEA, but can help to ensure that the plan is robust and coherent. Table 12.6 shows how this can be done. A similar process could be undertaken for the objectives at Stage A.

Table 12.6 Compatibility matrix (illustrative only)

2. Maintain the Green Belt in area X	✘			
3. Develop cycle network in area X	✓	?		
4. Develop cycle network in area Y (adjacent to X)	−	−	✓	
	1. Provide 2000 new homes in area X	2. Maintain the Green Belt in area X	3. Develop a cycle network in area Y	...

Key:

✓	Compatible
?	Uncertain link
✘	Incompatible
−	No link

Incompatibility between objectives (X above)	Revision of objective(s) to eliminate incompatibility	. . . or why objectives should not be revised
New housing likely to go into the Green Belt in X	Either revise 1 or 2 or (if possible) amend 1 to ensure that new houses don't go into Green Belt	
. . .		

(Based on Levett-Therivel 2004.)

D. Consultation on the draft plan and environmental report

The findings of A–C should be published in an Environmental Report alongside the draft plan, and made available to the public and statutory consultees. The Environmental Report should discuss:

- The baseline situation, including any problems
- Links to other PPPs and environmental objectives
- Environmental and sustainability problems
- How these problems were taken into account when choosing the preferred options and developing the plan (sub-components)
- Other options considered, and why these were rejected
- Significant environmental impacts of the plan as a whole and of key plan sub-components
- Proposed mitigation measures, e.g. changes to the plan, suggestions for changes to other plans
- Links to project EIA, design guidance, etc.
- Proposed monitoring arrangements.

After the consultation responses have been received, they must be "taken into account" in the final plan. Once the final plan has been agreed, it must be published alongside a statement that explains how the authority has taken the findings of the SEA and the consultation responses into account, and "the reasons for choosing the plan … as adopted, in the light of the other reasonable alternatives dealt with". The statement should also confirm the monitoring measures that will be carried out.

E. Monitoring the implementation of the plan

Finally, the authority must monitor the significant environmental impacts of the plan's implementation. Table 12.7 shows a possible format for this.

Table 12.7 Possible monitoring framework

What to monitor (indicator)	Where do monitoring data come from?	How often?	When should action be considered?	What could be done if a problem is identified?

(*Source*: Levett-Therivel 2004.)

12.6 Summary

The next few years will see a rapid evolution of SEA practice in Europe and internationally. Environmental databases will be set up and improved, impact prediction and assessment techniques will evolve, and strategic-level consultation will flourish. What is less certain is what effect this will have on decision-making: it should become more transparent, robust and sustainability-oriented, but will it do so in practice? And, will improved SEA lead to more concise, more effective EIA?

Notes

1. See Section 11.3 for a discussion of cumulative impacts and Chapter 9, Section 9.7 for a UK case study.
2. Tomes could be written on the difference between assessment and appraisal, sustainability whatever-one-calls-it and environmental whatever one calls it (WOCI). Here, appraisal is seen as a short form of SEA, sustainability assessment as an expansion of environmental assessment.
3. The UK comprises England, Wales, Scotland and Northern Ireland. Each of these will regulate for plans and programmes that fall within their responsibilities, and develop appropriate guidance.

References

CEAA (Canadian Environmental Assessment Agency) 1999. *Cabinet Directive on the Environmental Assessment of Policy, Plan and Programme Proposals*. Hull, Quebec: CEAA. www.ceaa-acee.gc.ca/016/index_e.htm.

CEQ (Council on Environmental Quality) 1978. *Regulations for implementing the procedural provisions of the National Environmental Policy Act*, 40 Code of Federal Regulations 1500–1508.

CEQ 1997. *National Environmental Policy Act – A Study of its Effectiveness after Twenty-Five Years*. Washington, DC: CEQ.

CCW (Countryside Council for Wales) and others 2004. *Strategic Environmental Assessment and Climate Change: Guidance for Practitioners*. Bangor: CCW.

DEFRA (Department for Environment, Food and Rural Affairs) 2003. *Policy Appraisal and the Environment*. London: HMSO. defra.gov.uk/environment/economics/appraisal/index .htm.

DfT (Department for Transport) 2004. *Strategic Environmental Assessment Guidance for Transport Plans and Programmes*. London: DfT.

DoE (Department of Environment) 1990. *This Common Inheritance: Britain's Environmental Strategy*, CM 1200. London: HMSO.

DoE 1991. *Policy Appraisal and the Environment*. London: HMSO.

DoE 1992. *Development Plans and Regional Planning Guidance, Policy Planning Guidance Note no. 12*. London: HMSO.

DoE 1993. *Environmental Appraisal of Development Plans: A Good Practice Guide*. London: DoE.

DoE 1994. *Environmental Appraisal in Government Departments*. London: HMSO.

English Nature, Royal Society for the Protection of Birds and others (2004) *Strategic Environmental Assessment and Biodiversity: Guidance for Practitioners*. Oxford: Southwest Ecological Surveys, Levett-Therivel and Oxford Brookes University.

EC (European Commission) 2001. *Directive 2001/42/EC on the assessment of the effects of certain plans and programmes on the environment*. Brussels: European Commission. http://europa.eu.int/comm/environment/eia/full-legal-text/0142_en.pdf.

EC 2003. *Implementation of Directive 2001/42 on the assessment of the effects of certain plans and programmes on the environment*. Brussels: European Commission. http://europa.eu.int/comm/environment/eia/030923_sea_guidance.pdf.

FEARO (Federal Environmental Assessment Review Office) 1993. *The Environmental Assessment Process for Policy and Program Proposals*. Hull, Quebec: FEARO.

Kleinschmidt, V. & D. Wagner (eds) 1998. *Strategic Environmental Assessment in Europe: Fourth European Workshop on Environmental Impact Assessment.* Kluwer Academic Publishers: Dordrecht, The Netherlands.

Levett-Therivel sustainability consultants (2004) *Templates for Strategic Environmental Assessment Reports,* prepared for Countryside Council for Wales. Oxford: Levett-Therivel.

Marsden, S. & S. Dovers 2002. *Strategic Environmental Assessment in Australasia.* Sydney: Federation Press.

MftE (Ministry for the Environment) 1991. Resource Management Act. Wellington, New Zealand. www.mfe.govt.nz/laws/rma.

MftE 2000. *Guide to Using Section 32 of the Resource Management Act.* Wellington, New Zealand: MftE.

ODPM (Office of the Deputy Prime Minister) 2003. *The Strategic Environmental Assessment Directive: Guidance for Local Planning Authorities.* London: ODPM. http://www.odpm.gov.uk.

ODPM 2004. SEA Directive: Guidance for Planning Authorities: Pilot Studies. London: ODPM. http://www.odpm.gov.uk.

Partidario, M. R. & R. Clark (eds) 2000. *Perspectives on Strategic Environmental Assessment,* 29–43. Lewis Publishers: Boca Raton.

Sadler, B. & R. Verheem 1996. SEA: *Status, Challenges and Future Directions,* Report 53. The Hague, Netherlands: Ministry of Housing, Spatial Planning and the Environment.

Scottish Executive 2003. *Environmental Assessment of Development Plans, Interim Planning Advice.* Edinburgh: Scottish Executive. http://www.scotland.gov.uk/library5/planning/eadp-00.asp.

Smith, S. P. & W. R. Sheate 2001. Sustainability appraisal of English regional plans: incorporating the requirements of the EU Strategic Environmental Assessment Directive. *Impact Assessment and Project Appraisal* **29**, 263–76.

State of California 1986. *The California Environmental Quality Act.* Sacramento: Office of Planning and Research.

Sustainable Development Unit 2003. *Integrated Policy Appraisal,* http://www.sustainable-development.gov.uk/sdig/integrating/index.htm.

Therivel, R. 1995. Environmental Appraisal of Development Plans: Current Status. *Planning Practice and Research* **10**(2), 223–34.

Therivel, R. 1996. "Environmental Appraisal of Development Plans: Status in Late 1995", *Report,* March, 14–16.

Therivel, R. 2004. *Strategic Environmental Assessment in Action.* London: Earthscan.

Therivel, R. & P. Minas 2002. Ensuring effective sustainability appraisal. *Impact Assessment and Project Appraisal* **19**(2), 81–91.

Therivel, R. & M. R. Partidario (eds) 1996. *The Practice of Strategic Environmental Assessment.* London: Earthscan.

Therivel, R., E. Wilson, S. Thompson, D. Heaney, D. Pritchard 1992. *Strategic Environmental Assessment.* London: Earthscan.

Wood, C. 1991. EIA of Policies, Plans and Programmes. *EIA Newsletter* 5, 2–3.

Wood, C. 1988. EIA in plan-making. In *Environmental Impact Assessment: Theory and Practice,* P. Wathern (ed.). London: Unwin Hyman.

Appendix 1
The text of Council Directive 97/11/EC of 3 March 1997 amending Directive 85/337/EEC on the assessment of the effects of certain public and private projects on the environment[1]

Article 1

1. This Directive shall apply to the assessment of the environmental effects of those public and private projects which are likely to have significant effects on the environment.
2. For the purposes of this Directive:
    "project" means:

    −   the execution of construction works or of other installations or schemes,
    −   other interventions in the natural surroundings and landscape including those involving the extraction of mineral resources;

    "developer" means:
    the applicant for authorization for a private project or the public authority which initiates a project;
    "development consent" means:
    the decision of the competent authority or authorities which entitles the developer to proceed with the project.
3. The competent authority or authorities shall be that or those which the Member States designate as responsible for performing the duties arising from this Directive.
4. Projects serving national defence purposes are not covered by this Directive.
5. This Directive shall not apply to projects the details of which are adopted by a specific act of national legislation, since the objectives of this Directive, including that of supplying information, are achieved through the legislative process.

---

[1] Consolidated version.

Article 2

1. *Member States shall adopt all measures necessary to ensure that, before consent is given, projects likely to have significant effects on the environment by virtue, inter alia, of their nature, size or location are made subject to a requirement for development consent and an assessment with regard to their effects. These projects are defined in Article 4.*
2. The environmental impact assessment may be integrated into the existing procedures for consent to projects in the Member States, or, failing this, into other procedures or into procedures to be established to comply with the aims of this Directive.
3. *Without prejudice to Article 7, Member States may, in exceptional cases, exempt a specific project in whole or in part from the provisions laid down in this Directive.*
    In this event, the Member States shall:

    (a) consider whether another form of assessment would be appropriate and whether the information thus collected should be made available to the public;
    (b) make available to the public concerned the information relating to the exemption and the reasons for granting it;
    (c) inform the Commission, prior to granting consent, of the reasons justifying the exemption granted, and provide it with the information made available, *where applicable*, to their own nationals.

    The Commission shall immediately forward the documents received to the other Member States.

    The Commission shall report annually to the Council on the application of this paragraph.
4. *Member States may provide for a single procedure in order to fulfil the requirements of this Directive and the requirements of Council Directive 96/61/EC of 24 September 1996 on integrated pollution prevention and control.*[2]

Article 3

The environmental impact assessment shall identify, describe and assess in an appropriate manner, in the light of each individual case and in accordance with Articles 4 to 11, the direct and indirect effects of a project on the following factors:

- *human beings, fauna and flora;*
- *soil, water, air, climate and the landscape;*
- *material assets and the cultural heritage;*
- *the interaction between the factors mentioned in the first, second and third indents.*

Article 4

1. *Subject to Article 2(3), projects listed in Annex I shall be made subject to an assessment in accordance with Articles 5 to 10.*

---

[2] OJ No L257, 10.10.1996, p. 26.

2.   *Subject to Article 2(3), for projects listed in Annex II, the Member States shall determine through:*

   (a)   *a case-by-case examination, or*
   (b)   *thresholds or criteria set by the Member States*

   whether the project shall be made subject to an assessment in accordance with Articles 5 to 10.
   Member States may decide to apply both procedures referred to in (a) and (b).

3.   *When a case-by-case examination is carried out or thresholds or criteria are set for the purpose of paragraph 2, the relevant selection criteria set out in Annex III shall be taken into account.*

4.   *Member States shall ensure that the determination made by the competent authorities under paragraph 2 is made available to the public.*

Article 5

1.   *In the case of projects which, pursuant to Article 4, must be subjected to an environmental impact assessment in accordance with Articles 5 to 10, Member States shall adopt the necessary measures to ensure that the developer supplies in an appropriate form the information specified in Annex IV inasmuch as:*

   (a)   *the Member States consider that the information is relevant to a given stage of the consent procedure and to the specific characteristics of a particular project or type of project and of the environmental features likely to be affected;*
   (b)   *the Member States consider that a developer may reasonably be required to compile this information having regard inter alia to current knowledge and methods of assessment.*

2.   *Member States shall take the necessary measures to ensure that, if the developer so requests before submitting an application for development consent, the competent authority shall give an opinion on the information to be supplied by the developer in accordance with paragraph 1. The competent authority shall consult the developer and authorities referred to in Article 6(1) before it gives its opinion. The fact that the authority has given an opinion under this paragraph shall not preclude it from subsequently requiring the developer to submit further information.*

   Member States may require the competent authorities to give such an opinion, irrespective of whether the developer so requests.

3.   *The information to be provided by the developer in accordance with paragraph 1 shall include at least:*

   –   *a description of the project comprising information on the site, design and size of the project,*
   –   *a description of the measures envisaged in order to avoid, reduce and, if possible, remedy significant adverse effects,*
   –   *the data required to identify and assess the main effects which the project is likely to have on the environment,*
   –   *an outline of the main alternatives studied by the developer and an indication of the main reasons for his choice, taking into account the environmental effects,*
   –   *a non-technical summary of the information mentioned in the previous indents.*

4. *Member States shall, if necessary, ensure that any authorities holding relevant information, with particular reference to Article 3, shall make this information available to the developer.*

Article 6

1. *Member States shall take the measures necessary to ensure that the authorities likely to be concerned by the project by reason of their specific environmental responsibilities are given an opportunity to express their opinion on the information supplied by the developer and on the request for development consent. To this end, Member States shall designate the authorities to be consulted, either in general terms or on a case-by-case basis. The information gathered pursuant to Article 5 shall be forwarded to those authorities. Detailed arrangements for consultation shall be laid down by the Member States.*
2. *Member States shall ensure that any request for development consent and any information gathered pursuant to Article 5 are made available to the public within a reasonable time in order to give the public concerned the opportunity to express an opinion before the development consent is granted.*
3. The detailed arrangements for such information and consultation shall be determined by the Member States, which may in particular, depending on the particular characteristics of the projects or sites concerned:

   – determine the public concerned,
   – specify the places where the information can be consulted,
   – specify the way in which the public may be informed, for example by bill-posting within a certain radius, publication in local newspapers, organization of exhibitions with plans, drawings, tables, graphs, models,
   – determine the manner in which the public is to be consulted, for example, by written submissions, by public enquiry,
   – fix appropriate time limits for the various stages of the procedure in order to ensure that a decision is taken within a reasonable period.

Article 7

1. *Where a Member State is aware that a project is likely to have significant effects on the environment in another Member State or where a Member State likely to be significantly affected so requests, the Member State in whose territory the project is intended to be carried out shall send to the affected Member State as soon as possible and no later than when informing its own public, inter alia:*

   (a) *a description of the project, together with any available information on its possible transboundary impact;*
   (b) *information on the nature of the decision which may be taken,*

   and shall give the other Member State a reasonable time in which to indicate whether it wishes to participate in the Environmental Impact Assessment procedure, and may include the information referred to in paragraph 2.

2.   *If a Member State which receives information pursuant to paragraph 1 indicates that it intends to participate in the Environmental Impact Assessment procedure, the Member State in whose territory the project is intended to be carried out shall, if it has not already done so, send to the affected Member State the information gathered pursuant to Article 5 and relevant information regarding the said procedure, including the request for development consent.*

3.   *The Member States concerned, each insofar as it is concerned, shall also:*

(a)   *arrange for the information referred to in paragraphs 1 and 2 to be made available, within a reasonable time, to the authorities referred to in Article 6(1) and the public concerned in the territory of the Member State likely to be significantly affected; and*

(b)   *ensure that those authorities and the public concerned are given an opportunity, before development consent for the project is granted, to forward their opinion within a reasonable time on the information supplied to the competent authority in the Member State in whose territory the project is intended to be carried out.*

4.   *The Member States concerned shall enter into consultations regarding, inter alia, the potential transboundary effects of the project and the measures envisaged to reduce or eliminate such effects and shall agree on a reasonable time frame for the duration of the consultation period.*

5.   *The detailed arrangements for implementing the provisions of this Article may be determined by the Member States concerned.*

Article 8

The results of consultations and the information gathered pursuant to Articles 5, 6 and 7 must be taken into consideration in the development consent procedure.

Article 9

1.   *When a decision to grant or refuse development consent has been taken, the competent authority or authorities shall inform the public thereof in accordance with the appropriate procedures and shall make available to the public the following information:*

—   *the content of the decision and any conditions attached thereto,*
—   *the main reasons and considerations on which the decision is based,*
—   *a description, where necessary, of the main measures to avoid, reduce and, if possible, offset the major adverse effects.*

2.   *The competent authority or authorities shall inform any Member State which has been consulted pursuant to Article 7, forwarding to it the information referred to in paragraph 1.*

Article 10

The provisions of this Directive shall not affect the obligation on the competent authorities to respect the limitations imposed by national regulations and administrative provisions and accepted legal practices with regard to commercial and industrial confidentiality, including intellectual property, and the safeguarding the public interest.

Where Article 7 applies, the transmission of information to another Member State and the receipt of information by another Member State shall be subject to the limitations in force in the Member State in which the project is proposed.

Article 11

1. The Member States and the Commission shall exchange information on the experience gained in applying this Directive.
2. *In particular, Member States shall inform the Commission of any criteria and/or thresholds adopted for the selection of the projects in question, in accordance with Article 4(2).*
3. *Five years after the entry into force of this Directive, the Commission shall send the European Parliament and the Council a report on the application and effectiveness of Directive 85/337/EEC as amended by this Directive. The report shall be based on the exchange of information provided for by Article 11(1) and (2).*
4. *On the basis of this report, the Commission shall, where appropriate, submit to the Council additional proposals with a view to ensuring further coordination in the application of this Directive.*

Article 12

1. *Member States shall bring into force the laws, regulations and administrative provisions necessary to comply with this Directive by 14 March 1999 at the latest. They shall forthwith inform the Commission thereof.*

   When Member States adopt these provisions, they shall contain a reference to this Directive or shall be accompanied by such reference at the time of their official publication. The procedure for such reference shall be adopted by Member States.
2. *If a request for development consent is submitted to a competent authority before the end of the time limit laid down in paragraph 1, the provisions of Directive 85/337/EEC prior to these amendments shall continue to apply.*

Article 13

This Directive shall enter into force on the twentieth day following that of its publication in the Official Journal of the European Communities.

Article 14

This Directive is addressed to the Member States.
Done at Brussels, 3 March 1997.
For the Council
The President
M. DE BOER

Annex I: Projects subject to Article 4(1)

1. Crude-oil refineries (excluding undertakings manufacturing only lubricants from crude oil) and installations for the gasification and liquefaction of 500 tonnes or more of coal or bituminous shale per day.

2. Thermal power stations and other combustion installations with a heat output of 300 megawatts or more, and

    Nuclear power stations and other nuclear reactors including the dismantling or decommissioning of such power stations or reactors[1] (except research installations for the production and conversion of fissionable and fertile materials, whose maximum power does not exceed 1 kilowatt continuous thermal load).

3. (a) Installations for the reprocessing of irradiated nuclear fuel.
    (b) Installations designed:

    - for the production or enrichment of nuclear fuel,
    - for the processing of irradiated nuclear fuel or high-level radioactive waste,
    - for the final disposal of irradiated nuclear fuel,
    - solely for the final disposal of radioactive waste,
    - solely for the storage (planned for more than 10 years) of irradiated nuclear fuels or radioactive waste in a different site than the production site.

4. Integrated works for the initial smelting of cast-iron and steel;
    Installations for the production of non-ferrous crude metals from ore, concentrates or secondary raw materials by metallurgical, chemical or electrolytic processes.

5. Installations for the extraction of asbestos and for the processing and transformation of asbestos and products containing asbestos: for asbestos-cement products, with an annual production of more than 20 000 tonnes of finished products, for friction material, with an annual production of more than 50 tonnes of finished products, and for other uses of asbestos, utilization of more than 200 tonnes per year.

6. Integrated chemical installations, i.e. those installations for the manufacture on an industrial scale of substances using chemical conversion processes, in which several units are juxtaposed and are functionally linked to one another and which are:

    (i) for the production of basic organic chemicals;
    (ii) for the production of basic inorganic chemicals;

---

[1] Nuclear power stations and other nuclear reactors cease to be such an installation when all nuclear fuel and other radioactively contaminated elements have been removed permanently from the installation site.

    (iii) for the production of phosphorous-, nitrogen- or potassium-based fertilizers (simple or compound fertilizers);

    (iv) for the production of basic plant health products and of biocides;

    (v) for the production of basic pharmaceutical products using a chemical or biological process;

    (vi) for the production of explosives.

7. (a) Construction of lines for long-distance railway traffic and of airports[2] with a basic runway length of 2 100 m or more;

    (b) Construction of motorways and express roads;[3]

    (c) Construction of a new road of four or more lanes, or realignment and/or widening of an existing road of two lanes or less so as to provide four or more lanes, where such new road, or realigned and/or widened section of road would be 10 km or more in a continuous length.

8. (a) Inland waterways and ports for inland-waterway traffic which permit the passage of vessels of over 1 350 tonnes;

    (b) Trading ports, piers for loading and unloading connected to land and outside ports (excluding ferry piers) which can take vessels of over 1 350 tonnes.

9. Waste disposal installations for the incineration, chemical treatment as defined in Annex IIA to Directive 75/442/EEC[4] under heading D9, or landfill of hazardous waste (i.e. waste to which Directive 91/689/EEC[5] applies).

10. Waste disposal installations for the incineration or chemical treatment as defined in Annex IIA to Directive 75/442/EEC under heading D9 of non-hazardous waste with a capacity exceeding 100 tonnes per day.

11. Groundwater abstraction or artificial groundwater recharge schemes where the annual volume of water abstracted or recharged is equivalent to or exceeds 10 million cubic metres.

12. (a) Works for the transfer of water resources between river basins where this transfer aims at preventing possible shortages of water and where the amount of water transferred exceeds 100 million cubic metres/year;

    (b) In all other cases, works for the transfer of water resources between river basins where the multi-annual average flow of the basin of abstraction exceeds 2 000 million cubic metres/year and where the amount of water transferred exceeds 5% of this flow.

In both cases transfers of piped drinking water are excluded.

13. Waste water treatment plants with a capacity exceeding 150 000 population equivalent as defined in Article 2 point (6) of Directive 91/271/EEC.[6]

---

[2] For the purposes of this Directive, "airport" means airports which comply with the definition in the 1944 Chicago Convention setting up the International Civil Aviation Organization (Annex 14).

[3] For the purposes of the Directive, "express road" means a road which complies with the definition in the European Agreement on Main International Traffic Arteries of 15 November 1975.

[4] OJ No L 194, 25. 7. 1975, p. 39. Directive as last amended by Commission Decision 94/3/EC (OJ No L 5, 7. 1. 1994, p. 15).

[5] OJ No L 377, 31. 12. 1991, p. 20. Directive as last amended by Directive 94/31/EC (OJ No L 168, 2. 7. 1994, p. 28).

[6] OJ No L 135, 30. 5. 1991, p. 40. Directive as last amended by the 1994 Act of Accession.

14. Extraction of petroleum and natural gas for commercial purposes where the amount extracted exceeds 500 tonnes/day in the case of petroleum and 500 000 m³/day in the case of gas.

15. Dams and other installations designed for the holding back or permanent storage of water, where a new or additional amount of water held back or stored exceeds 10 million cubic metres.

16. Pipelines for the transport of gas, oil or chemicals with a diameter of more than 800 mm and a length of more than 40 km.

17. Installations for the intensive rearing of poultry or pigs with more than:

    (a) 85 000 places for broilers, 60 000 places for hens;
    (b) 3 000 places for production pigs (over 30 kg); or
    (c) 900 places for sows.

18. Industrial plants for the

    (a) production of pulp from timber or similar fibrous materials;
    (b) production of paper and board with a production capacity exceeding 200 tonnes per day.

19. Quarries and open-cast mining where the surface of the site exceeds 25 hectares, or peat extraction, where the surface of the site exceeds 150 hectares.

20. Construction of overhead electrical power lines with a voltage of 220 kV or more and a length of more than 15 km.

21. Installations for storage of petroleum, petrochemical, or chemical products with a capacity of 200 000 tonnes or more.

Annex II: Projects subject to Article 4(2)

1. **Agriculture, silviculture and aquaculture**

   (a) Projects for the restructuring of rural land holdings;
   (b) Projects for the use of uncultivated land or semi-natural areas for intensive agricultural purposes;
   (c) Water management projects for agriculture, including irrigation and land drainage projects;
   (d) Initial afforestation and deforestation for the purposes of conversion to another type of land use;
   (e) Intensive livestock installations (projects not included in Annex I);
   (f) Intensive fish farming;
   (g) Reclamation of land from the sea.

2. **Extractive industry**

   (a) Quarries, open-cast mining and peat extraction (projects not included in Annex I);
   (b) Underground mining;
   (c) Extraction of minerals by marine or fluvial dredging;
   (d) Deep drillings, in particular:

   – geothermal drilling,
   – drilling for the storage of nuclear waste material,
   – drilling for water supplies, with the exception of drillings for investigating the stability of the soil;

   (e) Surface industrial installations for the extraction of coal, petroleum, natural gas and ores, as well as bituminous shale.

3. **Energy industry**

   (a) Industrial installations for the production of electricity, steam and hot water (projects not included in Annex I);
   (b) Industrial installations for carrying gas, steam and hot water; transmission of electrical energy by overhead cables (projects not included in Annex I);
   (c) Surface storage of natural gas;
   (d) Underground storage of combustible gases;
   (e) Surface storage of fossil fuels;
   (f) Industrial briquetting of coal and lignite;
   (g) Installations for the processing and storage of radioactive waste (unless include in Annex I);

(h) Installations for hydroelectric energy production;
(i) Installations for the harnessing of wind power for energy production (wind farms).

4. **Production and processing of metals**

(a) Installations for the production of pig iron or steel (primary or secondary fusion including continuous casting);
(b) Installations for the processing of ferrous metals:
    (i)   hot-rolling mills;
    (ii) smitheries with hammers;
    (iii) application of protective fused metal coats;
(c) Ferrous metal foundries;
(d) Installations for the smelting, including the alloyage, of non-ferrous metals excluding precious metals, including recovered products (refining, foundry casting etc.);
(e) Installations for surface treatment of metals and plastic materials using an electrolytic or chemical process;
(f) Manufacture and assembly of motor vehicles and manufacture of motor-vehicle engines;
(g) Shipyards;
(h) Installations for the construction and repair of aircraft;
(i) Manufacture of railway equipment;
(j) Swaging by explosives;
(k) Installations for the roasting and sintering of metallic ores.

5. **Mineral industry**

(a) Coke ovens (dry coal distillation);
(b) Installations for the manufacture of cement;
(c) Installations for the production of asbestos and the manufacture of asbestos products (projects not included in Annex I);
(d) Installations for the manufacture of glass including glass fibre;
(e) Installations for smelting mineral substances including the production of mineral fibres;
(f) Manufacture of ceramic products by burning, in particular roofing tiles, bricks, refractory bricks, tiles, stoneware or porcelain.

6. **Chemical industry (Projects not included in Annex I)**

(a) Treatment of intermediate products and production of chemicals;
(b) Production of pesticides and pharmaceutical products, paint and varnishes, elastomers and peroxides;
(c) Storage facilities for petroleum, petrochemical and chemical products.

7. **Food industry**

(a) Manufacture of vegetable and animal oils and fats;
(b) Packing and canning of animal and vegetable products;
(c) Manufacture of dairy products;
(d) Brewing and malting;
(e) Confectionery and syrup manufacture;

(f) Installations for the slaughter of animals;

(g) Industrial starch manufacturing installations;

(h) Fish-meal and fish-oil factories;

(i) Sugar factories.

8. **Textile, leather, wood and paper industries**

(a) Industrial plants for the production of paper and board (projects not included in Annex I);

(b) Plants for the pretreatment (operations such as washing, bleaching, mercerization) or dyeing of fibres or textiles;

(c) Plants for the tanning of hides and skins;

(d) Cellulose-processing and production installations.

9. **Rubber industry**

Manufacture and treatment of elastomer-based products.

10. **Infrastructure projects**

(a) Industrial estate development projects;

(b) Urban development projects, including the construction of shopping centres and car parks;

(c) Construction of railways and intermodal transshipment facilities, and of intermodal terminals (projects not included in Annex I);

(d) Construction of airfields (projects not included in Annex I);

(e) Construction of roads, harbours and port installations, including fishing harbours (projects not included in Annex I);

(f) Inland-waterway construction not included in Annex I, canalization and flood relief works;

(g) Dams and other installations designed to hold water or store it on a long-term basis (projects not included in Annex I);

(h) Tramways, elevated and underground railways, suspended lines or similar lines of a particular type, used exclusively or mainly for passenger transport;

(i) Oil and gas pipeline installations (projects not included in Annex I);

(j) Installations of long-distance aqueducts;

(k) Coastal work to combat erosion and maritime works capable of altering the coast through the construction, for example, of dykes, moles, jetties and other sea defence works, excluding the maintenance and reconstruction of such works;

(l) Groundwater abstraction and artificial groundwater recharge schemes not included in Annex I;

(m) Works for the transfer of water resources between river basins not included in Annex I.

11. **Other projects**

(a) Permanent racing and test tracks for motorized vehicles;

(b) Installations for the disposal of waste (projects not included in Annex I);

(c) Waste-water treatment plants (projects not included in Annex I);

(d) Sludge-deposition sites;

(e) Storage of scrap iron, including scrap vehicles;

(f) Test benches for engines, turbines or reactors;

(g) Installations for the manufacture of artificial mineral fibres;

(h) Installations for the recovery or destruction of explosive substances;

(i) Knackers' yards.

12. **Tourism and leisure**

(a) Ski-runs, ski-lifts and cable-cars and associated developments;

(b) Marinas;

(c) Holiday villages and hotel complexes outside urban areas and associated developments;

(d) Permanent camp sites and caravan sites;

(e) Theme parks.

13. Any change or extension of projects listed in Annex I or Annex II, already authorized, executed or in the process of being executed, which may have significant adverse effects on the environment;

Projects in Annex I, undertaken exclusively or mainly for the development and testing of new methods or products and not used for more than two years.

Annex III: Selection criteria referred to in Article 4(3)

1. **Characteristics of projects**
    The characteristics of projects must be considered having regard, in particular, to:

    - the size of the project,
    - the cumulation with other projects,
    - the use of natural resources,
    - the production of waste,
    - pollution and nuisances,
    - the risk of accidents, having regard in particular to substances or technologies used.

2. **Location of projects**
    The environmental sensitivity of geographical areas likely to be affected by projects must be considered, having regard, in particular, to:

    - the existing land use,
    - the relative abundance, quality and regenerative capacity of natural resources in the area,
    - the absorption capacity of the natural environment, paying particular attention to the following areas:

        (a) wetlands;
        (b) coastal zones;
        (c) mountain and forest areas;
        (d) nature reserves and parks;
        (e) areas classified or protected under Member States' legislation; special protection areas designated by Member States pursuant to Directive 79/409/EEC and 92/43/EEC;
        (f) areas in which the environmental quality standards laid down in Community legislation have already been exceeded;
        (g) densely populated areas;
        (h) landscapes of historical, cultural or archaeological significance.

3. **Characteristics of the potential impact**
    The potential significant effects of projects must be considered in relation to criteria set out under 1 and 2 above, and having regard in particular to:

    - the extent of the impact (geographical area and size of the affected population),
    - the transfrontier nature of the impact,
    - the magnitude and complexity of the impact,
    - the probability of the impact,
    - the duration, frequency and reversibility of the impact.

Annex IV: Information referred to in Article 5(1)

1. Description of the project, including in particular:
   - a description of the physical characteristics of the whole project and the land-use requirements during the construction and operational phases,
   - a description of the main characteristics of the production processes, for instance, nature and quantity of the materials used,
   - an estimate, by type and quantity, of expected residues and emissions (water, air and soil pollution, noise, vibration, light, heat, radiation, etc.) resulting from the operation of the proposed project.

2. An outline of the main alternatives studied by the developer and an indication of the main reasons for this choice, taking into account the environmental effects.

3. A description of the aspects of the environment likely to be significantly affected by the proposed project, including, in particular, population, fauna, flora, soil, water, air, climatic factors, material assets, including the architectural and archaeological heritage, landscape and the inter-relationship between the above factors.

4. A description[7] of the likely significant effects of the proposed project on the environment resulting from:
   - the existence of the project,
   - the use of natural resources,
   - the emission of pollutants, the creation of nuisances and the elimination of waste, and the description by the developer of the forecasting methods used to assess the effects on the environment.

5. A description of the measures envisaged to prevent, reduce and where possible offset any significant adverse effects on the environment.

6. A non-technical summary of the information provided under the above headings.

7. An indication of any difficulties (technical deficiencies or lack of know-how) encountered by the developer in compiling the required information.

[7] This description should cover the direct effects and any indirect, secondary, cumulative, short, medium and long-term, permanent and temporary, positive and negative effects of the project.

Appendix 2
Directive 2001/42/EC-SEA of the European Parliament and of the Council
of 27 June 2001 on the assessment of the effects of certain plans and programmes on the environment

THE EUROPEAN PARLIAMENT AND THE COUNCIL OF THE EUROPEAN UNION

Having regard to the Treaty establishing the European Community, and in particular Article 175(1) thereof,

Having regard to the proposal from the Commission,[1]

Having regard to the opinion of the Economic and Social Committee,[2]

Having regard to the opinion of the Committee of the Regions,[3]

Acting in accordance with the procedure laid down in Article 251 of the Treaty,[4] in the light of the joint text approved by the Conciliation Committee on 21 March 2001,

Whereas:

(1) Article 174 of the Treaty provides that Community policy on the environment is to contribute to, *inter alia*, the preservation, protection and improvement of the quality of the environment, the protection of human health and the prudent and rational utilisation of natural resources and that it is to be based on the precautionary principle. Article 6 of the Treaty provides that environmental protection requirements are to be integrated into the definition of Community policies and activities, in particular with a view to promoting sustainable development.

(2) The Fifth Environment Action Programme: Towards sustainability – A European Community programme of policy and action in relation to the environment and

---

[1] OJ C 129, 25.4.1997, p. 14 and OJ C 83, 25.3.1999, p. 13.

[2] OJ C 287, 22.9.1997, p. 101.

[3] OJ C 64, 27.2.1998, p. 63 and OJ C 374, 23.12.1999, p. 9.

[4] Opinion of the European Parliament of 20 October 1998 (OJ C 341, 9.11.1998, p. 18), confirmed on 16 September 1999 (OJ C 54, 25.2.2000, p. 76), Council Common Position of 30 March 2000 (OJ C 137, 16.5.2000, p. 11) and Decision of the European Parliament of 6 September 2000 (OJ C 135, 7.5.2001, p. 155). Decision of the European Parliament of 31 May 2001 and Decision of the Council of 5 June 2001.

sustainable development,[5] supplemented by Council Decision No 2179/98/EC[6] on its review, affirms the importance of assessing the likely environmental effects of plans and programmes.

(3) The Convention on Biological Diversity requires Parties to integrate as far as possible and as appropriate the conservation and sustainable use of biological diversity into relevant sectoral or cross-sectoral plans and programmes.

(4) Environmental assessment is an important tool for integrating environmental considerations into the preparation and adoption of certain plans and programmes which are likely to have significant effects on the environment in the Member States, because it ensures that such effects of implementing plans and programmes are taken into account during their preparation and before their adoption.

(5) The adoption of environmental assessment procedures at the planning and programming level should benefit undertakings by providing a more consistent framework in which to operate by the inclusion of the relevant environmental information into decision making. The inclusion of a wider set of factors in decision making should contribute to more sustainable and effective solutions.

(6) The different environmental assessment systems operating within Member States should contain a set of common procedural requirements necessary to contribute to a high level of protection of the environment.

(7) The United Nations/Economic Commission for Europe Convention on Environmental Impact Assessment in a Transboundary Context of 25 February 1991, which applies to both Member States and other States, encourages the parties to the Convention to apply its principles to plans and programmes as well; at the second meeting of the Parties to the Convention in Sofia on 26 and 27 February 2001, it was decided to prepare a legally binding protocol on strategic environmental assessment which would supplement the existing provisions on environmental impact assessment in a transboundary context, with a view to its possible adoption on the occasion of the 5th Ministerial Conference "Environment for Europe" at an extraordinary meeting of the Parties to the Conventions, scheduled for May 2003 in Kiev, Ukraine. The systems operating within the Community for environmental assessment of plans and programmes should ensure that there are adequate transboundary consultations where the implementation of a plan or programme being prepared in one Member State is likely to have significant effects on the environment of another Member State. The information on plans and programmes having significant effects on the environment of other States should be forwarded on a reciprocal and equivalent basis within an appropriate legal framework between Member States and these other States.

(8) Action is therefore required at Community level to lay down a minimum environmental assessment framework, which would set out the broad principles of the environmental assessment system and leave the details to the Member States, having regard to the principle of subsidiarity. Action by the Community should not go beyond what is necessary to achieve the objectives set out in the Treaty.

(9) The Directive is of a procedural nature, and its requirements should either be integrated into existing procedures in Member States or incorporated in specifically

[5] OJ C 138, 17.5.1993, p. 5.
[6] OJ L 275, 10.10.1998, p. 1.

established procedures. With a view to avoiding duplication of the assessment, Member States should take account, where appropriate, of the fact that assessments will be carried out at different levels of a hierarchy of plans and programmes.

(10) All plans and programmes which are prepared for a number of sectors and which set a framework for future development consent of projects listed in Annexes I and II to Council Directive 85/337/EEC of 27 June 1985 on the assessment of the effects of certain public and private projects on the environment,[7] and all plans and programmes which have been determined to require assessment pursuant to Council Directive 92/43/EEC of 21 May 1992 on the conservation of natural habitats and of wild flora and fauna,[8] are likely to have significant effects on the environment, and should as a rule be made subject to systematic environmental assessment. When they determine the use of small areas at local level or are minor modifications to the above plans or programmes, they should be assessed only where Member States determine that they are likely to have significant effects on the environment.

(11) Other plans and programmes which set the framework for future development consent of projects may not have significant effects on the environment in all cases and should be assessed only where Member States determine that they are likely to have such effects.

(12) When Member States make such determinations, they should take into account the relevant criteria set out in this Directive.

(13) Some plans or programmes are not subject to this Directive because of their particular characteristics.

(14) Where an assessment is required by this Directive, an environmental report should be prepared containing relevant information as set out in this Directive, identifying, describing and evaluating the likely significant environmental effects of implementing the plan or programme, and reasonable alternatives taking into account the objectives and the geographical scope of the plan or programme; Member States should communicate to the Commission any measures they take concerning the quality of environmental reports.

(15) In order to contribute to more transparent decision making and with the aim of ensuring that the information supplied for the assessment is comprehensive and reliable, it is necessary to provide that authorities with relevant environmental responsibilities and the public are to be consulted during the assessment of plans and programmes, and that appropriate time frames are set, allowing sufficient time for consultations, including the expression of opinion.

(16) Where the implementation of a plan or programme prepared in one Member State is likely to have a significant effect on the environment of other Member States, provision should be made for the Member States concerned to enter into consultations and for the relevant authorities and the public to be informed and enabled to express their opinion.

(17) The environmental report and the opinions expressed by the relevant authorities and the public, as well as the results of any transboundary consultation, should

[7] OJ L 175, 5.7.1985, p. 40. Directive as amended by Directive 97/11/EC (OJ L 73, 14.3.1997, p. 5).
[8] OJ L 206, 22.7.1992, p. 7. Directive as last amended by Directive 97/62/EC (OJ L 305, 8.11.1997, p. 42).

be taken into account during the preparation of the plan or programme and before its adoption or submission to the legislative procedure.

(18) Member States should ensure that, when a plan or programme is adopted, the relevant authorities and the public are informed and relevant information is made available to them.

(19) Where the obligation to carry out assessments of the effects on the environment arises simultaneously from this Directive and other Community legislation, such as Council Directive 79/409/EEC of 2 April 1979 on the conservation of wild birds,[9] Directive 92/43/EEC, or Directive 2000/60/EC of the European Parliament and the Council of 23 October 2000 establishing a framework for Community action in the field of water policy,[10] in order to avoid duplication of the assessment, Member States may provide for coordinated or joint procedures fulfilling the requirements of the relevant Community legislation.

(20) A first report on the application and effectiveness of this Directive should be carried out by the Commission five years after its entry into force, and at seven-year intervals thereafter. With a view to further integrating environmental protection requirements, and taking into account the experience acquired, the first report should, if appropriate, be accompanied by proposals for amendment of this Directive, in particular as regards the possibility of extending its scope to other areas/sectors and other types of plans and programmes.

HAVE ADOPTED THIS DIRECTIVE:

Article 1. Objectives

The objective of this Directive is to provide for a high level of protection of the environment and to contribute to the integration of environmental considerations into the preparation and adoption of plans and programmes with a view to promoting sustainable development, by ensuring that, in accordance with this Directive, an environmental assessment is carried out of certain plans and programmes which are likely to have significant effects on the environment.

Article 2. Definitions

For the purposes of this Directive:

(a) "plans and programmes" shall mean plans and programmes, including those co-financed by the European Community, as well as any modifications to them:

- which are subject to preparation and/or adoption by an authority at national, regional or local level or which are prepared by an authority for adoption, through a legislative procedure by Parliament or Government, and
- which are required by legislative, regulatory or administrative provisions;

---

[9] OJ L 103, 25.4.1979, p. 1. Directive as last amended by Directive 97/49/EC (OJ L 223, 13.8.1997, p. 9).
[10] OJ L 327, 22.12.2000, p. 1.

(b) "environmental assessment" shall mean the preparation of an environmental report, the carrying out of consultations, the taking into account of the environmental report and the results of the consultations in decision-making and the provision of information on the decision in accordance with Articles 4 to 9;

(c) "environmental report" shall mean the part of the plan or programme documentation containing the information required in Article 5 and Annex I;

(d) "The public" shall mean one or more natural or legal persons and, in accordance with national legislation or practice, their associations, organisations or groups.

Article 3. Scope

1. An environmental assessment, in accordance with Articles 4 to 9, shall be carried out for plans and programmes referred to in paragraphs 2 to 4 which are likely to have significant environmental effects.

2. Subject to paragraph 3, an environmental assessment shall be carried out for all plans and programmes,

    (a) which are prepared for agriculture, forestry, fisheries, energy, industry, transport, waste management, water management, telecommunications, tourism, town and country planning or land use and which set the framework for future development consent of projects listed in Annexes I and II to Directive 85/337/EC, or

    (b) which, in view of the likely effects on sites, have been determined to require an assessment pursuant to Article 6 or 7 of Directive 92/43/EEC.

3. Plans and programmes referred to in paragraph 2 which determine the use of small areas at local level and minor modifications to plans and programmes referred to in paragraph 2 shall require an environmental assessment only where the Member States determine that they are likely to have significant environmental effects.

4. Member States shall determine whether plans and programmes, other than those referred to in paragraph 2, which set the framework for future development consent of projects, are likely to have significant environmental effects.

5. Member States shall determine whether plans or programmes referred to in paragraphs 3 and 4 are likely to have significant environmental effects either through case-by-case examination or by specifying types of plans and programmes or by combining both approaches. For this purpose Member States shall in all cases take into account relevant criteria set out in Annex II, in order to ensure that plans and programmes with likely significant effects on the environment are covered by this Directive.

6. In the case-by-case examination and in specifying types of plans and programmes in accordance with paragraph 5, the authorities referred to in Article 6(3) shall be consulted.

7. Member States shall ensure that their conclusions pursuant to paragraph 5, including the reasons for not requiring an environmental assessment pursuant to Articles 4 to 9, are made available to the public.

8. The following plans and programmes are not subject to this Directive:

   • plans and programmes the sole purpose of which is to serve national defence or civil emergency,
   • financial or budget plans and programmes.

9. This Directive does not apply to plans and programmes co-financed under the current respective programming periods[1] for Council Regulations (EC) No 1260/1999[2] and (EC) No 1257/1999.[3]

Article 4. General obligations

1. The environmental assessment referred to in Article 3 shall be carried out during the preparation of a plan or programme and before its adoption or submission to the legislative procedure.
2. The requirements of this Directive shall either be integrated into existing procedures in Member States for the adoption of plans and programmes or incorporated in procedures established to comply with this Directive.
3. Where plans and programmes form part of a hierarchy, Member States shall, with a view to avoiding duplication of the assessment, take into account the fact that the assessment will be carried out, in accordance with this Directive, at different levels of the hierarchy. For the purpose of, *inter alia*, avoiding duplication of assessment, Member States shall apply Article 5(2) and (3).

Article 5. Environmental report

1. Where an environmental assessment is required under Article 3(1), an environmental report shall be prepared in which the likely significant effects on the environment of implementing the plan or programme, and reasonable alternatives taking into account the objectives and the geographical scope of the plan or programme, are identified, described and evaluated. The information to be given for this purpose is referred to in Annex I.
2. The environmental report prepared pursuant to paragraph 1 shall include the information that may reasonably be required taking into account current knowledge and methods of assessment, the contents and level of detail in the plan or programme, its stage in the decision-making process and the extent to which certain matters are more appropriately assessed at different levels in that process in order to avoid duplication of the assessment.
3. Relevant information available on environmental effects of the plans and programmes and obtained at other levels of decision-making or through other

[1] The 2000–2006 programming period for Council Regulation (EC) No 1260/1999 and the 2000–2006 and 2000–2007 programming periods for Council Regulation (EC) No 1257/1999.
[2] Council Regulation (EC) No 1260/1999 of 21 June 1999 laying down general provisions on the Structural Funds (OJ L 161, 26.6.1999, p. 1).
[3] Council Regulation (EC) No 1257/1999 of 17 May 1999 on support for rural development from the European Agricultural Guidance and Guarantee Fund (EAGGF) and amending and repealing certain regulations (OJ L 160, 26.6.1999, p. 80).

Community legislation may be used for providing the information referred to in Annex I.

4.   The authorities referred to in Article 6(3) shall be consulted when deciding on the scope and level of detail of the information which must be included in the environmental report.

Article 6. Consultations

1.   The draft plan or programme and the environmental report prepared in accordance with Article 5 shall be made available to the authorities referred to in paragraph 3 of this Article and the public.
2.   The authorities referred to in paragraph 3 and the public referred to in paragraph 4 shall be given an early and effective opportunity within appropriate time frames to express their opinion on the draft plan or programme and the accompanying environmental report before the adoption of the plan or programme or its submission to the legislative procedure.
3.   Member States shall designate the authorities to be consulted which, by reason of their specific environmental responsibilities, are likely to be concerned by the environmental effects of implementing plans and programmes.
4.   Member States shall identify the public for the purposes of paragraph 2, including the public affected or likely to be affected by, or having an interest in, the decision-making subject to this Directive, including relevant non-governmental organisations, such as those promoting environmental protection and other organisations concerned.
5.   The detailed arrangements for the information and consultation of the authorities and the public shall be determined by the Member States.

Article 7. Transboundary consultations

1.   Where a Member State considers that the implementation of a plan or programme being prepared in relation to its territory is likely to have significant effects on the environment in another Member State, or where a Member State likely to be significantly affected so requests, the Member State in whose territory the plan or programme is being prepared shall, before its adoption or submission to the legislative procedure, forward a copy of the draft plan or programme and the relevant environmental report to the other Member State.
2.   Where a Member State is sent a copy of a draft plan or programme and an environmental report under paragraph 1, it shall indicate to the other Member State whether it wishes to enter into consultations before the adoption of the plan or programme or its submission to the legislative procedure and, if it so indicates, the Member States concerned shall enter into consultations concerning the likely transboundary environmental effects of implementing the plan or programme and the measures envisaged to reduce or eliminate such effects.

Where such consultations take place, the Member States concerned shall agree on detailed arrangements to ensure that the authorities referred to in Article 6(3) and the public referred to in Article 6(4) in the Member State likely to be significantly affected are informed and given an opportunity to forward their opinion within a reasonable time-frame.

3. Where Member States are required under this Article to enter into consultations, they shall agree, at the beginning of such consultations, on a reasonable time-frame for the duration of the consultations.

Article 8. Decision making

The environmental report prepared pursuant to Article 5, the opinions expressed pursuant to Article 6 and the results of any transboundary consultations entered into pursuant to Article 7 shall be taken into account during the preparation of the plan or programme and before its adoption or submission to the legislative procedure.

Article 9. Information on the decision

1. Member States shall ensure that, when a plan or programme is adopted, the authorities referred to in Article 6(3), the public and any Member State consulted under Article 7 are informed and the following items are made available to those so informed:

    (a) the plan or programme as adopted;
    (b) a statement summarising how environmental considerations have been integrated into the plan or programme and how the environmental report prepared pursuant to Article 5, the opinions expressed pursuant to Article 6 and the results of consultations entered into pursuant to Article 7 have been taken into account in accordance with Article 8 and the reasons for choosing the plan or programme as adopted, in the light of the other reasonable alternatives dealt with, and
    (c) the measures decided concerning monitoring in accordance with Article 10.

2. The detailed arrangements concerning the information referred to in paragraph 1 shall be determined by the Member States.

Article 10. Monitoring

1. Member States shall monitor the significant environmental effects of the implementation of plans and programmes in order, *inter alia*, to identify at an early stage unforeseen adverse effects, and to be able to undertake appropriate remedial action.

2. In order to comply with paragraph 1, existing monitoring arrangements may be used if appropriate, with a view to avoiding duplication of monitoring.

Article 11. Relationship with other Community legislation

1. An environmental assessment carried out under this Directive shall be without prejudice to any requirements under Directive 85/337/EEC and to any other Community law requirements.

2.   For plans and programmes for which the obligation to carry out assessments of the effects on the environment arises simultaneously from this Directive and other Community legislation, Member States may provide for coordinated or joint procedures fulfilling the requirements of the relevant Community legislation in order, *inter alia*, to avoid duplication of assessment.

3.   For plans and programmes co-financed by the European Community, the environmental assessment in accordance with this Directive shall be carried out in conformity with the specific provisions in relevant Community legislation.

Article 12. Information, reporting and review

1.   Member States and the Commission shall exchange information on the experience gained in applying this Directive.

2.   Member States shall ensure that environmental reports are of a sufficient quality to meet the requirements of this Directive and shall communicate to the Commission any measures they take concerning the quality of these reports.

3.   Before 21 July 2006 the Commission shall send a first report on the application and effectiveness of this Directive to the European Parliament and to the Council.

   With a view further to integrating environmental protection requirements, in accordance with Article 6 of the Treaty, and taking into account the experience acquired in the application of this Directive in the Member States, such a report will be accompanied by proposals for amendment of this Directive, if appropriate. In particular, the Commission will consider the possibility of extending the scope of this Directive to other areas/sectors and other types of plans and programmes.

   A new evaluation report shall follow at seven-year intervals.

4.   The Commission shall report on the relationship between this Directive and Regulations (EC) No 1260/1999 and (EC) No 1257/1999 well ahead of the expiry of the programming periods provided for in those Regulations, with a view to ensuring a coherent approach with regard to this Directive and subsequent Community Regulations.

Article 13. Implementation of the Directive

1.   Member States shall bring into force the laws, regulations and administrative provisions necessary to comply with this Directive before 21 July 2004. They shall forthwith inform the Commission thereof.

2.   When Member States adopt the measures, they shall contain a reference to this Directive or shall be accompanied by such reference on the occasion of their official publication. The methods of making such reference shall be laid down by Member States.

3.   The obligation referred to in Article 4(1) shall apply to the plans and programmes of which the first formal preparatory act is subsequent to the date referred to in paragraph 1. Plans and programmes of which the first formal preparatory act is before that date and which are adopted or submitted to the legislative procedure more than 24 months thereafter, shall be made subject to the obligation

referred to in Article 4(1) unless Member States decide on a case-by-case basis that this is not feasible and inform the public of their decision.

4.   Before 21 July 2004, Member States shall communicate to the Commission, in addition to the measures referred to in paragraph 1, separate information on the types of plans and programmes which, in accordance with Article 3, would be subject to an environmental assessment pursuant to this Directive. The Commission shall make this information available to the Member States. The information will be updated on a regular basis.

Article 14. Entry into force

This Directive shall enter into force on the day of its publication in the *Official Journal of the European Communities*.

Article 15. Addressees

This Directive is addressed to the Member States.
Done at Luxembourg, 27 June 2001.
For the European Parliament, The President, N. FONTAINE
For the Council, The President, B. ROSENGREN

Annex I: Information referred to in Article 5(1)

The information to be provided under Article 5(1), subject to Article 5(2) and (3), is the following:

(a) an outline of the contents, main objectives of the plan or programme and relationship with other relevant plans and programmes;

(b) the relevant aspects of the current state of the environment and the likely evolution thereof without implementation of the plan or programme;

(c) the environmental characteristics of areas likely to be significantly affected;

(d) any existing environmental problems which are relevant to the plan or programme including, in particular, those relating to any areas of a particular environmental importance, such as areas designated pursuant to Directives 79/409/EEC and 92/43/EEC;

(e) the environmental protection objectives, established at international, Community or Member State level, which are relevant to the plan or programme and the way those objectives and any environmental considerations have been taken into account during its preparation;

(f) the likely significant effects[1] on the environment, including on issues such as biodiversity, population, human health, fauna, flora, soil, water, air, climatic factors, material assets, cultural heritage including architectural and archaeological heritage, landscape and the interrelationship between the above factors;

(g) the measures envisaged to prevent, reduce and as fully as possible offset any significant adverse affects on the environment of implementing the plan or programme;

(h) an outline of the reasons for selecting the alternatives dealt with, and a description of how the assessment was undertaken including any difficulties (such as technical deficiencies or a lack of know-how) encountered in compiling the required information;

(i) a description of the measures envisaged concerning monitoring in accordance with Article 10;

(j) a non-technical summary of the information provided under the above headings.

---

[1] These effects should include secondary, cumulative, synergistic, short, medium and long-term, permanent and temporary, positive and negative effects.

Annex II: Criteria for determining the likely significance of effects referred to in Article 3(5)

1. The characteristics of plans and programmes, having regard, in particular, to

   - the degree to which the plan or programme sets a framework for projects and other activities, either with regard to the location, nature, size and operating conditions or by allocating resources,
   - the degree to which the plan or programme influences other plans and programmes including those in a hierarchy,
   - the relevance of the plan or programme for the integration of environmental considerations in particular with a view to promoting sustainable development,
   - environmental problems relevant to the plan or programme,
   - the relevance of the plan or programme for the implementation of Community legislation on the environment (e.g. plans and programmes linked to waste-management or water protection).

2. Characteristics of the effects and of the area likely to be affected, having regard, in particular, to

   - the probability, duration, frequency and reversibility of the effects,
   - the cumulative nature of the effects,
   - the transboundary nature of the effects,
   - the risks to human health or the environment (e.g. due to accidents),
   - the magnitude and spatial extent of the effects (geographical area and size of the population likely to be affected),
   - the value and vulnerability of the area likely to be affected due to:

     - special natural characteristics or cultural heritage,
     - exceeded environmental quality standards or limit values,
     - intensive land-use,

   - the effects on areas or landscapes which have a recognised national, Community or international protection status.

Appendix 3
The Lee and Colley review package

The Lee and Colley method reviews EISs under four main topics, each of which is examined under a number of sub-headings:

(i) Description of the development, the local environment and the baseline conditions:

- Description of the development
- Site description
- Residuals
- Baseline conditions

(ii) Identification and evaluation of key impacts:

- Identification of impacts
- Prediction of impact magnitudes
- Assessment of impact significance

(iii) Alternatives and mitigation:

- Alternatives
- Mitigation
- Commitment to mitigation

(iv) Communication of results:

- Presentation
- Balance
- Non-technical summary

In outline, the content and quality of the environmental statement is reviewed under each of the subheads, using a sliding scale of assessment symbols A–F:

Grade A   indicates that the work has generally been well performed with no important omissions.

Grade B   is generally satisfactory and complete with only minor omissions and inadequacies.

Grade C   is regarded as just satisfactory despite some omissions or inadequacies.

Grade D   indicates that parts are well attempted but, on the whole, just unsatisfactory because of omissions or inadequacies.

Grade E   is not satisfactory, revealing significant omissions or inadequacies.

Grade F   is very unsatisfactory with important task(s) poorly done or not attempted.

Having analysed each sub-head, aggregated scores are given to the four review areas, and a final summary grade is attached to the whole statement.

Appendix 4
Environmental impact statement review package (IAU, Oxford Brookes University)

Using the review packages

The IAU review package was developed for a research project into the changing quality of EISs which was funded by the DoE, the Scottish and Welsh Offices in 1995/96. the package is a robust mechanism for systematically reviewing EISs. The full review package has now been updated to combine the requirements of the 1999 EIA Regulations, the DoE checklist, a review package developed by Manchester University, an EU review checklist as well as notions of best practice developed by the IAU. The package is divided into 8 sections and within each section are a number of individual review criterion. In all, the package assesses the quality of an EIS against 92 criteria, some of which are not necessarily relevant to all projects. Each criterion is graded on the basis of the quality of the material provided and each section is then awarded an overall grade. From the grades given to each section an overall grade for the EIS is arrived at. The IAU review grades are based upon the grading system developed by Manchester University for their review package. These grades are:

A = indicates that the work has generally been well performed with no important omissions;
B = is generally satisfactory and complete with only minor omissions and inadequacies;
C = is regarded as just satisfactory despite some omissions or inadequacies;
D = indicates that parts are well attempted but, on the whole, just unsatisfactory because of omissions or inadequacies;
E = is not satisfactory, revealing significant omissions or inadequacies;
F = is very unsatisfactory with important task(s) poorly done or not attempted.

These grades can be used to test an EIS's compliance with the relevant Regulations, with the pass/fail mark lying between grades "C" and "D". By using this grading system the reviewer can more readily identify the aspects of the EIS that need completing and because the grades are well established the competent authority can confidently justify any requests for further information. The assessment of EIS quality against these grades is rather like the marking of an academic essay in that while the activity – i.e. review – is carried out independently, objectively and systematically, the attributing of individual grades to individual criterion is inherently subjective. One way of reducing the subjectivity of the review is for the EIS to be assessed by two independent reviewers on the basis of a "double blind" approach. Here each reviewer assesses the

EIS against the criteria and grades the EIS on the basis of "A" to "F" for each criterion and for the ES as a whole. The reviewers then compare results and agree grades.

In arriving at overall grades, from all of the individual grades, a decision must be made over whether, for example, an "A" grade for one area outweighs a "D" grade for another area. This will depend entirely on perspective, as an individual reviewer may consider some aspects to be more important than others and so it is not a simple matter of counting up all of the "A", "B" and "Cs" and giving an overall grade based on the most common or average grade. In some cases a clear "F" grade for one of the minimum regulatory requirements (e.g. Non Technical Summary) could be seen as resulting in an overall fail for the EIS because of the importance of that particular aspect. Other areas (e.g. consideration of alternatives) may be seen as less crucial where that aspect is not of particular relevance to the project in question. An "F" grade for one such criteria, may not, in such cases, prevent an EIS being attributed a "C" grade, or above, overall. Attributing the overall grade for an EIS through this process requires the reviewer to come to a judgement on the weight to be given to the individual review areas and is rather like attributing weight to planning considerations.

The success of EIS review relies a great deal on the experience of the reviewer and their ability to make a judgement on the quality of the EIS as a whole, based upon the systematic assessment of its parts. In reviewing the EIS a reviewer should come to a view on the information provided based upon a balance between:

- what it "must" contain;
- what it could contain; and
- what it can be reasonably expected to contain.

Oxford Brookes University

Impacts Assessment Unit

Environmental Impact Statement Review Package

Name of Project:

EIS Submitted by:

Date Submitted:

Review Grades

A = Relevant tasks well performed, no important tasks left incomplete.

B = Generally satisfactory and complete, only minor omissions and inadequacies.

C = Can be considered just satisfactory despite omissions and/or inadequacies.

D = Parts are well attempted but must, as a whole, be considered just unsatisfactory because of omissions and/or inadequacies.

E = Not satisfactory, significant omissions or inadequacies.

F = Very unsatisfactory, important task(s) poorly done or not attempted.

NA = Not applicable in the context of the EIS or the project.

1. DESCRIPTION OF THE DEVELOPMENT

Criterion	Review Grade	Comments
Principal features of the project		
1.1 Explains the purpose(s) and objectives of the development.		
1.2 Indicates the nature and status of the decision(s) for which the environmental information has been prepared.		
1.3 Gives the estimated duration of the construction, operational and, where appropriate, decommissioning phase, and the programme within these phases.		
1.4 Provides a description of the development comprising information on the site, design and size of the development.[1]		
1.5 Provides diagrams, plans or maps and photographs to aid the description of the development.		
1.6 Indicates the physical presence or appearance of the completed development within the receiving environment.		
1.7 Describes the methods of construction.		
1.8 Describes the nature and methods of production or other types of activity involved in the operation of the project.		
1.9 Describes any additional services (water, electricity, emergency services etc.) and developments required as a consequence of the project.		
1.10 Describes the project's potential for accidents, hazards and emergencies.		
Land requirements		
1.11 Defines the land area taken up by the development and/or construction site and any associated arrangements, auxiliary facilities and landscaping areas, and shows their location clearly on a map. For a linear project, describes the land corridor, vertical and horizontal alignment and need for tunnelling and earthworks.		
1.12 Describes the uses to which this land will be put, and demarcates the different land use areas.		
1.13 Describes the reinstatement and after-use of landtake during construction.		
Project inputs		
1.14 Describes the nature and quantities of materials needed during the construction and operational phases.		

[1] Schedule 4 Part II Criterion (1999 EIA Regulations)

Criterion	Review Grade	Comments
1.15 Estimates the number of workers and visitors entering the project site during both construction and operation.		
1.16 Describes their access to the site and likely means of transport.		
1.17 Indicates the means of transporting materials and products to and from the site during construction and operation, and the number of movements involved.		
Residues and emissions		
1.18 Estimates the types and quantities of waste matter, energy (noise, vibration, light, heat, radiation etc.) and residual materials generated during construction and operation of the project, and rate at which these will be produced.		
1.19 Indicates how these wastes and residual materials are expected to be handled/treated prior to release/ disposal, and the routes by which they will eventually be disposed of to the environment.		
1.20 Identifies any special or hazardous wastes (defined as . . .) which will be produced, and describes the methods for their disposal as regards their likely main environmental impacts.		
1.21 Indicates the methods by which the quantities of residuals and wastes were estimated. Acknowledges any uncertainty, and gives ranges or confidence limits where appropriate.		

Overall Grade for Section 1 =
Comments

2. DESCRIPTION OF THE ENVIRONMENT

Criterion	Review Grade	Comments
Description of the area occupied by and surrounding the project		
2.1 Indicates the area expected to be significantly affected by the various aspects of the project with the aid of suitable maps. Explains the time over which these impacts are likely to occur.		
2.2 Describes the land uses on the site(s) and in surrounding areas.		

Criterion	Review Grade	Comments
2.3 Defines the affected environment broadly enough to include any potentially significant effects occurring away from the immediate areas of construction and operation. These may be caused by, for example, the dispersion of pollutants, infrastructural requirements of the project, traffic etc.		
Baseline conditions		
2.4 Identifies and describes the components of the affected environment potentially affected by the project.		
2.5 The methods used to investigate the affected environment are appropriate to the size and complexity of the assessment task. Uncertainty is indicated.		
2.6 Predicts the likely future environmental conditions in the absence of the project. Identifies variability in natural systems and human use.		
2.7 Uses existing technical data sources, including records and studies carried out for environmental agencies and for special interest groups.		
2.8 Reviews local, regional and national plans and policies, and other data collected as necessary to predict future environmental conditions. Where the proposal does not conform to these plans and policies, the departure is justified.		
2.9 Local, regional and national agencies holding information on baseline environmental conditions have been approached.		

Overall Grade for Section 2 =
Comments

3. SCOPING, CONSULTATION, AND IMPACT IDENTIFICATION

Criterion	Review Grade	Comments
Scoping and consultation		
3.1 There has been a genuine attempt to contact the general public, relevant public agencies, relevant experts and special interest groups to appraise them of the project and its implication. Lists the groups approached.		

Criterion	Review Grade	Comments
3.2 Statutory consultees have been contacted. Lists the consultees approached.		
3.3 Identifies valued environmental attributes on the basis of this consultation.		
3.4 Identifies all project activities with significant impacts on valued environmental attributes. Identifies and selects key impacts for more intense investigation. Describes and justifies the scoping methods used.		
3.5 Includes a copy or summary of the main comments from consultees and the public, and measures taken to respond to these comments.		
Impact identification		
3.6 **Provides the data required to identify the main effects which the development is likely to have on the environment.**[1]		
3.7 Considers direct and indirect/secondary effects of constructing, operating and, where relevant, after-use or decommissioning of the project (including positive and negative effects). Considers whether effects will arise as a result of "consequential" development.		
3.8 Investigates the above types of impacts in so far as they affect: human beings, flora, fauna, soil, water, air, climate, landscape, interactions between the above, material assets, cultural heritage.		
3.9 Also noise, land use, historic heritage, communities.		
3.10 If any of the above are not of concern in relation to the specific project and its location, this is clearly stated.		
3.11 Identifies impacts using a systematic methodology such as project specific checklists, matrices, panels of experts, extensive consultations, etc. Describes the methods/approaches used and the rationale for using them.		
3.12 The investigation of each type of impact is appropriate to its importance for the decision, avoiding unnecessary information and concentrating on the key issues.		
3.13 Considers impacts which may not themselves be significant but which may contribute incrementally to a significant effect.		

Criterion	Review Grade	Comments
3.14 Considers impacts which might arise from non-standard operating conditions, accidents and emergencies.		
3.15 If the nature of the project is such that accidents are possible which might cause severe damage within the surrounding environment, an assessment of the probability and likely consequences of such events is carried out and the main findings reported.		

Overall Grade for Section 3 =
Comments

4. PREDICTION AND EVALUATION OF IMPACTS

Criterion	Review Grade	Comments
Prediction of magnitude of impacts		
4.1 Describes impacts in terms of the nature and magnitude of the change occurring and the nature, location, number, value, sensitivity of the affected receptors.		
4.2 Predicts the timescale over which the effects will occur, so that it is clear whether impacts are short, medium or long term, temporary or permanent, reversible or irreversible.		
4.3 Where possible, expresses impact predictions in quantitative terms. Qualitative descriptions, where necessary, are as fully defined as possible.		
4.4 Describes the likelihood of impacts occurring, and the level of uncertainty attached to the results.		
Methods and data		
4.5 **Provides the data required to assess the main effects which the development is likely to have on the environment.**[1]		
4.6 The methods used to predict the nature, size and scale of impacts are described, and are appropriate to the size and importance of the projected disturbance.		
4.7 The data used to estimate the size and scale of the main impacts are sufficient for the task, clearly described, and their sources clearly identified. Any gaps in the data are indicated and accounted for.		

Criterion	Review Grade	Comments
Evaluation of impact significance		
4.8 Discusses the significance of effects in terms of the impact on the local community (including distribution of impacts) and on the protection of environmental resources.		
4.9 Discusses the available standards, assumptions and value systems which can be used to assess significance.		
4.10 Where there are no generally accepted standards or criteria for the evaluation of significance, alternative approaches are discussed and, if so, a clear distinction is made between fact, assumption and professional judgement.		
4.11 Discusses the significance of effects taking into account the appropriate national and international standards or norms, where these are available. Otherwise the magnitude, location and duration of the effects are discussed in conjunction with the value, sensitivity and rarity of the resource.		
4.12 Differentiates project-generated impacts from other changes resulting from non-project activities and variables.		
4.13 Includes a clear indication of which impacts may be significant and which may not and provides justification for this distinction.		

Overall Grade for Section 4 =
Comments

5. ALTERNATIVES

Criterion	Review Grade	Comments
5.1 *Provides an outline of the main alternatives studied and gives an indication of the main reasons for their choice, taking into account the environmental effects.*[1]		
5.2 Considers the "no action" alternative, alternative processes, scales, layouts, designs and operating conditions where available at an early stage of project planning, and investigates their main environmental advantages and disadvantages.		
5.3 If unexpectedly severe adverse impacts are identified during the course of the investigation, which are difficult to mitigate, alternatives rejected in the earlier planning phases are re-appraised.		

Criterion	Review Grade	Comments
5.4 The alternatives are realistic and genuine.		
5.5 Compares the alternatives' main environmental impacts clearly and objectively with those of the proposed project and with the likely future environmental conditions without the project.		

Overall Grade for Section 5 =
Comments

6. MITIGATION AND MONITORING

Criterion	Review Grade	Comments
Description of mitigation measure		
6.1 **Provides a description of the measures envisaged in order to avoid, reduce and, if possible, remedy significant adverse effects.**[1]		
6.2 Mitigation measures considered include modification of project design, construction and operation, the replacement of facilities/resources, and the creation of new resources, as well as "end-of-pipe" technologies for pollution control.		
6.3 Describes the reasons for choosing the particular type of mitigation, and the other options available.		
6.4 Explains the extent to which the mitigation methods will be effective. Where the effectiveness is uncertain, or where mitigation may not work, this is made clear and data are introduced to justify the acceptance of these assumptions.		
6.5 Indicates the significance of any residual or unmitigated impacts remaining after mitigation, and justifies why these impacts should not be mitigated.		
Commitment to mitigation and monitoring		
6.6 Gives details of how the mitigation measures will be implemented and function over the time span for which they are necessary.		
6.7 Proposes monitoring arrangements for all significant impacts, especially where uncertainty exists, to check the environmental impact resulting from the implementation of the project and its conformity with the predictions made.		
6.8 The scale of any proposed monitoring arrangements corresponds to the potential scale and significance of deviations from expected impacts.		

Criterion	Review Grade	Comments
Environmental effects of mitigation		
6.9 Investigates and describes any adverse environmental effects of mitigation measures.		
6.10 Considers the potential for conflict between the benefits of mitigation measures and their adverse impacts.		

Overall Grade for Section 6 =
Comments

7. NON-TECHNICAL SUMMARY

Criterion	Review Grade	Comments
7.1 *There is a non-technical summary of the information provided under paragraphs 1 to 4 of Part 2 of Schedule 4.*[1]		
7.2 The non-technical summary contains at least a brief description of the project and the environment, an account of the main mitigation measures to be undertaken by the developer, and a description of any remaining or residual impacts.	·	
7.3 The summary avoids technical terms, lists of data and detailed explanations of scientific reasoning.		
7.4 The summary presents the main findings of the assessment and covers all the main issues raised in the information.		
7.5 The summary includes a brief explanation of the overall approach to the assessment.		
7.6 The summary indicates the confidence which can be placed in the results.		

Overall Grade for Section 7 =
Comments

8. ORGANISATION AND PRESENTATION OF INFORMATION

Criterion	Review Grade	Comments
Organisation of the information		
8.1 Logically arranges the information in sections.		
8.2 Identifies the location of information in a table or list of contents.		
8.3 There are chapter or section summaries outlining the main findings of each phase of the investigation.		

Criterion	Review Grade	Comments
8.4 When information from external sources has been introduced, a full reference to the source is included.		

Presentation of information

8.5 Mentions the relevant EIA legislation, name of the developer, name of competent authority(ies), name of organisation preparing the EIS, and name, address and contact number of a contact person.		
8.6 Includes an introduction briefly describing the project, the aims of the assessment, and the methods used.		
8.7 The statement is presented as an integrated whole. Data presented in appendices are fully discussed in the main body of the text.		
8.8 Offers information and analysis to support all conclusions drawn.		
8.9 Presents information so as to be comprehensible to the non-specialist. Uses maps, tables, graphical material and other devices as appropriate. Avoids unnecessarily technical or obscure language.		
8.10 Discusses all the important data and results in an integrated fashion.		
8.11 Avoids superfluous information (i.e. information not needed for the decision).		
8.12 Presents the information in a concise form with a consistent terminology and logical links between different sections.		
8.13 Gives prominence and emphasis to severe adverse impacts, substantial environmental benefits, and controversial issues.		
8.14 Defines technical terms, acronyms and initials.		
8.15 The information is objective, and does not lobby for any particular point of view. Adverse impacts are not disguised by euphemisms or platitudes.		

Difficulties compiling the information

8.16 Indicates any gaps in the required data and explains the means used to deal with them in the assessment.		
8.17 Acknowledges and explains any difficulties in assembling or analysing the data needed to predict impacts, and any basis for questioning assumptions, data or information.		

Overall Grade for Section 8 =
Comments

COLLATION SHEET

Minimum Requirements of Schedule 4 Part II (1999 EIA Regulations)

Criterion	Overall Grade	Areas where more information required
1) A description of the development comprising information on the site, design and size of the development.		
2) A description of the measures envisaged in order to avoid, reduce and, if possible, remedy significant adverse effects.		
3) The data required to identify and assess the main effects which the development is likely to have on the environment.		
4) An outline of the main alternatives studied and an indication of the main reasons for their choice, taking into account the environmental effects.		
5) A non-technical summary of the information provided under 1 to 4 above.		
Overall Grade (A–F):		

List of Information that is required to complete the EIS

IAU Best Practice Requirements

Criterion	Overall Grade	Areas where more information required
Description of the development		
Description of the environment		
Scoping, consultation, and impact identification		
Prediction and evaluation of impacts		
Alternatives		
Mitigation and Monitoring		
Non-Technical Summary		
Organisation and Presentation of Information		
Overall Grade (A–F):		

Comments

Appendix 5
Key EIA journals and websites

Key journals

Environmental Impact Assessment Review
Elsevier Inc., New York, USA
www.elsevier.com

The Environmentalist
Magazine of the Institute of Environmental Management and Assessment (IEMA),
Lincoln, UK
www.iema.net

Impact Assessment and Project Appraisal
Journal of the International Association for Impact Assessment
Beech Tree Publishing, Guildford, UK

Journal of Environmental Assessment Policy and Management
Imperial College Press, London, UK
www.env.ic.ac.uk/jeapm/Welcome.html

Journal of Environmental Management
Academic Press, London, UK

Journal of Environmental Planning and Management
Carfax Publishing Company, London, UK
www.tandf.co.uk/journals

Journal of Environmental Policy and Planning
Carfax Publishing Company, London, UK
www.tandf.co.uk/journals

Journal of European Environmental Policy
Wiley, Chichester, UK
www.interscience.wiley.com

Journal of Planning and Environmental Law
Sweet and Maxwell, London, UK

Land Use Policy
Elsevier Science

Local Environment
Carfax Publishing Company, London, UK
www.tandf.co.uk/journals

*Review of European Community and International
Environmental Law*
Blackwell Publishing, Oxford, UK
www.blackwellpublishing.com

Sustainable Development
Wiley, Chichester, UK

Town Planning Review
Liverpool University Press, Liverpool, UK
www.liverpool-unipress.co.uk/journals

Key websites

Australia and New Zealand

Australian EIA Network
www.deh.gov.au/assessments/eianet/index.html

Environment Australia
www.deh.gov.au

New Zealand Ministry for the Environment
www.mfe.govt.nz

Parliamentary Commissioner for the Environment
www.pce.govt.nz

USA and Canada

Canadian Environmental Assessment Agency
www.ceaa-acee.gc.ca

Environment Canada
www.ec.gc.ca

Government of British Columbia, Environmental Assessment Office
www.eao.gov.bc.ca

*Nova Scotia Department of Environment & Labor, Environmental
Assessment Branch, Canada*
www.gov.ns.ca/enla/ess/ea

Quebec Association for Impact Assessment, Canada
www.cam.org/~aqei

US Council on Environmental Quality
www.ceq.eh.doe.gov

US Environmental Protection Agency
www.epa.gov

United Kingdom

Department for Environment, Food and Rural Affairs (DEFRA)
www.defra.gov.uk/environment

Department for International Development
www.dfid.gov.uk

EIA Unit, University of Aberystwyth
www.aber.ac.uk/~eiawww

Environment Agency
www.environment-agency.gov.uk

Environmental Policy and Management Group, Imperial College London
www.env.ic.ac.uk/research/epmg

Impacts Assessment Unit (IAU), Oxford Brookes University
www.brookes.ac.uk/iau

Institute of Environmental Management and Assessment
www.iema.net

Manchester EIA Centre, University of Manchester
www.art.man.ac.uk/eia

Office of the Deputy Prime Minister (ODPM)
www.odpm.gov.uk

Scottish Executive
www.scotland.gov.uk

UK Sustainable Development Unit
www.sustainable-development.gov.uk

Other European Union and Northern Europe

Arctic EIA Centre
http://finnbarents.urova.fi/aria

Centre for Planning and EIA, Oslo, Norway
www.nibr.no

EIA Centre – Milan, Italy
www.centrovia.it

European Commission – Environment Section
www.europa.eu.int/comm/environment/eia/home.htm

European Environment Agency – Copenhagen, Denmark
www.eea.eu.int

Eurostat (EU statistical information)
www.eurostat.eu.int

Finnish Environment Institute
www.environment.fi

Ministry of the Environment, France
www.environnement.gouv.fr

Netherlands Commission for EIA
www.eia.nl

Nordic Network for Environmental Assessment and Sustainable Regional Development
www.nordregio.se

Stockholm Environment Institute
www.sei.se

Swedish EIA Centre
www-mkb.slu.se/engelsk/indexe.htm

Central and Eastern Europe

Environmental Assessment in Countries in Transition
www.personal.ceu.hu/departs/envsci/eianetwork

Institute of Environmental Protection, Poland
www.ios.edu.pl

Ministry of Environment, Republic of Lithuania
www.am.lt

Ministry of Environmental Protection, Natural Resources and Forestry, Poland
www.mos.gov.pl

Regional Environmental Centre for Central and Eastern Europe
www.rec.org

Middle East

EIA Unit, Suez Canal University, Egypt
www.eiaunit.cjb.net

Israel Ministry of the Environment
www.environment.gov.il

Lebanon Ministry of Environment
www.moe.gov.lb

Asia

Hong Kong Environmental Protection Department
www.epd.gov.hk/eia

Ministry of Environment, Republic of Korea
www.me.go.kr

Ministry of the Environment, Japan
www.env.go.jp/en/index.html

Africa

Department of Environmental Affairs and Tourism, South Africa
www.environment.gov.za

International Organizations

Asian Development Bank
www.adb.org

European Bank for Reconstruction and Development
www.ebrd.com

International Association for Impact Assessment (IAIA)
www.iaia.org

Organisation for Economic Co-operation and Development (OECD)
www.oecd.org

United Nations Economic Commission for Europe (UNECE)
www.unece.org/env/eia

United Nations Environment Programme (UNEP)
www.unep.ch

World Bank
http://lnweb18.worldbank.org/ESSD/envext.nsf/47ByDocName/Environmental
Assessment

Other lists of EIA websites

EIA – Preliminary Index of Useful Internet Web Sites
Compiled for IAIA by the Canadian International Development Agency
www.iaia.org/eialist.html

Author index

ADB (Asian Development Bank) 313
Ahmad, B. 330–1
Anderson, F. R. et al. 29
Angelsen, A. et al. 293
ANZECC (Australia and New Zealand
    Environment and Conservation
    Council) 18, 307
Arnstein, S. R. 161
Arts, J. 188, 204
Atkinson, N. & R. Ainsworth 180–1
Au, E. & G. Sanvícens 190
Audit Commission 158

Baglo, M. A. 296–7
Bailey, P. et al. 331
Balram, S. et al. 329
Barde, J. P. & D. W. Pearce 143, 145
Barker, A. & C. Wood 94
Baseline Environmental Consulting 189
Bateman, I. 143
Baxter, W. et al. 326
BBC (British Broadcasting
    Corporation) 286
Beanlands, G. 293
Beanlands, G. E. & P. Duinker 192
Bear, D. 29, 35
Beattie, R. 13, 135
Berkes, F. 192
Bird, A. 229
Bisset, R. 118, 147, 193
Bisset, R. & P. Tomlinson 186
Blandford, C. Associates 193
Boulding, K. 9
Bourdillon, N. 140
Bowers, J. 142
Bowles, R. T. 196
Braun, C. 222
Braybrooke, C. & D. Lindblom 14
Breakell, M. & J. Glasson 8
Breese, G. et al. 16
Bregman, J. I. & K. M. Mackenthun
    108, 130

Briffett, C. 293, 295
Brito, E. & I. Verocai 293–4, 299
Buckley, R. 186, 193
Burdge, R. 329
Buxton, R. 181

California Resources Agency 188
Calow, P. 7
Canter, L. W. 29, 35, 88, 158
Carley, M. J. & E. S. Bustelo 7
Cashmore, M. et al. 227
Catlow, J. & C. G. Thirlwall 21, 39
CCW (Countryside Council for Wales)
    et al. 353
CEAA (Canadian Environmental
    Assessment Agency) 293, 306, 307,
    326, 346
CEC (Commission of the European
    Communities) 7, 9, 11–12, 16, 23,
    37, 40, 42, 44–9, 56, 94, 100, 135,
    141, 149, 175, 181, 188, 205, 218,
    229, 293, 296, 303, 328
CEPA (Commonwealth Environmental
    Protection Agency) 21, 308, 326,
    329–30
CEQ (Council on Environmental
    Quality) 31, 34, 88, 93, 149, 326, 345
Chadwick, A. 329
Chadwick, A. & J. Glasson 192, 201–2
Chico, I. 293
China 301
Clark, B. 160
Clark, B. D. 22
Clark, B. D. & R. G. H. Turnbull 38, 42
Clark, B. D. et al. 39, 112
Clark, R. & D. Richards 293
CLEIAA (Capacity Development and
    Linkages for Environmental Impact
    Assessment in Africa) 293
Cleland, D. I. & H. Kerzner 89
Coles, T. et al. 22, 233–4
Cooper, L. M. & W. R. Sheate 270

Countryside Agency et al.   120
CPRE (Council for the Protection of Rural
    England)   39, 61, 80, 322
Culhane, P. J.   185

Dagg, S. et al.   328
Dallas, W. G.   159
d'Almeida, K.   293–4, 296
Dasgupta, A. K. & D. W. Pearce   142
Davis, S.   331
Davison, J. B. R.   225
De Jongh, P. E.   135, 148
Dee, N. et al.   115–17
DEFRA (Department for Environment,
    Food and Rural Affairs)   338, 351
DETR (Department of Environment,
    Transport and the Regions)
    5, 12–13, 22, 60, 62, 64–5, 72, 75–6,
    81, 91–2, 126, 128, 135, 151–2, 157,
    212, 219, 228, 234
DFID (Department for International
    Development)   313
DfT (Department for Transport)   78,
    94, 353
Dickman, M.   192
Dipper, B. et al.   201
Dixon, J. & T. Fookes   293
Dobry, G.   39
DoE (Department of the Environment)
    3, 18, 23, 39, 60, 62–3, 72, 92–3,
    141, 143, 145, 167, 169, 174, 177–8,
    212, 218, 221, 224–8, 232, 234–6,
    351–2
DoE et al.   10–11
DoE/DoT   248, 250, 252
DoEn (Department of Energy)   195–6
DoT (Department of Transport)   78, 227,
    248–52
DTI   276, 278–9, 281–2

Eastman, C.   92, 94
EBRD (European Bank for Reconstruction
    and Development)   312
EC (European Commission)   90, 93,
    108–9, 334–5, 348–9
Ecotech Research and Consulting
    Ltd   193
Edwards, W. & J. R. Newman   147
EIB (European Investment Bank)   313
Elsom, D.   133
ENDS (Environmental Data Services)
    57–8
English Nature   128
English Nature et al.   353
Environment Agency   60, 93
Environment Australia   308

EPD (Environmental Protection
    Department)   190
ESRC (Economic and Social Research
    Council)   56
Essex Planning Officers'
    Association   227, 322
ETSU (Energy Technology Support Unit)
    193, 262
Etzioni, A.   14
European Bank for Reconstruction and
    Development   37

Faludi, A.   13
FEARO (Federal Environmental Assessment
    Review Office)   306, 346
Ferrary, C.   222
Finsterbusch, K.   7
Fisher, D.   301
Flyberg, B.   7
Forester, J.   14
Fortlage, C.   61, 88–9, 94, 105, 149, 151
Friend, J. K. & A. Hickling   136
Friend, J. K. & W. N. Jessop   13, 136
Frost, R.   99–100, 190
Frost, R. & D. Wenham   214
Fuller, K.   92, 175, 234

Gibson, R.   306–7, 314
Ginger, C. & P. Mohai   171
Glasson, J.   138, 190–1, 231, 329
Glasson, J. & D. Heaney   197
Glasson, J. & G. Bellanger   46, 50
Glasson, J. & N. N. B. Salvador   36, 294
Glasson, J. et al.   130–1, 195–7, 199–200,
    267, 270, 333
Golden, J. et al.   135
Goodland, R. & J. R. Mercier   314
Goodland, R. & V. Edmundson   293
Goodland, R. et al.   293
Gosling, J.   218
Government of New South Wales   93
Green, H. et al.   129, 147
Greene, G. et al.   187

Habermas, J.   14
Hall, E.   35, 225
Hancock, T.   159
Hanley, N. D. & C. Splash   14
Hansen, P. E. & S. E. Jorgensen   130
Hart, S. L.   22
Harvey, N.   293, 307
Healey, P.   14
Health Canada   330
Health & Safety Commission   7
Highland Council   269–70
Hill, M.   146, 149

HMG (Her Majesty's Government) 12
HMSO (Her Majesty's Stationery Office)
38, 141
Holling, C. S. 137, 187
Holmberg, J. & R. Sandbrook 10
Holstein, T. 332, 333
House of Lords 40–1, 159, 262
Huggett, D. 256
Hughes, J. & C. Wood 220
Hui, S. Y. M. & M. W. Ho 190
Hunter, C. & H. Green 267
Hyder Consulting 326
Hydro-Québec 163

IAIA (International Association for Impact
Assessment) 7, 187, 329, 332
IAU (Impact Assessment Unit) 218–19
IEA (Institute of Environmental
Assessment) 190, 212, 217
IEMA (Institute of Environmental
Management and Assessment)
158, 308
Iglesias, S. 299
INEM (International Network for
Environmental Management) 336
Institute of Environmental Assessment
and Landscape Institute 128
IOS (International Organization for
Standardization) 335
IWM (Institute of Waste
Management) 262

Jahiel, A. R. 301
Jendroska, J. & J. Sommer 302
Joao, E. 123
Jones, C. E. 221, 223, 225, 228, 232–3,
235–6
Jones, C. E. & C. Wood 179
Jones, C. E. et al. 22, 94, 220, 235
Journal of American Institute of
Planners 130

Kaiser, E. et al. 14
Kakonge, J. O. 293–4
Kennedy, W. V. 13
Kenyan, R. C. 171
Kirkby, J. et al. 10
Kleinschmidt, V. & D. Wagner 348
Kobus, D. & N. Lee 223, 227–8, 232,
235–6
Kreuser, P. & R. Hammersley 228
Kristensen, P. et al. 130

Land Use Consultants 268
Lane, P. and Associates 326–7
Lawrence, D. 13

Lee, N. 100, 133–4, 145
Lee, N. & C. George 293
Lee, N. & C. M. Wood 41–2, 187
Lee, N. & D. Brown 223, 226–8, 234
Lee, N. & R. Colley 175, 223, 227–8
Lee, N. & R. Dancey 225, 227
Lee, N. et al. 221, 223, 228, 232,
234–5
Legore, S. 32
Leopold, L. B. et al. 111
Leu, W.-S. et al. 221, 235
Levett-Therivel 355, 357–63
Lewis, J. A. 130
Lichfield, N. 14, 21, 143
Lichfield, N. et al. 141, 143
Lindblom, E. C. E. 14
Loewenstein, L. K. 130

McCormick, J. 157
McDonald, G. T. & L. Brown 332
McDonic, G. 175
McHarg, I. 118
Maclaren, V. W. & J. B. Whitney
137
McLoughlin, J. B. 13
McNab, A. 160, 167
Mandelker, D. R. 29
Mao, W. & P. Hills 299, 301
Marsden, S. & S. Dovers 308, 347
Marshall, R. 194
Marstrand, P. K. 99
Mathieson, A. & G. Wall 267
Mercier, J. R. 313
MftE (Ministry for the
Environment) 346–7
Mills, J. 99, 193, 225
Mollison, K. 160
Moreira, I. V. 36
Morgan, R. 293
Morris, P. & D. Thurling 133
Morris, P. & R. Therivel xii, 100,
108, 149
Morrison-Saunders, A. 188
Morrison-Saunders, A. et al. 159, 204
Mulvihill, P. R. & D. C. Baker 91
Munn, R. E. 3, 108, 129, 135

NAFW (National Assembly for Wales)
60, 62
NAO (National Audit Office) 251
Nelson, P. 225
Netherlands Commission for
Environmental Impact
Assessment 293
Newton, J. 329
Nuclear Electric 202–3

ODPM (Office of the Deputy Prime Minister)   75, 91–2, 94, 97, 102–3, 106, 108, 126, 128, 149, 166, 169, 180, 221–2, 349, 352–6, 359
Odum, E. P. et al.   117
Odum, W.   325
OECD   303
Okaru, V. & A. Barannik   293–4
O'Riordan, T.   8, 10, 13, 321
O'Riordan, T. & W. R. D. Sewell   42
Orloff, N.   29
Ortolano, L.   301
Overseas Development Administration   37

Padgett, R. & L. K. Kriwoken   308
Parker, B. C. & R. V. Howard   110
Parkin, J.   138, 147
Parkinson, P.   332
Partidario, M. R.   175
Partidario, M. R. & R. Clark   341
Pearce, D. & A. Markandya   145
Pearce, D. et al.   142
Pearce, D. W.   10, 12, 142
Peterson, E. et al.   325
Petts, J.   14, 161, 261–5, 267, 328
Petts, J. & G. Eduljee   88
Petts, J. & P. Hills   40, 212
Piper, J. M.   271–5, 326
Planning Service (Northern Ireland)   60
Preston, D. & B. Bedford   21
Pritchard, G. et al.   221, 223, 225, 229, 234–5
Project Appraisal   331–2

Radcliff, A. & G. Edward-Jones   213, 221, 225, 234
Rau, J. G. & D. C. Wooten   108, 130, 133, 135
Read, R.   165
Redclift, M.   10
Regional Environmental Centre for Central and Eastern Europe   293
Reid, D.   10
ReliefWeb   284, 286
Rendel Planning   101
Richey, J. S. et al.   147
Rodriguez-Bachiller, A.   106
Rodriguez-Bachiller, A. with J. Glasson   100, 133–4, 149, 175, 337
Roe, D. et al.   293
Ross, W. A.   161, 178
RSPB (Royal Society for the Protection of Birds)   277
Rzeszot, U. A.   293, 295

SACTRA (Standing Advisory Committee on Trunk Road Assessment)   78
Sadler, B.   7, 36, 186, 211, 221, 231, 323–4
Sadler, B. & R. Verheem   340
Salter, J. R.   181
Sassaman, R. W.   109–10
Scanlon, J. & M. Dyson   308
Scholten, J.   305
Scottish Executive   353
Scottish Executive Rural Affairs Department   60
Scott-Samuel, A. et al.   330
SDD (Scottish Development Department)   38
SEDD (Scottish Executive Development Department)   60, 62, 81
SEERA   13
Sheate, W. R.   61, 222, 241, 243, 246–7
Sheate, W. R. & M. Sullivan   253
Sinclair, A. J. & P. Fitzpatrick   307
Sippe, R.   231
Skolimowski, P.   10
Skutsch, M. M. & R. T. N. Flowerdew   117
Smith, S. P. & W. R. Sheate   356
Snary, C.   262–6
SNH (Scottish Natural Heritage)   269
Sorensen, J. C.   118–19
Sorensen, J. C. & M. L. Moss   108
Stakhiv, E.   21
State of California   36, 90, 345
Steinemann, A.   95
Stevens Committee   39
Stover, L. V.   117
Sustainable Development Unit   352
Suter, G. W.   130
Swaffield, S.   42

Tarling, J. P.   234–5
Taylor, L. & R. Quigley   330
Therivel, R.   7, 23, 133, 340, 352, 355
Therivel, R. & M. R. Partidario   7, 341, 348
Therivel, R. & P. Minas   352–4
Therivel, R. et al.   7, 344
Thomas, I.   293, 307
Tomlinson, E. & S. F. Atkinson   191
Tomlinson, P.   41–2, 136, 173, 175
Turner, R. K. & D. W. Pearce   10
Turner, T.   31, 158, 338

UN World Commission on Environment and Development   9–10
UNECE (United Nations Economic Commission for Europe)   4, 7, 328
UNEP (United Nations Environment Programme)   37, 103–5, 313

UNEP et al.   284–5
US Environmental Protection Agency   130

Vanclay, F.   7, 138, 329
Vanclay, F. & D. Bronstein   7
Von Neuman, J. & O. Morgenstern   146
Voogd, J. H.   136–7
VROM   129, 134–5

WA EPA (Western Australian
    Environmental Protection
    Authority)   138, 310–11
Waldeck, S. et al.   308
Wang, Y. et al.   300–1
Wathern, P.   22, 108
Weaver, A. B. et al.   88–9
Weiss, E. H.   170
Welles, H.   293
Wende, W.   91, 167
West Yorkshire County Council   120
Westman, W. E.   130, 162–3
Weston, J.   13, 57, 158, 177–80, 182,
    213, 219, 221, 225, 228,
    233–6, 262

Weston, J. & R. Smith   254, 256–61
Weston, J. et al.   158, 177–80, 182
White House   35
White, P. R. et al.   14
WHO   140
Williams, G. & A. Hill   161
Williams, R. H.   41–2
Willis, K. G. & N. A. Powe   145
Winpenny, J. T.   141, 143, 145
Wiszniewska, B. et al.   302–3
WO (Welsh Office)   81
Wood, C.   42, 212, 214–15, 221–2, 229,
    293, 307–8, 342, 348
Wood, C. & C. Jones   167, 177, 223,
    232–3
Wood, C. & J. Bailey   308
Wood, C. & N. Lee   100
Wood, G.   133, 194
Wood, G. & C. Bellanger   212,
    214–17
World Bank   37, 293, 295,
    313, 322

Zambellas, L.   225–6

Subject index

accuracy of predictions 192–4
affected parties, role in EIA 55–6
afforestation and EIA 48, 61, 80
Africa, EIA in 294, 297–9
Agenda 21 10–11
alternatives
  consideration of in EIA 49, 93–6, 248,
    250–2
  presentation and comparison of 96
  types of 94–5
  UK regulatory requirements 93–4
Annex I and II projects 42–3, 45, 47–8,
  60, 62, 65–70, 90, 303, 372–8
appropriate assessment, example of 254–61
Asia, EIA in 294–5, 299–301
Asian Development Bank, EIA
  guidelines 313
auditing
  definitions of 191–2
  environmental impact auditing 186
  environmental management
    auditing 186
  examples of 192–4, 201–2
  findings from 192–4, 201–2
  importance of in EIA 185–7
  of predictive techniques 194
  problems with 192
  types of 191–2
Australia, EIA system 188, 296,
  307–12, 326
Austria, EIA system 49

baseline data 100, 102–7
baseline, environmental 100, 102–7
baseline studies, see baseline data
Belgium, EIA system 46, 48
benefits of EIA 235–6
Benin, EIA system 294, 296–7

California
  EIA system 36, 90, 345–6
  monitoring procedures in 188–9

Canada, EIA system 93, 305–7, 326, 346
carrying capacity 333
CEC, see Commission of the European
  Communities
checklists, for impact identification 96,
  108–10
China, EIA system 299–301
Commission of the European
  Communities
  Action Programmes on the
    Environment 11, 23, 40, 42
  complaints to 181, 241–4, 247
  Directive 85/337 on EIA 40–4
  EIS review criteria 175–6
  Member States' EIA systems 46–50
  SEA Directive, see strategic
    environmental assessment (SEA)
  views on EIA 190, 243–4, 322
community impact evaluation 143
competent authority
  challenging decision of 179–81
  role of in EIA 44–7, 177, 179, 218,
    220, 228, 235
  see also local planning authorities
    and EIA
consequential impacts 254
consultation
  post-EIA submission 227–8
  pre-EIA submission 220–2
  with statutory consultees 167–8
  see also public participation
consultees 56, 76–7, 106, 167–9, 221,
  225–6, 227–9, 235–6
  see also statutory consultees, role of
    in EIA
contact group approach, as method of
  public participation 261–7
contingent valuation 144–5
cost–benefit analysis 8, 14, 39, 141–5
costs of EIA 233–5
Council on Environmental Quality
  28–36, 93, 325, 345

Countries in Transition, EIA in 289
cumulative effects assessment, case study
    of 270–5
cumulative impacts, assessment of 49,
    325–8, 343

decision-making
    decision-making theory and EIA 13–14
    effect of EIA on 231–3
    and EIA 8, 13–14, 77, 176–81, 229–33
    judicial review of in EIA cases 179–81
Delphi technique 129, 135, 147–9, 329
Denmark, EIA system 46, 49
Department of the Environment
    guidance on EIA 62, 65, 72–6, 81, 91,
        94, 149–50
    views on EIA 39, 41, 322
developers, role in EIA 8, 54–5
developing countries, implementation of
    EIA in 295
development plans, SEA of 352–5
Directive 97/11/EC
    main features of 44–5
    review of 47–50, 181
    text of 366–81
distributional impacts 21, 115, 143
drainage projects and EIA 61, 78–9

Eastern Europe, EIA in 295, 301–3
EC Directive 85/337
    amendments to 44–5
    compliance with in UK 80–1
    differences in implementation between
        Member States 28, 46–50
    implementation of in UK 58–81
    legislative history 40–2
    procedures 42–4
    reviews of 46–7
eco-management and auditing 334–6
EIA, see environmental impact assessment
EIA commissions 46, 174, 303–5
EIS, see environmental impact statements
electricity supply industry and EIA
    case studies of 241–7, 275–84
    and problems of project
        definition 241–7
    UK EIA regulations 62, 79–80
energy projects and EIA, see electricity
    supply industry and EIA
environment, dimensions of 18–19
environmental appraisal 351–2
environmental audit/auditing 14, 186,
    334–7
environmental baseline
    establishment of 100, 102–5
    sources and presentation of data 105–7

environmental capital 10
environmental consultants 57–8
environmental data, types and sources
    of 105–7
environmental evaluation system 115–17
environmental impact assessment
    changing perspectives on 13–14
    costs and benefits of 233–6
    current issues in 21–3
    and decision-making 8, 13–14, 77,
        176–81, 229–33
    definitions of 3–4
    in developing countries 295
    development of in UK 37–40
    directive, see EC Directive 85/337
    effectiveness of 211, 323–5
    future changes in 321–5
    guidance, see guidance on EIA, from
        UK government
    implementation of in UK 58–81
    improvements to 323–4
    journals 408–9
    main stages in the process 4–6
    and major projects 15–17
    managing the process 87–9
    methods 22
    origins and development of
        in the European Union 36
        in the UK 37–40
        in the USA 28–36
        worldwide 36–7
    as a process 4–6
    participants 22, 54–8
    and planning permission 77
    of policies, plans and programmes,
        see strategic environmental
        assessment (SEA)
    possible changes in 321–34
    procedures, see regulations on EIA
        in the UK
    and project authorization 176–8, 229
    and public inquiries 178–9
    purposes of 8–13
    quality of 22–3
    regulations, see regulations on EIA
        in the UK
    relationship to other environmental
        management decision tools 14,
        35, 229
    scope of 21
    screening, see screening
    and sustainable development 8–13
    in the UK, see United Kingdom
    theoretical context of 13–14
    types of projects requiring 42–3, 45,
        47–8, 60, 62, 64–70, 89–93

environmental impact assessment
  (*Continued*)
  uncertainty in   135–7
  websites   409–12
  worldwide spread of   36–7
  worldwide status of   290–6
  and the land-use planning system   38
  see also prediction
environmental impact auditing   186
  see also auditing
environmental impact design   332–3
environmental impact statements
  clarity of   170–2
  collections of   217
  contents and scope of   6, 33, 43–4,
    71–6, 168–70, 220–2
  factors affecting quality of   226–7
  non-technical summary   6, 169
  numbers prepared in EU Member
    States   48
  numbers prepared in the UK   3, 212–17
  numbers prepared in the USA   34
  presentation of   172
  quality of   22–3, 222–7
  review criteria   175–6, 393–407
  review of   172, 175–6, 228, 305
environmental impacts, nature of   19–21
environmental injustice   35
environmental interest groups, growth of
  in the UK   56
environmental management
  auditing   186
environmental management systems   14,
  57, 334–6
  Eco-Management and Audit
    Scheme   335–7
  implementation of   336
  links with EIA   336
  standards and regulations   335–6
environmental standards   140–1
European Bank for Reconstruction and
  Development, EIA procedures
  312–13
European Court of Justice   181
European Union
  Directive on SEA   348–50
  divergence in EIA practice within   28,
    46–50
  EIA systems in   46–50
  see also Commission of the European
    Communities
evaluation
  methods   137–8
  monetary valuation methods   141–5
  multi-criteria methods   145–9
  of significance in EIA, *see* significance
  in the EIA process   137–41

use of weighting   145–9
expert systems   333

facilitators, role of in EIA   57–8
Finland, EIA system   49
forecasting, *see* prediction
forestry projects and EIA   61, 80
France, EIA system   46, 49

geographical information systems
  106–7, 133
Germany, EIA system   46, 49
global impacts, consideration of
  in EIA   35
goals achievement matrix   146–7, 149
Greece, EIA system   46, 49
guidance on EIA, from UK government
  60, 62, 72–6, 81, 91, 94, 149–50

Habitats Directive and EIA   254–61, 348
health impact assessment   330–1
highway projects and EIA
  case studies of   248–61
  regulations   78
Hong Kong, EIA in   190, 294, 344

impact identification
  checklists   96, 108–10
  comparison of methods   122
  matrices   96, 109–15
  methods   96, 107–21
  networks   118–21
  quantitative methods of   115–18
impact magnitude   111–12, 128
impact prediction, *see* prediction
impact significance, *see* significance
impacts
  consequential   253
  distributional   21, 115, 143
  identification of, *see* impact
    identification
  indirect   20, 253
  mitigation of, *see* mitigation
  nature of   19–21
  reversibility of   20
  significance of, *see* significance
  socio-economic, *see* socio-economic
    impacts
importance weighting, *see* weighting, use
  of in EIA methods
indicative criteria and thresholds
  65–70, 90
indirect impacts, treatment of   253
integrated environmental
  assessment   331–2
international funding institutions, and
  EIA   37, 312–13

Ireland, EIA system 46
Italy, EIA system 46

journals 408–9
judicial review, in EIA cases 179–81

land drainage projects, EIA
 regulations 61, 78–9
land-use planning system, limitations
 of 38
Lee and Colley, EIS review criteria
 175–6, 228, 394
Leopold matrix 111–14, 146
life cycle assessment 14
local government, views on EIA 322
local planning authorities and EIA 56,
 63–4, 70, 72, 76–7, 175, 177, 225–6,
 228–9, 231–3

magnitude of impacts 111–12, 128
major projects
 characteristics of 15–17
 and EIA 15
 life cycle of 16–17
management, of the EIA process 87–9
Manual of Environmental Appraisal
 78, 332
mathematical models, for impact
 prediction 130–2
matrices/matrix methods, for impact
 identification 109–15
methods
 of assessment 22
 of impact identification, *see* impact
 identification
 of prediction, *see* prediction
mitigation
 case study of 267–70
 in the EIA process 152–3
 types of 149–52
monetary valuation, techniques 141–5
monitoring
 in California 188–9
 case study of 194–203
 elements of 187
 in Hong Kong 190
 importance of in EIA 185–7
 requirement for in EIA 49, 188
 of socio-economic impacts 194–203
 UK examples 190–1, 194–203
multi-criteria methods, of
 evaluation 145–9

National Environmental Policy Act
 (NEPA) 21, 28–36
 comparison with EC Directive on
 EIA 42

EIA procedures 31–4, 344
 legal interpretation of 29–31
 legislative history 29
 recent developments in 34–6
Netherlands, EIA system 46, 48–9, 93,
 188, 303–5
networks, for impact identification
 118–20
New Zealand, EIA system 296, 326, 332,
 346–7
non-technical summary, importance of 6
North Sea oil and gas, and development of
 EIA 38–9

offshore wind energy, SEA case
 study 275–84
overlay maps 96, 118, 120

participants in EIA, role of 22, 54–8
Peru, EIA system 297–9
physical models 132
planning balance sheet 143
planning gain 151
planning permission and EIA 77
Poland, EIA system 295, 301–3
policies, plans and programmes, EIA of,
 see strategic environmental
 assessment (SEA)
policy appraisal, *see* strategic
 environmental assessment (SEA)
Portugal, EIA system 46, 49
power stations and EIA, *see* electricity
 supply industry and EIA
prediction
 accuracy of 192–4
 choice of methods 133–5
 dimensions of 126–9
 methods and models 129–35
 and uncertainty 135–7
pressure groups, views on EIA 322–3
probability of impacts 129–30,
 135–7, 171
project authorization and EIA 176–8, 229
project description 96–100
project design and EIA 332–3
project EIA
 improving its effectiveness 323–34
 limitations of 341–2
project screening, *see* screening
projects
 definition of in EIA 241–7
 dimensions of 96–9
 types of requiring EIA, *see* screening
public consultation in EIA 49, 76–7,
 167–8
 see also public participation
public inquiries and EIA 178–9

public participation   49, 157–67,
    328–9
  advantages and disadvantages of   158–9
  improving its effectiveness   328–9
  methods   162–5, 261–7
  new approaches to   261–7
  post-EIA submission   228–9
  requirements for effective
      participation   159–65
  types of   162–3, 165
  UK procedures for   166–7
  see also public consultation in EIA

quality of life assessment   121–2
quality of life capital, *see* quality of life
    assessment
quantitative methods   115–18

refugees, environmental impact of
    284–6
regulations on EIA in the UK   58–81
  eco-management and audit   335–6
  electricity   61, 79
  forestry   80
  highways   61, 78
  land drainage   78–9
  overview of   58–62
  pipeline works   80
  Town and Country Planning (AEE)
      Regulations
      Procedures   62–5, 70–2, 76–7
      projects covered by   60, 62, 64–70
regulators, role of in EIA   56
review criteria, for environmental
    impact statements   175–6,
    393–407
review, of environmental impact
    statements   172, 175–6, 228, 305
  see also review criteria, for
      environmental impact statements
risk assessment   7
roads, EIA regulations   61, 78

Schedule 1 and 2 projects, *see* Annex I and
    II projects
scoping   91–3
  and pre-submission consultation   92
  requirement for   72, 76, 93
  in UK practice   92, 220–2
Scotland, EIA regulations   61, 78
screening   89–92
  thresholds   48, 65–70, 90–2
  in UK practice   218–20
SEA, *see* strategic environmental
    assessment (SEA)
sensitivity analysis   137

significance
  assessment of   137–49
  determinants of   138
  see also evaluation
Sizewell B power station, monitoring of
    socio-economic impacts   194–203
social impact assessment (SIA)   7
socio-economic impact assessment   7
socio-economic impacts   20, 38, 128, 138,
    194–203, 329–30
  inclusion of in EIA   7
  inclusion of in SEA   343
  monitoring of   194–203
  prediction of   130–3
Sorensen network   118–19
South America, EIA in   294, 297–9
Spain, EIA system   46, 49
specified information, required in
    EIA   43–5, 71–2, 96–7
statutory consultees, role of in EIA   56,
    76–7, 167–8
strategic environmental assessment
    (SEA)   7, 23
  in Canada   346
  case study of   275–84
  definitions of   341–2
  EU Directive on   23, 347–50, 382–92
  implementation of in the UK   352–62
  methods   279–82
  need for   341–4
  in New Zealand   346–7
  of offshore wind energy
      development   275–84
  problems with   280, 282–3, 344
  and sustainable development   343–4
  in the European Union   347–50
  in the UK   351–62
  in the USA   345–6
  of UK development plans   352–6
  UK government guidance on   352–6
  UNECE SEA protocol   350–1
  worldwide spread of   344–50
sustainability, *see* sustainable development
sustainability appraisal   351–3, 354–6
sustainability indicators   13
sustainable development   8–13
Sweden, EIA system   46, 49

threshold of concern   96, 108–10
thresholds, *see* screening
tourism, impacts of   267–70
Town and Country Planning (AEE)
    Regulations
  consultation arrangements   76–7
  contents of the EIA   70–5
  projects covered by   60, 62, 64–70

trans-boundary impacts 45, 50
transmission lines, and project definition
   in EIA 61, 241–7
transport projects and EIA 61, 78,
   248–61

uncertainty, dealing with in EIA
   135–7
United Kingdom
   development of EIA in 37–40
   early EIA initiatives 38–9
   EIA regulations, overview of
     58–62
   implementation of EC Directive
     85/337 58–81
United Nations, SEA Protocol
   350–1

United States of America
   federal EIA system 28–36
   recent trends in EIA 34–6
   SEA in 345–6
   state-level EIA systems 36
   see also National Environmental Policy
     Act (NEPA)

waste disposal schemes, EIA case
   study 261–7
websites 409–12
weighted matrix 96, 112, 115
weighting, use of in EIA methods 96, 112,
   115, 117, 145–7, 149
Western Australia, EIA in 308–12
wind power projects and EIA 275–84
World Bank, EIA procedures 313

eBooks

eBooks – at www.eBookstore.tandf.co.uk

A library at your fingertips!

eBooks are electronic versions of printed books. You can store them on your PC/laptop or browse them online.

They have advantages for anyone needing rapid access to a wide variety of published, copyright information.

eBooks can help your research by enabling you to bookmark chapters, annotate text and use instant searches to find specific words or phrases. Several eBook files would fit on even a small laptop or PDA.

NEW: Save money by eSubscribing: cheap, online access to any eBook for as long as you need it.

Annual subscription packages

We now offer special low-cost bulk subscriptions to packages of eBooks in certain subject areas. These are available to libraries or to individuals.

For more information please contact webmaster.ebooks@tandf.co.uk

We're continually developing the eBook concept, so keep up to date by visiting the website.

www.eBookstore.tandf.co.uk